COMPARATIVE INVERTEBRATE NEUROCHEMISTRY

Edited by G.G. Lunt, Department of Biochemistry, University of Bath, and R.W. Olsen, Department of Pharmacology, University of California, Los Angeles

Neurochemistry is a rapidly advancing field of research and workers on mammalian and other vertebrate systems, as well as those interested in insects and other invertebrates, are turning their attention to invertebrates to show how neural transmission occurs in 'simple' systems. Hence the subject is attracting interest from many neurophysiologists and neuropharmacologists as well as from zoologists. It is also of relevance in applied fields such as the pesticide industry, since it is possible to produce pesticides which interfere with nervous transmitter systems of major insect pests, in a similar way to how drugs act in other systems.

This book presents a review of invertebrate neurochemistry at a level suitable for the advanced student or researcher in zoology, neuroscience or pharmacology. Most of the book is concerned with insects, but other invertebrate systems are also covered and comparisons are made with mammalian neurochemistry. The book is organised around the major neurotransmitter classes and in each case both 'pure' and applied aspects are covered.

Comparative Invertebrate Neurochemistry

Edited by G.G. LUNT and R.W. OLSEN

CORNELL UNIVERSITY PRESS
Ithaca, New York

First published 1988 by Cornell University Press

Library of Congress Cataloging-in-Publication Data

Comparative invertebrate neurochemistry.

    Includes bibliographies and index.
    1. Invertebrates — Physiology. 2. Nervous system — Invertebrates. 3. Neurochemistry. I. Lunt, George G. II. Olsen, Richard W.
QL364.C66   1988   592'.0188   87-27446
ISBN 0-8014-2177-2

Printed and bound in Great Britain

Contents

List of Contributors

David J. Beadle, School of Biological Sciences and Environmental Health, Thames Polytechnic, London SE18 6PF, England.

Ian R. Duce, Department of Zoology, University of Nottingham, University Park, Nottingham NG7 2RD, England.

Amira T. Eldefrawi and Mohyee E. Eldefrawi, Department of Pharmacology and Experimental Therapeutics, University of Maryland School of Medicine, Baltimore, Maryland 21201, USA.

George C. Lunt, Department of Biochemistry, University of Bath, Claverton Down, Bath BA2 7AY, England.

Richard W. Olsen, Department of Pharmacology, UCLA School of Medicine, Los Angeles, California 90024, USA.

Nicholas Platt, Department of Anatomy and Neurobiology, Washington University School of Medicine, 660 South Euclid Avenue, St Louis, Missouri 63110, USA.

Timothy N. Robinson, Astra Neuroscience Research Unit, 1 Wakefield Street, London WC1 1PJ, England

Stuart E. Reynolds, School of Biological Sciences, University of Bath, Claverton Down, Bath BA2 7AY, England.

Peter F.T. Vaughan, Department of Biochemistry, University of Glasgow, Glasgow G12 8QQ, Scotland.

Eliahu Zlotkin, Institue of Life Sciences, The Hebrew University of Jerusalem, 91904 Jerusalem, Israel.

Preface

The attractions of invertebrate nervous systems have long been appreciated by neurophysiologists. Indeed some of the milestones in our understanding of nervous systems have their foundations in experiments done on invertebrate preparations, typified by the role of the squid axon in dissecting the events that constitute the action potential. More recently we have seen how the relatively simple nervous system of *Aplysia* has permitted new insights into the molecular mechanisms of memory and learning.

Neurochemists, however, have not been enthusiastic about invertebrate tissues as their experimental material. Much of the biochemical information on invertebrate nervous systems that has accrued has been incidental, almost as a by-product of what were primarily physiological investigations. Fortunately the field is changing, and research groups are making a positive choice to turn to invertebrate tissues.

Two important factors have contributed to this. First, the study of analogous systems in invertebrates and vertebrates can tell us much about the evolution of nervous systems. The application of the techniques of molecular genetics to the study of such molecules as receptors and ion channels can provide detailed information about their composition that, in turn, allows us to better understand their function. By extending such studies to the invertebrates we should be able to understand how such systems have developed. Secondly, invertebrate pests are responsible for enormous losses of agricultural crops and are major vectors of disease in man. The pesticide industry has in the past adopted a predominantly empirical approach to the control of pests. However, it is increasingly apparent that such an approach is no longer tenable. The industry needs new targets at which to direct more specific, safer, control agents. It is generally agreed that the nervous system is probably the most vulnerable target at which to aim such agents, and consequently there has been a major effort to promote research aimed at identifying potential pest-specific sites.

It is against this background of increasing emphasis on comparative studies that we have assembled this volume. We have sought to review our current understanding of the major neurotransmitter systems in invertebrates and to highlight their relationship to the generally much better characterised vertebrate systems. Our authors have all worked with invertebrate systems for many years and have been in part responsible for the current upsurge of interest. In addition to considering the major transmitter systems we have included two chapters that cover slightly different areas. Cell culture, and more particularly neuronal cell culture, is a relatively new discipline that has had a great impact on biology. Cultured cells offer possibilities that the natural tissue cannot. Invertebrate neuronal culture is just beginning to realise its potential and is being enthusiastically taken up by a number of groups. Neurotoxins have provided neuro-scientists with some of the most specific and potent research tools that we have; how could we study nicotinic acetylcholine receptors without α-bungarotoxin or the Na^+ channel without tetrodotoxin? There are toxins that differentiate between vertebrate and invertebrate targets: not only could such toxins provide new probes for the comparative neurochemist, but they may also give valuable leads to the pesticide chemist. The chapter on invertebrate neurotoxins constitutes one of the most comprehensive accounts of these fascinating agents.

A book of this kind depends entirely on the efforts of colleagues, most of whom would rather be doing neurochemistry than writing a chapter for a book! We are most grateful to each of them for contributing to our text. Our thanks must also go to our publisher, who has waited so patiently for the final manuscript.

G.G. Lunt
Bath
R.W. Olsen
Los Angeles

1

Acetylcholine

Amira T. Eldefrawi and Mohyee E. Eldefrawi

INTRODUCTION

It is estimated that 97 per cent of animal species (1 200 000) are invertebrates and 78 per cent are insects. Yet most biochemical research on
cholinergic transmission has been conducted on vertebrates. There are
obvious advantages to studying the invertebrate nervous system. It has
simpler cellular organisation and clearly identifiable neurones, which allow
for the investigation and understanding of the basic principles of neurotransmission and neuronal integration of signals. Also, the availability of
anatomical, physiological and pharmacological data on certain neurones,
whose regulation of certain behaviours is known (such as learning in the
marine snail *Aplysia* (Kandel, 1979) and in the American cockroach
Periplaneta americana (Sattelle *et al.*, 1983), and food aversion learning in
the terrestrial mollusc *Limax maximus* (Kelly, 1981)) provides information on the basic mechanisms that underlie these behaviours. Furthermore, the genetics of an insect, *Drosophila*, are the best studied of all
animal species, which makes possible the systematic and direct generation
of single gene mutants, selection of the relevant ones and identification and
isolation of the altered gene and its product. The mutation can be mapped
accurately in the chromosome and cloned using recombinant DNA technology. The availability of behavioural mutants in *Drosophila* may help
elucidate the nature of the macromolecular components involved in neurochemical mechanisms and their role in normal behaviour.

Acetylcholine (ACh) acts as a neurotransmitter throughout the Animal
Kingdom. It is also present in single-cell animals and even in plants (Jaffe,
1970) with cholinesterase (Fluck and Jaffe, 1975), though its role there is
still unknown. In arthropods, ACh is the transmitter of messages from
sensory neurones to the central nervous system (CNS) and within the CNS,
but not from motor neurones to skeletal muscles where the transmitter is
mostly glutamate, as shown in insects (see Sattelle, 1980), lobster (Barker
et al., 1972) and crabs (Florey, 1973). On the other hand, ACh is a motor

1

excitatory transmitter on to body wall muscles in annelids (leeches and earthworms) and possibly in velvet worms (see Gardner and Walker, 1982), as it is in vertebrate motor neurones. ACh is localised in certain neurones in the CNS, as shown in locust brain using anti-ACh antibodies (Geffard *et al.*, 1985).

The simultaneous measurement of 17 putative transmitters and metabolites in insect tissues revealed a concentration of ACh ranging from 7.6 nmol/mg protein in housefly brain to 25 and 50 nmol/mg protein in locust brain and thoracic ganglion, respectively (Clarke and Donnellan, 1982). By comparison, the fleshfly brain and cotton leafworm nerve cord had 17.7 and 12.4 nmol/mg protein, respectively. However, ACh is also widely distributed in tissues that are not cholinergic (Sastry and Sadavong-vivad, 1979), such as locust muscle (54 nmol/mg protein (Clarke and Donnellan, 1982)), where it may be involved in metabolic processes. Compared with other putative neurotransmitters in the CNS, the concentration of ACh is generally lower than those of glutamate, aspartate, proline, glycine and alanine (e.g. 476, 135, 399, 133, 94 nmol/mg protein, respectively, in locust brain). However, some of the high concentrations of these amino acids may be due to their significant involvement in amino acid metabolism.

ACh may coexist in the same neurone with other transmitters such as serotonin and octopamine, as shown in *Aplysia* ganglia (Brownstein *et al.*, 1974), or with peptides such as neurotensin, vasoactive intestinal polypeptide (VIP) (Said, 1984) or enkephalins in vertebrate peripheral nervous system (see O'Donohue *et al.*, 1985). Such cotransmitters may influence the same or different receptors in the same or different neurones. An excellent example is the effect of VIP on the affinity of muscarinic receptors for ACh (Lundberg *et al.*, 1982).

The arrival of an action potential from the axon induces ACh release by exocytosis of vesicles in which it is stored. ACh diffuses to the postsynaptic membrane in a few microseconds, and binds at the postsynaptic membrane to ACh receptor (AChR) proteins, inducing changes in their conformations. There are two major types of AChR. The nicotinic receptor is activated by ACh or nicotine and inhibited by *d*-tubocurarine. It is a single protein that includes a cationic channel and traverses the membrane. The muscarinic receptor is activated by ACh and muscarine and is inhibited by atropine and quinuclidinyl benzilate (QNB). It is located in the extracellular portion of the membrane bilayer and is conformationally coupled to enzymes, such as adenylate cyclase and phosphodiesterase, and regulatory proteins, such as nucleotide binding protein. Activation of the nicotinic AChR results in a rapid change in membrane conductance, and the reaction is usually complete in a few milliseconds. On the other hand, activation of the cuscarinic AChR results in a much slower reaction, taking hundreds of milliseconds to seconds, because several proteins and second messengers are involved in producing the effect. Cholinesterases (ChE) in synaptic membranes and basal laminae hydrolyse ACh, thereby reducing its concentration in the synaptic cleft. Choline is then taken up mostly by a high-affinity Na^+-dependent carrier mechanism in the presynaptic membrane. Choline acetyltransferase (ChAT) synthesises ACh in the neuronal cytoplasm from choline and acetyl-CoA. ACh is then packaged into new vesicles at a concentration of 520 mM with ATP at 170 mM (Wagner *et al.*, 1978), possibly via an energy-requiring packaging mechanism.

Our knowledge of cholinergic transmission in vertebrates and invertebrates continues to grow. The fastest growth during the past ten years, however, occurred in the area of AChRs. Accordingly, we shall devote a major portion of this chapter to emphasising these latest developments.

ACETYLCHOLINE RELEASE

In the nerve terminal, ACh is present in vesicles, which also contain ATP, GTP, Ca^{2+} and possibly UTP (see Winkler *et al.*, 1983). Depolarisation of the nerve terminal causes opening of Ca^{2+} channels and entry into the cytoplasm of Ca^{2+}, which binds to molecules that reduce an energy barrier between nerve terminal and vesicular membranes at the active zone, allowing membrane fusion and the vesicular membrane to open and discharge its transmitter quanta (5000–10000 ACh molecules/quantum) by exocytosis within 100–200 μs at a Ca^{2+} concentration of 1–10 μM (see Reichardt and Kelly, 1983; Silinsky, 1985). Thus, most of the average of 1 ms between the arrival of the action potential at the nerve terminal and the opening of channels in the postsynaptic membrane is due to a lag in Ca^{2+} channel opening. The ACh release process is inhibited by botulinum toxin (Gundersen *et al.*, 1982) and activated by dendrotoxin (from mamba venom) (Harvey and Karlsson, 1984). Injection of ACh or AChE into the presynapse of *Aplysia* has not significantly changed the size of individual ACh quanta released (Tauc and Baux, 1982), which supports the vesicular hypothesis. ACh also leaves the nerve terminal in a non-quantal fashion at a slow rate, which is unaffected by nerve stimulation (Miledi *et al.*, 1983) and whose physiologic significance is still obscure.

By using mild homogenisation, a Ficoll high-density gradient of low osmotic pressure and a microflotation technique, well sealed nerve ending membranes (i.e. synaptosomes) were isolated from insect CNS. They were found to contain synaptic vesicles, AChR, ChAT and a high-affinity Na^+- and Cl^--dependent carrier system for choline, and also released ACh (Breer, 1983). The isolated insect nerve endings are different from those of the vertebrates in rarely showing postsynaptic attachments, possibly due to the infrequent appearance of membrane thickenings, in frequently containing glycogen deposits, in the high representation of cholinergic nerve

3

endings and a predominance of nicotinic AChR, in the absence of ganglio-sides, and in the low molar ratio of cholesterol to phospholipids (Breer, 1981a; Breer and Jeserich, 1984). Exposure of locust nerve synaptosomes that had been preincubated with [³H]choline to a high K^+ medium evoked release of [³H]ACh (Breer, 1982), which was dependent upon extracellular Ca^{2+} and was inhibited by Mg^{2+} (Breer and Knipper, 1985) as in the verte-brates (Miledi et al., 1983). Also, similarly, muscarinic agonists inhibited ACh release, suggesting that the insect presynaptic membrane carries muscarinic receptors which act as feedback regulators.

CHOLINE ACETYLTRANSFERASE

Synthesis of ACh in cholinergic neurones occurs via ChAT, which is believed to be mostly in the cytoplasm, with some possibly membrane bound (Benishin and Carroll, 1981). This enzyme is a much more specific marker of cholinergic neurones than AChE, which is widely distributed in tissues as shown in Drosophila brain (Hall and Kankel, 1976). ChAT was detected in insects, earthworm and Nereis CNS (Marsden et al., 1981), and leech CNS and body wall muscles (Perkins and Cottrell, 1972) as well as in the parasitic nematodes, platyhelminths (see Mansour, 1979) and planar-ians (Erzen and Brzin, 1979). In the nematode Ascaris lumbricoides, ChAT was localised within identified motor neurones (Johnson and Stretton, 1985).

Insect heads are a very rich source of ChAT, with specific activity in the crude homogenate of Drosophila of 0.028 μmol/min/mg, compared with 43 μmol/min/mg protein for the ChAT purified from Drosophila (Driskell et al., 1978), and 45 μmol of ACh synthesised/min/mg protein for ChAT purified from rat brain (Strauss and Nirenberg, 1985). The K_m of ChAT for choline and acetyl-CoA from Drosophila is 90 μM and 47 μM, respect-ively, in contrast to the snail (Helix aspersa) ChAT, in which it is 370 μM and 51 μM, respectively; the American cockroach, in which it is 550 μM and 16 μM respectively; the horseshoe crab (Limulus polyphemus), in which it is 2700 μM and 68 μM, respectively (Emson et al., 1974); and the tobacco hornworm (Manduca sexta), in which it is 1100 μM and 400 μM, respectively (Lester and Gilbert, 1985). The drug specificities of the three enzymes are similar, but the isoelectric points are 5.3 and 5.0 for the horseshoe crab and cockroach enzymes, respectively.

Although multiple forms of ChAT, with different molecular weights and isoelectric points, were reported in mammalian brains (Malthe-Sorenssen, 1976), inhibition of proteolysis has yielded only one form. The molecular weight of ChAT from Drosophila head and Manduca sexta larval brain are reported to be 69000 (Driskell et al., 1978) and 68000 (Lester and Gilbert, 1985), respectively, values which are in the range of 50000 (Potter *et al.*, 1968), 73000 (Hersh *et al.*, 1984) and 62000 (White and Wu, 1973) for rat, bovine and human brains, respectively. Isoelectric points of 5.1 (Driskell *et al.*, 1978), 5.3 and 5.0 (Emson *et al.*, 1974) are reported for the *Drosophila*, horseshoe crab and the cockroach ChAT, respectively. However, antibodies that inhibited *Drosophila* brain ChAT did not interact with the mammalian brain ChAT (Crawford *et al.*, 1982), suggesting differences between them. Most reports suggest ChAT is made of one subunit. Others report two non-identical subunits in squid brain ChAT of 51000 and 69000 daltons (Husain and Mautner, 1973) or 27000 and 56000 daltons (Polsky and Shuster, 1976) and in rat brain ChAT of 28000–29000 daltons and 50000–51000 daltons (Badamchian and Carroll, 1985).

Invertebrate ChAT has different drug specificities from those of the vertebrates. *Manduca* ChAT is not inhibited by the specific inhibitor 4-(1-naphthylvinyl)-pyridine, but is inhibited by non-specific sulphhydryl inhibitors such as iodoacetamide and 5,5-dithio-bis(2-nitrobenzoic acid) (Lester and Gilbert, 1985). *o*-Bromoacetophenone, *o,p*-dibromoacetophenone and *o*-bromo-2'-acetonaphthone are more inhibitory on the mouse ChAT than on that of the housefly. On the other hand, *N,N*-dialkyl-2-phenylaziridinium ions are inactive on the mouse enzyme but have some inhibitory activity on that of the housefly (Yu and Booth, 1971). Other differences between invertebrates and vertebrates include unusually low substrate affinities for *Aplysia* ChAT (Giller and Schwartz, 1971), unusual temperature coefficients for snail ChAT (Cottrell and Powell, 1970) and different thermal stabilities for squid ChAT (Prempeh *et al.*, 1972).

After synthesis, ACh is transported into the storage vesicles, and this process is enhanced by increased ACh release. A bicarbonate-stimulated Ca^{2+}- or Mg^{2+}-ATPase in the vesicles drives active uptake of ACh (Parsons *et al.*, 1982), possibly linked to an internally acidic proton gradient generated by ATPase (Anderson *et al.*, 1983). It is likely that a transporter for ACh exists in vesicle membranes which is different from ATPase (Michaelson and Angel, 1981). It is suggested that there is a small, active, highly labile fraction of vesicles that are the source of released ACh (Jope and Johnson, 1986).

CHOLINESTERASES

ChEs are very efficient hydrolases, which introduce a water molecule when hydrolysing ACh into acetate and choline (see Eldefrawi, 1984). They have been found with ACh in plants (Jaffe, 1970) and many animal species, including protozoa (Seeman and Houlihan, 1951) and sponges (Lentz, 1966), and may even be present in non-cholinergic neurones (Eckenstein and Sofroniew, 1983).

5

There are two major kinds of ChEs in vertebrates. Acetylcholinesterase (AChE) (true or specific ChE), whose preferred substrate is ACh, is inhibited by excess substrate, is reactivated by oximes, is selectively inhibited by BW284C51 (1:5 bis(4-allyldimethylammoniumphenyl)-pentan-3-one dibromide) and is selectively resistant to iso-OMPA (*N,N'*-diisopropyl phosphorodiamidic anhydride). The second kind is pseudocholinesterase, also called non-specific ChE, butyryl cholinesterase (BuChE), or, in a few cases, propionylcholinesterase (PrChE), whose preferred substrates are BuCh and PrCh, respectively. The selectivity of drugs for AChE and BuChE differs among animal species. For example, the ratio of I_{50} (the concentration that inhibits 50 per cent of enzyme activity) of iso-OMPA for ACheE to BuChE varies from 56 for human enzymes to 13 200 for dog enzymes (Aldridge, 1953). The two enzymes are believed to be produced by different genes (Harris, 1975) and they lack immunological cross-reactivity (Vigny *et al.*, 1978), yet there is close homology in their molecular forms.

Most ChEs of insects can be characterised as AChE, as shown for the enzyme of leafhoppers, housefly heads, wax moth, cockroach nerve cord (Frontali *et al.*, 1971), confused flour beetle *Tribolium confusum* (Chaudhary *et al.*, 1966), the tobacco hornworm, *Manduca sexta* (Prescott *et al.*, 1977) and the aphid, *Aphis citricola* (Manulis *et al.*, 1981). On the other hand, the pear aphid *Toxoptera piricola* (Sakai, 1967), pea aphid *Macrosiphum pisi* and carpet beetle larvae (Chadwick, 1963) are more like BuChE. The honeybee enzyme is more active in hydrolysing acetyl-β-methylcholine than ACh (Wolfe and Smallman, 1956), and thus is different from that in housefly heads. Mite ChEs are either AChE as in *Tetranychus urticae* or BuChE as in *T. kanzawai* (Sakai, 1967) or PrChE as in *T. telarius* (Voss and Matsumura, 1965). The *T. telarius* ChE has a preference for ACh over BuCh or PrCh, but is not inhibited by excess substrate (Dauterman and Mehrotra, 1963).

ChEs in arthropod muscles occur in very small amounts because the motorneurones are non-cholinergic, but in annelids and also in the phylum Onychophora, which is situated phylogenetically between annelids and arthropods, the neurones are cholinergic and contain high concentrations of ChEs (Florey and Florey, 1965). In crayfish muscles, the enzyme is identified as AChE though it is not inhibited by excess substrate (Spielholz and van der Kloot, 1973). However, the lobster axon enzyme, like those of the housefly head and bovine erythrocyte, has a low K_m for acetylthiocholine, is inhibited by excess substrate, has low activity towards butyrylthiocholine, and is insensitive to di-isopropyl phosphorodiaminofluoride (Soeda *et al.*, 1975). One characteristic that distinguishes the lobster enzyme over housefly and mammalian brain AChEs is its 100-fold higher sensitivity to the local anaesthetic, procaine. ChEs of *Planaria torva* have different sensitivities to inhibitors from ChEs of vertebrates (Erzen and

6

Brzin, 1979). ChE was also detected in earthworms, *Nereis* (Marsden *et al.*, 1981) and leeches (Nistri *et al.*, 1978) as well as in parasitic nematodes and platyhelminths (see Mansour, 1979).

AChEs, isolated from many tissues, exist in multiple molecular forms (i.e. isozymes), which may differ in molecular weight, size, glycoprotein and amino acid compositions, isoelectric point and possibly turnover number (Silver, 1974; Massoulie and Bon, 1982; MacPhee-Quigley *et al.*, 1985). The vertebrate AChE occurs as aggregates of collagen-tailed molecules and tailless or globular molecules. These have different abilities to undergo intermolecular interactions (e.g. hydrophobic or ionic), which probably also determine their localisation in the cell (e.g. mostly globular forms in plasma membranes or collagen-tailed forms in basal lamina) (see Massoulie and Bon, 1982). Two different molecular-weight forms of AChE were extracted from *Pieris brassicae* heads: a larger hydrophilic form (7.3 S) and a smaller hydrophobic form (6.5 S) that aggregates in the absence of detergent. They have similar substrate and inhibitor specificities, but, unlike the vertebrate AChE forms, the hydrophobic one is not converted into the hydrophilic component by mild pronase treatment, suggesting that *Pieris* brain AChE does not apparently contain collagen-tailed AChE (Arpagaus and Toutant, 1985).

Four isozymes were detected in housefly heads, two in the American cockroach and one in the southern army worm (Eldefrawi *et al.*, 1970). In the bean aphid a single enzyme was detected, and in the green peach aphid several were found (Beranek, 1974). AChE isozymes may be multimers of a common catalytic subunit as shown for electric organ AChE (Bon and Massoulie, 1976) and as seen by the effects of different detergents and concentrations in eliminating certain enzyme species (Houk *et al.*, 1981). Nevertheless, different molecular species can still be isolated, as shown with the two AChE isozymes isolated from the mosquito *Culex tarsalis.* One has an isoelectric point of 5.5 and is the predominant species in head homogenates, and the other has a pI of 4.6 and is presumably extraganglionic (Houk *et al.*, 1981).

AChE is synthesised and secreted by both nerve and muscle cells (Oh *et al.*, 1977). It is transported within neurones both towards and away from nerve cell bodies, though much faster to the nerve ending (Skau and Brimijoin, 1978). Its presence has been detected in early developmental stages long before the differentiation of nerve and muscle tissue, such as in the uncleaved eggs of the sea urchin *Pseudocentrolus depressus* (Ozaki, 1974), and as early as the 8-cell stage in the ascidian *Ciona intestinalis* (Meedel and Whittaker, 1979). Primordial AChE was also detected in the teleoblast, the germinal bands and the germinal plate during the early phase of embryogenesis of the leech *Helobdella triserialis* (Fitzpatrick-McElligott and Stent, 1981). The fact that AChE precedes the formation of any nervous tissue suggests that it plays some role in development of the

7

germinal bands. It is similar to that of the vertebrates where in foetal rat (Kupfer and Koelle, 1951) and chick embryo (Mumenthaler and Engel, 1961) some AChE is present in the muscles prior to innervation by cholinergic motorneurones.

CHOLINE UPTAKE

Cholinergic nerve terminals possess a high-affinity, carrier-mediated choline transport system, which is detected in synaptosomes that have been lysed, emptied, then resealed so as to eliminate intracellular metabolism and sequestration of choline. [^{3}H]Choline is taken up by *Torpedo* electric organ and locust CNS preparations in a time-, concentration-, temperature- and Na^+-dependent manner with artificially imposed ion gradients as the driving force, primarily a transmembrane Na^+ gradient with some contribution by K^+ and Cl^- (Breer and Lueken, 1983). It is highly sensitive to hemicholinium-3. It is suggested that there is Na^+-linked cotransport, and two Na^+ ions are involved for uptake of one choline molecule. Neither ATP nor ATPase is directly involved in choline transport, though the latter maintains the necessary K^+ diffusion potential (Breer and Knipper, 1985).

The high-affinity uptake of the [^{3}H]choline transport system was also demonstrated in synaptosomal preparations from squid optic lobe (Dowdall and Whittaker, 1973), lobster *Homarus americanus* ventral nerve cord and *Limulus* CNS (Newkirk *et al.*, 1976, 1981), and visualised by autoradiography in *Drosophila* brain (Buchner and Rodrigues, 1983). The reported K_T values in μM for choline uptake are 1.0 for *Locusta*, 9.7 for *Cavia cobaya*, and 2.1 for *Loligo* and *Limulus*; and V_{max} values in pmol/min/mg protein are 15 in *Cavia*, 18 in *Limulus*, 145 in *Homarus*, 116 in *Locusta* and 280 in *Loligo* (see Breer and Jeserich, 1984). This compares with a K_T of 1.7 μM for the *Torpedo* electric organ choline uptake carrier, which was also purified further and reconstituted into liposomes (Ducis and Whittaker, 1985; Vyas and O'Regan, 1985).

ACETYLCHOLINE RECEPTORS

As in vertebrate CNS AChRs are localised in insect CNS postsynaptically as well as extrasynaptically, as shown in the cockroach *P. americana* sixth abdominal ganglion (Sattelle *et al.*, 1983), the fast coxal depressor motor neurone (David and Sattelle, 1984), cultured CNS neurones from embryonic cockroaches (Lees *et al.*, 1983) and isolated somata from *S. gregaria* thoracic ganglia (Suter and Usherwood, 1985). Also, as in vertebrates, some receptors are located on the presynaptic nerve terminal and modulate not only ACh but also other transmitters, such as neuromuscular synapse in *Schistocerca gregaria* (Usherwood and Cull-Candy, 1975), where the presynaptic receptors are nicotinic and muscarinic AChRs (Fulton, 1982).

Nicotinic receptors

A nicotinic AChR was the first neurotransmitter receptor to be purified, and its composition, structure, drug specificity and function were the first to be known in detail (see Changeux *et al.*, 1984).

Structure

The receptor isolated from *Torpedo* electric organ is glycosylated, and its molecular weight is approximately 295000 (Finer-Moore and Stroud, 1984). It is a pentamer of four different subunits: two α, one β, one γ and one δ, with molecular weights of 40000, 49000, 60000 and 65000, respectively (Figure 1). Each α-subunit carries an ACh-binding site. The subunit structure of calf (Noda *et al.*, 1983; Kubo *et al.*, 1985) and human (Shibahara, 1985) muscles are also similar. However, a novel polypeptide (ε) was found in calf muscle AChR and resembles the γ-subunit (Takai *et al.*, 1985). The primary structures of all four subunits of the *Torpedo* electric organ (Mishina *et al.*, 1984) and several of the subunits of vertebrate muscle receptors have been elucidated by cloning and sequencing complementary DNAs or genomic DNAs encoding these polypeptides (Shibahara *et al.*, 1985). The cloned cDNAs encoding each of the subunits were expressed to produce a normal functional receptor in oocytes of the frog *Xenopus* (Mishina *et al.*, 1984).

It is proposed that each subunit is rod shaped and crosses the membrane five times with the amino terminus located on the extracellular side and the carboxy terminus on the cytoplasmic side of the postsynaptic membrane (Young *et al.*, 1985). The membrane is crossed by four hydrophobic regions of each subunit which form α-helices; the fifth crossing occurs via an amphipathic region of a not very hydrophobic region alternating + and − charges. The ion channel of the AChR is formed by the homologous amphipathic helices of the five subunits, each of which contributes charged residues to the lining of a water-filled ionic channel. The channel is wide at the synaptic end and is narrow when extending through the membrane (Brisson and Unwin, 1985). Although there is a high degree of amino acid sequence homology between the *Torpedo* electric organ and calf muscle nicotinic AChR, and their channels have the same conductance, there is a 10-fold longer duration for the calf receptor, leading Sakmann *et al.* (1985) to suggest that the subunit may determine the channel closing step. It should be noted that the time course of the open channel is also affected by the composition of the lipid that surrounds the receptor.

Only two invertebrate nicotinic AChRs have been purified, and both are

9

Figure 1.1: A diagram of a nicotonic AChR of *Torpedo* electric organ showing its structure and subunits. The dashed line represents the ionic channel within the molecule

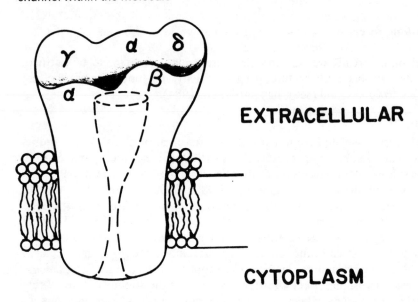

EXTRACELLULAR

CYTOPLASM

from the CNS: one from *Locusta migratoria* heads (Breer *et al.*, 1985) and the other from the cockroach *Periplaneta americana* nerve cord (Sattelle and Breer, 1985). The molecular weight of the cockroach membrane-bound receptor was estimated to be 108 000, using radiation inactivation and non-reducing conditions (Lummis *et al.*, 1984). The locust receptor was reported to be 250 000–300 000 daltons (Breer *et al.*, 1985), which is similar to the molecular weight of 285 000–290 000 for the vertebrate nicotinic AChR purified from *Torpedo* electric organ (Karlin, 1983; Noda *et al.*, 1983). The immunologic cross-reactivity between the isolated locust receptor and anti-*Torpedo* receptor antibodies confirms the identity of the isolated insect protein as a nicotinic AChR and suggests some structural homology between the two receptor types. This protein was apparently also reconstituted into planar bilayer and could transduce ACh-induced cation-specific permeability (Breer *et al.*, 1985). However, there are differences between the nicotinic AChR of locust CNS and that of *Torpedo*. The locust receptor appears to be heterogeneous with respect to its carbohydrate composition, has lower affinity for Con-A, and its α-bungarotoxin (α-BGT) binding is not inhibited by Con-A, suggesting different disposition of binding sites from those in the *Torpedo* receptor (Robinson *et al.*, 1984). Another difference is noted by the precipitation of the *Torpedo* receptor with certain polyclonal antibodies from an immunised sheep but not the nicotinic receptor of housefly brain (Eldefrawi and Eldefrawi, 1980). A significant finding is that both receptors isolated from the locust brain and ganglia and the cockroach nerve cord are reported to be aggregates (possibly four) of a single subunit of 65000 daltons. If true, it would suggest that the genes coding for the four subunits of the electric organ receptor may have originated from a common ancestral gene like that of the insect nicotinic receptor. The mRNA from locust nervous tissue, coding for α-BGT binding protein, was expressed in *Xenopus* oocytes (Breer and Benke, 1985), but no attempt was made to determine if these α-BGT-binding proteins were indeed functional receptors that cause agonist-induced changes in membrane potential.

In vertebrate brain, two nicotinic receptors have been identified; both bind α-BGT, but the function of only one is inhibited by it. The latter was purified from chick optic lobe and was also believed to be made up of a single subunit (Norman *et al.*, 1982). However, lately at least three subunits are believed to make up this receptor molecule: 48000, 56000 and 69000 daltons (Conti-Tronconi *et al.*, 1985). Its molecular weight is larger than the muscle nicotinic receptor and its smallest subunit is different from, but highly homologous to, the *Torpedo* receptor's α-subunit. Antibodies raised against the chick optic lobe receptor did not bind to peripheral AChR from chick muscle or *Torpedo* electric organ (Betz and Pfeiffer, 1984). The α-BGT binding protein in rat brain was also purified, and three subunits were identified of 49000, 53500 and 55000 daltons, with possibly a fourth (Kemp *et al.*, 1985). However, unlike the *Torpedo* receptor, whose smallest size subunit (40000 daltons) carries the ACh-binding site, in the rat brain protein it is the 55000-dalton subunit which does so, and in the chick optic lobe receptor it is the 56000-dalton subunit (Conti-Tronconi *et al.*, 1985).

Drug specificity

The vertebrate skeletal muscle and *Torpedo* electric organ receptors have two unequivalent ACh-binding sites on the two α-subunits. Drugs that bind to the ACh-binding or 'receptor' site and cause opening of the receptor's channel are agonists. The channel lifetime varies: suberyldicholine opens the channel of the vertebrate skeletal muscle receptor for 1.65 ms compared with 1 ms for ACh and 0.35 ms for carbamylcholine (Katz and Miledi, 1973). Inhibition of receptor function occurs through one of three general mechanisms. In *competitive antagonism*, the drug binds to the receptor site and prevents binding of ACh, and there is no opening of the receptor channel; thus it inhibits its binding of [^{3}H]ACh and [^{125}I]α-BGT binding to the receptor sites. Its effect is also manifested as a reduction in peak amplitude of postsynaptic current without affecting the lifetime or conductance of the single ionic channel. In *non-competitive antagonism* the drug binds to allosteric or 'channel' sites and inhibits binding of [^{3}H]perhydrohistrionicotoxin (H$_{12}$-HTX) or [^{3}H]phencyclidine (PCP) without inhibiting (and possibly potentiating) binding of [^{3}H]ACh. The

11

drug causes non-linearity in the current–voltage relationship and/or reduces channel lifetime, and its action is sensitive to temperature and the voltage across the membrane of which the AChR is a part. In *receptor desensitisation* the receptor becomes refractory to opening its channel, as a result of exposure to either an agonist for a prolonged period or to a high concentration for a shorter one, or as a result of binding of an antagonist to the 'channel sites'. This induces a conformational state of the receptor molecule that has a higher affinity for agonists. Desensitisation encompasses at least two kinetically identifiable states (Sakmann *et al.*, 1980; Aoshima *et al.*, 1981; El-Fakahany *et al.*, 1982). A detailed study of identified giant neurones from the parietal ganglia of the mollusc *Limnaea stagnalis*, whose channels are Cl^- selective, revealed that the AChR desensitised, but its onset was little influenced by wide changes in concentrations of Ca^{2+}, Na^+ or Cl^- in the medium (Bregestovski *et al.*, 1979). In another mollusc, *Limnaea heirosolima*, Na^+ even accelerated the onset of desensitisation (Ziskind and Werman, 1975). However, desensitisation of the electric eel and frog muscle receptor AChRs was affected by Ca^{2+} and Na^+ (Lester *et al.*, 1975). The embryonic dorsal unpaired medium (DUM) neurone cell body AChR desensitised in *P. americana* (Kerkut *et al.*, 1969) but not in *Schistocerca* (Goodman and Spitzer, 1979).

Certain drugs act as competitive as well as non-competitive inhibitors by binding to the receptor and channel sites, such as *d*-tubocurarine (Katz and Miledi, 1978). Others, such as the depolarising blocker decamethonium, act as agonists by binding to the ACh-binding site and activating the receptor to open its channel, as well as antagonists by binding to the channel site and inhibiting receptor function (Eldefrawi *et al.*, 1982). These 'channel sites' may bind a drug only if the ionic channel is open, only if it is closed (in resting conformation or with ACh bound) or in more than one of these receptor conformations. Examples of drugs that inhibit vertebrate muscle and *Torpedo* AChRs only in the open-channel conformation are gephyrotoxin (Souccar *et al.*, 1984), atropine (Adler *et al.*, 1978), neostigmine, pyridostigmine and physostigmine (Shaw *et al.*, 1985; Sherby *et al.*, 1985). Examples of drugs that inhibit only the closed-channel conformation are meproadifen (Aracava and Albuquerque, 1984) and imipramine (Eldefrawi *et al.*, 1981), and among those that inhibit either open- or closed-channel conformation are histrionicotoxin (Spivak *et al.*, 1982), PCP (Albuquerque *et al.*, 1980), amantadine (Tsai *et al.*, 1978), piperocaine (Tiedt *et al.*, 1979), and *d*-tubocurarine (Shaker *et al.*, 1982).

Similar allosteric sites are present in nicotinic AChRs of insects as shown by the voltage-dependent action on *P. americana* CNS neurones of histrionicotoxin (Sattelle and David, 1983), PCP (Sattelle *et al.*, 1985) and *d*-tubocurarine (David and Sattelle, 1984). Also, in *Aplysia* CNS, *d*-tubocurarine, procaine and hexamethonium are allosteric inhibitors of AChR-Na^+ (Ascher *et al.*, 1978; McCaman and Ono, 1982). The action of

12

amantadine on *P. americana* CNS is suggested to be on the receptor's open-channel conformation (Artola *et al.*, 1984), whereas *cis*-2-methyl-6-undecyl piperidine, an analogue of fire-ant venom, is proposed to act on the closed-receptor conformation at a site separate from the α-BGT-binding one on *P. americana* CNS D_f neurone (David *et al.*, 1984). Mecamylamine acts similarly on this neurone (David and Sattelle, 1984), though most of its action on frog sartorius muscle is on the receptor's open channel conformation (Varanda *et al.*, 1985).

Nicotinic AChRs in invertebrate CNS were identified by virtue of their specific [^{125}I]α-BGT binding (Table 1.1). The nicotinic AChRs of insect CNS are similar in their drug specificity to the vertebrate skeletal muscle type and the *Torpedo* electric organ receptors rather than the vertebrate CNS receptor. However, neither decamethonium, the potent inhibitor of vertebrate skeletal muscle AChR, nor hexamethonium, the potent inhibitor of vertebrate ganglionic AChR, has potent effects on [^{125}I]α-BGT binding to insect CNS nicotinic AChRs (Table 1.2). It should be noted that decamethonium binds to both the receptor and channel sites, causing depolarisation and blockade, respectively (Adams and Sakmann, 1978; Eldefrawi *et al.*, 1982). Thus, its inhibition of α-BGT binding would not reflect its total action on receptor function.

α-BGT inhibits nicotinic AChR function in the cockroach nerve cord and in the postsynaptic receptors (Sattelle *et al.*, 1983) as well as in the extrasynaptic ones on the cell body (David and Sattelle, 1984). The affinity of invertebrate nicotinic AChR for α-BGT is lower than the vertebrate muscle AChR, due mostly to faster dissociation, e.g. there is a half-time of 95 min in *Periplaneta* nerve cord (Lummis and Sattelle, 1985), 27.6 min in *Locusta* heads (Breer, 1981b), 30 min in housefly heads (Jones *et al.*,

Table 1.1: Characteristics of radiolabelled α-BGT binding to invertebrate CNS membranes

Species and tissue	Average K_d (nM)	Maximal conc. of binding sites (fmol/mg protein)	Reference
Musca domestica (heads)	3	2000	Eldefrawi and Eldefrawi (1980)
	6	23000	Jones *et al.* (1981)
Drosophila melanogaster (heads)	1.9	190	Dudai (1978)
	1.8	800	Rudloff (1978)
Periplaneta americana (nerve cord)	4.8	910	Lummis and Sattelle (1985)
Locusta migratoria (heads)	1.1	1775	Breer (1981b)
Acheta domesticus (last ganglion)	2.9	2329	Meyer and Reddy (1985)
Aplysia californica (ganglia)	0.8	24000	Shain *et al.* (1974)
Limulus polyphemus (brain)	3.5	6000	Thomas *et al.* (1978)

13

Table 1.2: Comparative pharmacology of radiolabelled α-BGT specific binding to invertebrate CNS membranes (i.e. nicotinic receptors)

Drug	[a]	Drosophila [b]	[c]	Periplaneta [d]	Locusta [e]	Acheta [f]	Aplysia [g]	Limulus [h]
					K_i or IC_{50} (M)			
Nicotine	8×10^{-7}	5×10^{-7}	5×10^{-7}	4×10^{-6}	4×10^{-8}	8×10^{-7}	—	7×10^{-6}
Carbamylcholine	1×10^{-4}	1×10^{-4}	—	—	1×10^{-4}	1×10^{-4}		6×10^{-4}
d-Tubocurarine	2×10^{-6}	3×10^{-6}	3×10^{-7}	9×10^{-7}	3×10^{-5}	9×10^{-7}	2×10^{-6}	2×10^{-6}
Decamethonium	9×10^{-4}	2×10^{-4}	—	1×10^{-4}	8×10^{-4}	—	2×10^{-4}	2×10^{-5}
Hexamethonium	—	—	7×10^{-5}	1×10^{-4}	8×10^{-4}	—	3×10^{-4}	3×10^{-4}
Atropine	5×10^{-5}	6×10^{-5}	2×10^{-5}	2×10^{-5}	1×10^{-5}	4×10^{-6}	3×10^{-5}	3×10^{-4}

[a] IC_{50} of 15 nM [^{125}I]α-BGT binding (Dudai, 1978)

[b] IC_{50} of 2.5 nM [^{125}I]α-BGT binding (Schmidt-Neilsen et al., 1977).

[c] IC_{50} of 0.1 nM [^{125}I]α-BGT binding (Rudloff, 1978)

[d] K_i obtained from inhibition of N-[^{3}H]propionylated α-BGT binding (Lummis and Sattelle, 1985).

[e] IC_{50} of 3.1 nM [^{125}I]α-BGT binding (Breer, 1981b).

[f] K_i values obtained from IC_{50} of 3.1 nM [^{125}I]α-BGT binding (Meyer and Reddy, 1985).

[g] IC_{50} of 3.4 nM [^{125}I]α-BGT binding (Shain et al., 1974).

[h] IC_{50} of 9 nM [^{125}I]α-BGT binding (Thomas et al., 1978).

1981), <10 min in *Limulus* nerve cord (Thomas *et al.*, 1978) and 30 min in *Aplysia* (Shain *et al.*, 1974). But no recovery from α-BGT inhibition was seen in *Drosophila* head up to 3 h (Schmidt-Nielsen *et al.*, 1977) nor in vertebrate skeletal muscle for days.

A few nicotinic AChRs, which are insensitive to α-BGT, are found in insect CNS, such as the embryonic dorsal unpaired median (DUM) neurones of *Schistocerca niteus* (Goodman and Spitzer, 1980). Neurones cultured from brain of *P. americana* embryos bound [^{125}I]α-BGT, and nicotine-induced depolarisation of the cell membrane was inhibited by α-BGT, except that a few cells could still depolarise after incubation with α-BGT (Lees *et al.*, 1983). α-BGT was also without effect on nicotinic ACh depolarisations in the gastric mill I muscles of the crab *Cancer pagurus*, and so was decamethonium. However, these were effectively blocked by vertebrate nicotinic ganglionic antagonists including dihydro-β-erythroidine, pempidine and mecamylamine (Marder and Paupardin-Tritsch, 1980). The AChRs of the leech *Hirudo medicinalis* dorsal muscle, while formally classifiable as nicotinic, possess several features that distinguish them from typical nicotinic AChRs of vertebrate skeletal muscles. Both carbamylcholine and succinylcholine acted as agonists, inducing muscle twitch. However, *d*-tubocurarine competitively antagonised carbamylcholine depolarisation, but not that induced by succinylcholine. On the other hand, α-BGT inhibited the succinylcholine-induced contraction without affecting the carbamylcholine-induced one. Whether a single or different receptors or binding sites are involved is not known (Ross and Triggle, 1972). In vertebrate CNS, most nicotinic responses are also not inhibited by α-BGT, such as in avian (Carbonetto *et al.*, 1978) and mammalian (Patrick and Stallcup, 1977) autonomic ganglion. However, only in the optic tectum of visual pathways of goldfish, toad, turtle and pigeon (Oswald and Freeman, 1981) and some sympathetic ganglia (Marshall, 1981) has inhibition of nicotinic receptor function been reported with α-BGT.

a-BGT may also bind to proteins other than AChR though with lower affinity, such as in the axon of the lobster *Homarus americanus* and the horseshoe crab *Limulus polyphemus* (Denburg *et al.*, 1972; Jones *et al.*, 1977), though not in axons of the blue crab *Callinectes sapidus* (Jones *et al.*, 1977).

The insect central nicotinic AChR is a molecular target for the toxic action of nicotine (Gepner *et al.*, 1978), which displaces [^{125}I]α-BGT binding to housefly brain and causes receptor desensitisation, as suggested by the time-dependent shift of the [^{125}I]α-BGT displacement curve that results from a change in receptor conformation to one of a higher affinity for agonists (Sherby *et al.*, 1985). The housefly brain receptor has a lower affinity for anatoxin-a, but much higher affinities than that of *Torpedo* for several nicotinoids, particularly (±)coniine and (−)nicotine and lobeline (Table 1.3), reaching over 1000-fold for the latter and suggesting that it

15

Table 1.3: Inhibitory constants of nicotinic chemicals for displacing the binding of [^{125}I]α-BGT to ACh receptors of housefly brain and *Torpedo* electric organ measured by ion exchange chromatography

	K_i (μM)		Relative affinities
Nicotinoid	Housefly	*Torpedo*	
Anatoxin-a	0.17	0.03	0.17
Cytisine	0.9	4.1	5
Nicotine	1.5	20.4	14
Nornicotine	2.1	6.8	3
Anabasine	4.5	40.9	9
Coniine	9.0	682	76
Lobeline	1.0	>1000	>1000
Mecamylamine	>1000	>1000	1

may be a powerful selective insecticide. However, it should be cautioned that, despite a poor effect on $[^{125}I]\alpha$-BGT binding to the *Torpedo* receptor, lobeline is an effective non-competitive blocker and inhibitor of $[^3H]H_{12}$-HTX binding to the *Torpedo* receptor (M. Eldefrawi, unpublished). Mecamylamine, like lobeline, is a competitive antagonist on mammalian ganglionic nicotinic receptors, but mecamylamine is a less potent allosteric inhibitor of both the *Torpedo* electric organ receptor (Varanda *et al.*, 1985) and housefly brain nicotinic receptors (Eldefrawi and Eldefrawi, 1987) (Table 1.3) as well as of those of the American cockroach nerve cord (Lummis and Sattelle, 1985) and *Drosophila* brain (Dudai, 1978).

The distinct difference in drug specificity of nicotinic receptors on insect CNS and the *Torpedo* electric organ is again demonstrated by the different effects charatoxins have on the two. Although they have little or no effect on $[^{125}I]\alpha$-BGT binding to the *Torpedo* receptor, several potentiate its binding to the honeybee brain receptor (Sherby *et al.*, 1986). Also, their potentiation of $[^3H]H_{12}$-HTX binding in the absence of ACh is not affected by *Naja* α-neurotoxin, which suggests that they bind to a site that is different from the ACh and H_{12}-HTX binding sites. The close analogue and insecticide nereistoxin bound to the ACh binding sites as well as the H_{12}-HTX binding sites, and acted similarly on the nicotinic receptors of *Torpedo* electric organ, honeybee brain and cockroach CNS (Sattelle *et al.*, 1985a; Sherby *et al.*, 1986).

Nicotinic receptors were also found in the body-wall longitudinal muscles of the earthworms *Pheretima communissima* (Ito *et al.*, 1970), *P. hawayana* (Chang, 1975) and *Lumbricus* (Gardner and Cashin, 1975; Rozhkova *et al.*, 1980) and in *Hirudo* (Walker *et al.*, 1968; Kuffler, 1978), as well as the gut of the polychaete *Syllis spongiphila* (Anderson and Mrose, 1976), the body-wall muscles of the nematode *Ascaris suum* (Rozhkova *et al.*, 1980) and the echiuroid *Urechis unicinctus* (Muneoka *et al.*, 1981).

In several cases, AChRs of muscles of invertebrates other than insects are also found to be very different in their drug sensitivities than the vertebrate receptors. The AChRs in trematodes are reported not to be inhibited by *d*-tubocurarine or atropine (Chance and Mansour, 1953). However, like the vertebrate skeletal muscle receptor, that in the body-wall muscle of the leech *Hirudo medicinalis* undergoes desensitisation (Walker *et al.*, 1968). Also the body-wall muscles of the worm-like marine invertebrate *Priapulus caudatus* are contracted by ACh and nicotine but not pilocarpine, yet atropine blocks the contractions and tubocurarine increases ACh responses (Mattisan *et al.*, 1974). Nicotine activates *Fasciola hepatica* muscles but not those of *Schistosoma mansoni*, yet both are paralysed by low concentrations of arecoline, whereas pilocarpine has no effect (Chance and Mansour, 1953; Barker *et al.*, 1966). Atropine and *d*-tubocurarine reverse the paralysis partially by cholinomimetic agents of

Fasciola muscles. The body-wall muscles of *Ascaris lumbricoides* are activated by nicotinic and not muscarinic drugs, are inhibited by mecamylamine and have a drug specificity like that of mammalian autonomic ganglia (Natoff, 1969). Also, ACh-activated responses in the stomatogastric ganglion of the crab *Cancer pagurus* are not inhibited by decamethonium, but rather by the vertebrate ganglionic blockers mecamylamine and hexamethonium (Marder and Paupardin-Tritsch, 1980).

The ganglia of the marine mollusc *Aplysia* sp. have the best identified central neurones that make synaptic connections with several identifiable postsynaptic neurones, and have well characterised receptors. The relatively large size of its neurones makes possible the measurement of ACh and choline in individual neurones (McCaman *et al.*, 1973). However, only small amounts of nerve tissue are available for biochemical studies of receptors. There are three kinds of ACh-receptors in *Aplysia* responding by increasing Na^+, Cl^- or K^+ conductances (Kehoe, 1972a, b), and the multicomponent responses to ACh arise from combinations of these three types of receptor. The drug sensitivities of these types are different (Kehoe *et al.*, 1976). The excitatory receptor that increases membrane permeability to Na^+ ($AChR-N^+$) is inhibited by *d*-tubocurarine and hexamethonium but not α-BGT, thus resembling the vertebrate central AChR. The inhibitory receptor that increases permeability to Cl^- ($AChR-Cl^-$) is inhibited by *d*-tubocurarine and high concentrations of α-BGT is insensitive to hexamethonium; however, the α-BGT action is reversible, unlike its action on the vertebrate skeletal muscle AChR. The inhibitory receptor that increases permeability to K^+ ($AChR-K^+$) is insensitive to *d*-tubocurarine, hexamethonium, atropine and α-BGT but is inhibited by tetraethylammonium, like the voltage-dependent K^+ channel. There are indications that receptors for two or more agonists or transmitters may share or control the same kind of ionic channel. α-BGT blocks the ACh-induced Cl^- conductance without affecting the ACh-induced Na and K conductances, but it also inhibits the histamine-induced Cl^- conductance (Ono and Salvaterra, 1981) (though not the GABA-induced Cl^- conductances (McCaman and Ono, 1982)).

Similar kinds of AChR-induced ionic permeabilities were reported in *Helix* neurones (Chad *et al.*, 1979), while the $AChR-Na^+$ and $AChR-Cl^-$ responses were reported in *Limnaea*, *Planobarius* (Zeimal and Vulfius, 1968) and *Navanax* (Levitan and Tauc, 1972).

Developmental neurobiology

Nerve cells influence the development of effector cells, and vice versa, through release of trophic factors, which could be an enzyme, ATP, etc. Little is known about these factors in invertebrates, but the study of synthesis and clustering of receptors gives clues to the influences of the neurone. The study of AChR development in the antennal lobes of *Manduca sexta* brain is ideal because they contain high concentrations of

18

cholinergic synapses, and the lobes and antennae develop *de novo* starting at the time of the metamorphic moult of the larva to the pupa. Also, by removing one antenna, the sensory input to its lobe is eliminated, while the other antenna and lobe can serve as control. The concentration of nicotinic AChR (detected by [^{125}I]α-BGT binding) in the antennal lobes rises gradually as the sensory axon grows into the lobes. Deafferentation (in the absence of sensory input) greatly reduces ChAT and AChE activities but has no effect on receptor numbers or distribution in the neuropil (Sanes *et al.*, 1977; Hildebrand *et al.*, 1979). Thus, neurones and AChRs complete much of their development normally without synaptic input.

Changes in receptor numbers may take place before synaptogenesis. In the giant interneurone 2 in the sixth abdominal ganglion of *P. americana*, chemosensitivity (which reflects receptor activity) appears at 40–45 per cent embryogenesis in both cell body and dendrites (Blagburn *et al.*, 1985). After 60–65 per cent embryogenesis, the dendrites become 1000 times more sensitive before onset of synapse formation, but at 80 per cent embryogenesis the sensitivity declines, possibly as a result of synapse maturation. Cell-body sensitivity increases during postembryonic development. In vertebrate muscle cultures, AChRs cluster before innervation, then nerves synapse anywhere (Frank and Fischbach, 1977), though *in vivo* AChR clusters form on myotubes only after innervation (Bevan and Steinbach, 1977). Also, during embryonic development of DUM neurones in the locust *S. niteus*, AChRs appear one week before detection of synaptic activity and are distributed over the whole surface of the cell (Goodman and Spitzer, 1980).

Genetics

Because we know more about the genetics of the fruitfly *Drosophila melanogaster* than about that of any other animal, and because of the availability of a variety of mutants, it has been a favourable organism for use in the genetic approach to neurobiological problems. ACh is a major transmitter in the CNS of *Drosophila melanogaster*. Its nicotinic AChR was identified (Hall and Teng, 1975; Dudai, 1978) and a single gene, designated *Cha*, was identified for ChAT (Greenspan, 1980) and another, *Ace*, for AChE, on the 3rd chromosome (Hall and Kankel, 1976). In a temperature-sensitive (*ts*) mutant, the phenotype can be switched on and off by temperature shifts at any desired time, thus permitting recovery of otherwise lethal mutations. One mutant, *shibire temperaturesensitive* (*shi^{ts}*) has reduced [^{3}H]ACh accumulation at the restrictive temperature of 37°C due to an abnormally rapid rate of release that is not blocked by inhibiting Na$^+$ channels with tetrodotoxin and does not result from decreased ACh synthesis (Wu *et al.*, 1983).

Characteristics of nicotinic AChR may change in mutants, as is found in a nicotine-resistant strain of *Drosophila*, where the receptor's isoelectric

19

point is reported to be 6.69 for the nicotine-resistant strain but 6.60 for the wild type (Hall *et al.*, 1978). The responsible gene is X-chromosome linked, and so is the major nicotine-resistant factor (Hall, 1980). AChR from hybrid females has an intermediate isoelectric point (Hall *et al.*, 1978). However, this change in receptor structure appears not to be reflected in a change in receptor function. The gene coding for AChE is different from, and not adjacent to, those which code for the muscarinic AChR (Haim *et al.*, 1979) or the nicotinic AChR, since a mutant with reduced AChE activity has normal α-BGT binding (Dudai, 1978).

Muscarinic receptors

Muscarinic receptors differ radically from nicotinic receptors in being smaller, without an ion channel as part of its structure, in their dependence on other proteins to produce cellular responses, and in the mediation of their action by second messengers. Most studies have been conducted on vertebrate muscarinic receptors; thus they are the ones whose structure, function and pharmacology are best known and will be detailed below before comparison with invertebrate receptors is made.

Mechanism of action

The action of the muscarinic AChR can be stimulatory or inhibitory. The receptor inhibits adenylate cyclase but activates guanylate cyclase and phosphodiesterase. It closes K^+ channels in vertebrate cortical neurones and opens K^+ channels in the heart. These apparent anomalies are better understood now since the different mechanisms of receptor action have been resolved.

The muscarinic receptor is embedded in the plasma membrane, and when it binds ACh it transmits the information through the membrane and into the cell by means of a family of nucleotide binding proteins (called G or N proteins) which are active only if they bind GTP (Figure 1.2). The GTP activates an 'amplifier enzyme' on the inner surface of the membranes, which converts highly phosphorylated precursor molecules into second messengers. Two major signal pathways are responsible for amplifying and expressing the receptor function. In one, the receptor activates an inhibitory G_i protein, which in turn inhibits adenylate cyclase and the production of c-AMP from ATP. Stimulatory G_s proteins are activated by other receptors (e.g. β-adrenergic receptor) and stimulate adenylate cyclase. Intracellularly, c-AMP activates A-kinase, which phosphorylates other proteins. In the other pathway, the receptor activates a G_a protein, activating phosphodiesterase, which splits the components of the membrane phosphatidylinositol 4,5-bisphosphate (PIP_2) to inositol trisphosphate (IP_3) and diacylglycerol (DG). The IP_3 release Ca^{2+} ions from non-mito-

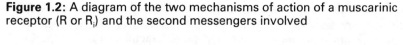

Figure 1.2: A diagram of the two mechanisms of action of a muscarinic receptor (R or R$_i$) and the second messengers involved

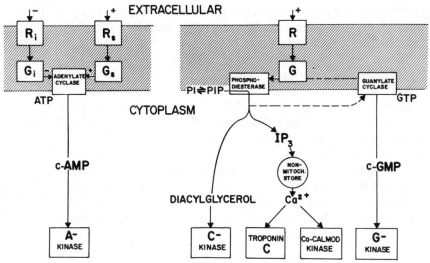

chondrial internal stores and later by influx from outside the cell. The third messenger Ca^{2+} then binds to calmodulin, activating a Ca^{2+}-calmodulin kinase or binds to troponin C, stimulating muscle contraction. DG activates C-kinase, which requires Ca^{2+} for its activity. The diglyceride is converted to phosphatidic acid by the addition of a phosphate group and then to the cytosine nucleotide derivative (CDP-DG). Inositol (I) and CDP-DG combine to produce phosphatidylinositol (PI). The latter is phosphorylated to PIP which regenerates PIP$_2$ (see Abdel-Latif *et al.*, 1985; Berridge, 1985). Another possible subpathway is activation of a guanylate cyclase by the G protein or phosphodiesterase, producing c-GMP which stimulates a G-kinase. These pathways are general for stimulation of Ca^{2+}-linked receptors.

Binding of the protein to the receptor increases dramatically the receptor's affinity for agonists. This is reversed by GTP binding to the G protein, and the agonist dissociates from the receptor. A solubilised receptor has only a low affinity for agonists (McMahon and Hosey, 1985), but its reconstitution into phospholipids to which is added a pure G protein or into lysed erythrocyte membrane containing a G protein recovers the receptor's high affinity and GTP effect (Florio and Sternweis, 1985). Long exposure to agonists causes muscarinic receptor desensitisation, as shown by their decreased capacity to inhibit adenylate cyclase in rat striatum (Olianas *et al.*, 1984) and ligand binding in neuroblastoma cells (clone N1E-115) (Feigenbaum and El-Fakahany, 1985).

21

myo-[2-^{3}H]Inositol was shown to be taken up and metabolised by pieces of the metathoracic ganglion of *Schistocerca gregaria* into both lipid- and acid-soluble products (Trimmer and Berridge, 1985). Although exogenous agonists had no effect on the levels of inositol phosphates, they were altered dramatically by atropine, suggesting the presence of a class of muscarinic-like AChRs that are linked to phosphatidylinositol metabolism.

In one study, two kinds of adenylate cyclase were detected in *Drosophila*, one sensitive and one insensitive to Ca^{2+}/calmodulin, with the former being less responsive to guanyl nucleotides. The two forms were distinguished genetically (Livingstone, 1985) and are similar to what was reported for mammalian brain adenylate cyclase. Furthermore, in *Drosophila* head membranes, the phosphorylation of at least seven proteins was enhanced in the presence of c-AMP, but a 56000-dalton c-AMP-binding phosphoprotein was dephosphorylated and its conformation changed. It was suggested that this protein is very similar to the regulatory subunit (RII) of a mammalian c-AMP-dependent protein kinase, and that its binding to c-AMP makes it very susceptible to the action of phosphatases.

Structure

Muscarinic receptors, like nicotinic receptors, are glycoproteins (Rauh *et al.*, 1986). There is substantial heterogeneity in the reported molecular sizes of muscarinic receptor subunits ranging from a broad band of subunits of 66000-75000 daltons (Dadi and Morris, 1984), 70000-75000 daltons (Ruess and Lieflander, 1979) and 80000-86000 daltons (Birdsall *et al.*, 1983). The purified muscarinic AChR from porcine brain is reported to have a major subunit of 70000 and several minor ones (Haga *et al.*, 1985). The porcine heart receptor is reported to be made of subunits of 75000 daltons (Herron and Schimerlik, 1983) or two subunits of 78000 and 14000 daltons (Peterson *et al.*, 1984). Only one subunit of 80000 daltons was obtained from human, rat, dog and *Drosophila* brains, dog and rat heart and guinea pig ileum muscle; the receptors had similar isoelectric points (5.9), and the antibody against the rat brain receptor cross-reacted with the other receptors (Venter *et al.*, 1984). The range of molecular weights reported for subunits suggests that species- and/or tissue-specific heterogeneity exists among these receptors, that there may be proteolytic activity in some preparations, or different degrees of removal of attached phospholipids or that there is substantial heterogeneity among muscarinic receptor structures (Hootman *et al.*, 1985).

Drug specificities

Although different muscarinic receptors have similar affinities for atropine or QNB, they differ in their affinities for agonists and the antagonist pirenzepine, which has a 50-fold higher affinity for vertebrate muscarinic receptors in cerebral cortex, hippocampus and autonomic ganglia than for heart, hindbrain, smooth muscle and parasynaptic receptors (Hammer *et al.*, 1980). This led to the hypothesis that there are subtypes of muscarinic receptors in vertebrates, M_1 and M_2, which have high and low affinities, respectively, for pirenzepine (see Birdsall and Hulme, 1983; Watson *et al.*, 1984). The M_1 receptor is suggested to be independent of a guanine nucleotide regulatory protein because GTP does not regulate [^{3}H]piren- zepine binding (Watson *et al.*, 1983); possibly, then, the M_1 receptor mediates phosphoinositide hydrolysis and Ca^{2+} mobilisation. The M_2 receptor is suggested to regulate adenylate cyclase because it predominates in rat heart (Hammer *et al.*, 1980). However, pirenzepine is less effective on receptors regulating phosphoinositide turnover than on those regulating c-AMP metabolism in embryonic chick heart cells, which also have a high affinity for pirenzepine. Thus, the muscarinic receptor in chick heart is predominantly M_1 (Brown *et al.*, 1985), in contrast to that of the rat heart.

An alternative hypothesis is that the difference between M_1 and M_2 receptors is a difference in materials that adhere to the receptor. Thus, subtypes reflect interconvertibility of binding sites, which is affected by the membrane environment and ligand-induced conformational changes (Schreiber and Sokolovsky, 1985). Chick heart receptors can exist *in vitro* in three agonist affinity states, which are interconverted by Gpp(NH)p and Mg^{2+}, involving interaction with one or more associated proteins (McMahon and Hosey, 1985). However, interconvertibility of low and high agonist affinities (Ikeda *et al.*, 1980) has not been shown to transmute permanently one receptor subtype to the other. Also, both M_1 and M_2 receptors possess multiple agonist affinities (Watson *et al.*, 1986), and most M_2 sites in rat brain have high affinities for [^{3}H]ACh (Keller *et al.*, 1985). Furthermore, the heterogeneity of affinities of brain muscarinic receptors for pirenzepine is maintained after solubilisation with digitonin (Luthin and Wolfe, 1985). Thus, the different affinities cannot be due to coupling with the G protein.

Gallamine has been shown to be a non-competitive allosteric inhibitor of muscarinic receptor binding of agonists and antagonists. Its effect is most marked on heart receptors, 40-fold more than that on lacrimal gland receptor, and may be absent in others (Stockton *et al.*, 1983).

Considering the complexity of the actions of muscarinic receptors and their drug specificities, almost all the biochemical studies of invertebrate muscarinic receptors have dealt only with binding of [^{3}H]QNB and the receptor's drug specificity. No studies have used [^{3}H]pirenzepine or searched for subtypes. There are indications that they may be present, since a Hill coefficient of less than 1 was reported for agonist inhibition of [^{3}H]QNB binding to the housefly brain receptor (Shaker and Eldefrawi, 1981), as reported for mammals.

Muscarinic responses were detected in the leech *Hirudo* CNS (Woodruff *et al.*, 1971) and body-wall muscles, gut and heart of earthworms *Lumbricus* (Kiefer, 1959), the gut of *Arenicola* (Wells, 1937), the stomatogastric ganglion of the crab *Cancer pagurus* (Marder and Paupardin-Tritsch, 1978) and the luminescent glands in the epithelium of *Chaetopterus* notopods (Anctil, 1981). These responses also control the fluorescence of elytra of the polynoid worm *Harmothoe lunulata* (Nicholas *et al.*, 1978).

A major difference in the invertebrate CNS receptors is that the majority of their AChRs are nicotinic rather than muscarinic as in mammalian brains. The density of muscarinic AChRs in insect brains ranges from 40 fmol/mg protein in the housefly (Shaker and Eldefrawi, 1981) to 1900 fmol/mg protein in the cricket *Acheta* (Table 1.4), compared with 1.7 pmol/mg protein in rat brain membranes (Aronstam *et al.*, 1979). However, these concentrations reflect not only density of receptors but degree of purity as well, since they are expressed per milligram of protein. In the CNS of the opisthobranch molluscs *Aplysia californica* and *Pleurobranchaea californica* and in homogenates of the nematode worm *Caenorhabditis elegans*, the concentrations of [^{3}H]QNB binding sites are also very low (Culotti and Klein, 1983; Murray *et al.*, 1985).

The affinities that invertebrate muscarinic receptors have for drugs, measured by competition with [^{3}H]QNB binding (Table 1.5), are generally similar to those of mammalian brain. For example, K_i values for rat brain binding of [^{3}H]QNB by scopolamine, atropine, pilocarpine and oxotremorine are 2×10^{-10} M, 3×10^{-9} M, 7×10^{-7} M and 7×10^{-7} M, respectively (Yamamura and Snyder, 1974). However, a higher affinity for the nicotinic antagonist *d*-tubocurarine is reported for the muscarinic receptors of housefly and *Drosophila* brains (Dudai and Ben-Barak, 1977; Shaker and Eldefrawi, 1981) and the *Caenorhabditis* receptor. The clear muscarinic pharmacology of the [^{3}H]QNB binding site in *Aplysia* ganglia (Murray *et al.*, 1985) does not necessarily contrast with the electrophysiological

Table 1.4: Characteristics of [^{3}H]QNB binding to invertebrate CNS membranes

Species and tissue	Average K_d (nM)	Maximal conc. of binding sites fmol/mg protein	Reference
Musca domestica (heads)	0.24	40	Shaker and Eldefrawi (1981)
Drosophila melanogaster (heads)	0.42	65	Haim *et al.* (1979)
Periplaneta americana (nerve cord)	8.0	138	Lummis and Sattelle (1985)
Locusta migratoria (heads)	0.77	116	Breer (1981b)
Acheta domesticus (ganglion)	7.5	1900	Meyer and Reddy (1985)
Aplysia californica (ganglion)	0.77	47	Murray *et al.* (1985)
Caenorhabditis elegans (body)	0.38	10	Culotti and Klein (1983)

Table 1.5: Comparative pharmacology of [³H]QNB specific binding to invertebrate CNS membranes (i.e. muscarinic receptors)

Drug	K_i or IC_{50} (M)						
	Drosophila[a]	Periplaneta[b]	Locusta[c]	Acheta[d]	Musca[e]	Aplysia[f]	Caenorhabditis[g]
Dexetimide	1×10^{-9}	5×10^{-9}	—	—	—	1×10^{-8}	—
Scopolamine	1×10^{-9}	5×10^{-9}	1×10^{-7}	—	7×10^{-9}	6×10^{-8}	2×10^{-9}
Atropine	4×10^{-9}	2×10^{-7}	4×10^{-7}	1×10^{-7}	3×10^{-9}	1×10^{-7}	2×10^{-10}
Pilocarpine	3×10^{-6}	1×10^{-5}	5×10^{-6}	—	3×10^{-5}	2×10^{-5}	—
Oxotremorine	9×10^{-6}	1×10^{-5}	—	6×10^{-5}	—	3×10^{-6}	1×10^{-5}
d-Tubocurarine	5×10^{-6}	5×10^{-6}	—	1×10^{-6}	7×10^{-6}	$>10^{-4}$	3×10^{-5}
Nicotine	—	$>10^{-3}$	$>10^{-2}$	1×10^{-4}	$>10^{-3}$	—	3×10^{-4}
Decamethonium	—	—	6×10^{-4}	—	3×10^{-4}	$>10^{-4}$	—

[a]K_i values (Haim et al., 1979).
[b]K_i values (Lummis and Sattelle, 1985).
[c]K_i values (Breer, 1981b).
[d]K_i values (Meyer and Reddy, 1985).
[e]K_i values (Shaker and Eldefrawi, 1981)
[f]IC_{50} of 1.4 nM [³H]QNB (Murray et al., 1985).
[g]K_i values (Culotti and Klein, 1983).

findings of apparent mixed nicotinic–muscarinic drug specificity of certain responses where atropine blocks nicotinic responses (Kehoe, 1972b). It should be noted that at high concentrations atropine inhibits the ionic channel of the vertebrate skeletal muscle nicotinic receptor (Adler *et al.*, 1978).

When vertebrate nicotinic or muscarinic receptors are exposed to high concentrations of agonists, two types of desensitisation change occur. One is acute desensitisation, which is rapid in onset, with recovery taking seconds to minutes. The other is chronic desensitisation, where the change takes hours to appear and recovery is slow and results from a reduction in receptor numbers (down-regulation) due to increased breakdown and possibly reduced synthesis. However, in an invertebrate, the nematode *Caenorhabditis*, the muscarinic receptor concentration was not altered by exposure to increased ACh levels (resulting from AChE inhibition) or by chronic administration of agonist (Culotti and Klein, 1983). Neither were there gross deficiencies in [^{3}H]QNB or [^{3}H]N-methylscopolamine binding in phenotypic revertants of AChE-deficient double mutants as well as a class of agonist-resistant mutants. Similarly, heterozygotes of a *Drosophila* mutant lacking AChE, which are viable and have about half or normal AChE activity, have the same concentration of [^{3}H]QNB binding sites as the normal strain (Haim *et al.*, 1979). It may be that invertebrates have not evolved the regulatory mechanism found in vertebrates which would reduce receptor numbers (down-regulation) upon excessive activation of the receptors, or that the change occurred but was too small to be detected (Culotti and Klein, 1983).

Nicotinic–musacarinic binding protein

In addition to the 'typical' nicotinic and muscarinic receptors identified in insect brains, a protein was discovered in housefly brain extract which bound nicotinic as well as muscarinic drugs but not α-BGT, with the highest affinity for decamethonium (K_D 0.15 μM) (Eldefrawi and O'Brien, 1970). Two protein forms were purified and separated by electrophoresis (Mansour *et al.*, 1977) with different pIs (4.7 and 4.9) (Harris *et al.*, 1981). Unlike the nicotinic receptor, they were solubilised without the aid of detergents and had a high content of hydrophilic amino acids. Although the identity and *in vivo* function of these proteins is still unknown, their high concentration in an insect brain, but not vertebrate brains, and their binding of almost all agonists and antagonists of nicotinic and muscarinic receptors, warrant further studies.

CONCLUDING REMARKS

It is obvious that although our basic understanding of the neurobiological role of ACh has not changed much since the days of Sir Henry Dale in the early 1940s, our knowledge of the molecular properties of the proteins and processes that are involved in cholinergic transmission has grown steadily. Probably the most impressive expansion in our understanding came from studies on cholinergic receptors. It is also this area of scientific investigation that promises future fruitful practical applications. Not only are muscarinic AChRs involved in memory and learning, but loss of cholinergic neurones in basal ganglia in mammalian brain is associated with Alzheimer's disease, which is one of the worst neurological diseases of old age. Investigations on the role of ACh in invertebrate model neuronal circuits are providing valuable information to help unravel the mysteries of the more complicated mammalian brain.

Our knowledge of the pharmacological properties of AChRs in different species should be extremely valuable. For example, infections by nematodes or roundworms that parasitise the gastrointestinal tract of mammals are usually treated by anticholinergic drugs that have a low therapeutic index and are toxic to the mammalian host as well. Knowledge of the special properties of AChRs of roundworms and nematodes and how they differ from those of their mammalian host should lead to the development of more selective therapeutics. Another area of potential benefit is in insect pest control. The majority of insecticides in use today (e.g. the organophosphate and carbamate anticholinesterases) are mostly indirect inhibitors of AChRs. They inhibit ChEs, allowing ACh to accumulate at cholinergic synapses, which leads to a transient stimulation followed by receptor desensitisation. Most of these chemicals are non-selective poisons and cause countless environmental problems. Differences in AChRs between vertebrates and insect pests can be utilised for the development of insect-selective and environmentally safe, effective control agents.

REFERENCES

Abdel-Latif, A.A., Smith, J.P. and Akhtar, R.A. (1985) Polyphosphoinositides and muscarinic cholinergic and α_1-adrenergic receptors in the iris smooth muscle. In: Bleasdale, J.E., Eichberg, J. and Hauser, G. (eds) *Inositol and phosphoinositides: metabolism and regulation*, pp. 275-98. Humana Press, Clifton, NJ

Adams, P.R. and Sakmann, B. (1978) Decamethonium both opens and blocks endplate channels. *Proc. Natl. Acad. Sci. USA 75*, 2994-8

Adler, M., Albuquerque, E.X. and Lebeda, F.J. (1978) Kinetic analysis of end plate currents altered by atropine and scopolamine. *Molec. Pharmacol. 14*, 514-29

Albuquerque, E.X., Tsai, M.-C., Aronstam, R.S., Witkop, B., Eldefrawi, A.T. and Eldefrawi, M.E. (1980) Phencyclidine interactions with ionic channels of the acetylcholine receptor and electrogenic membranes. *Proc. Natl. Acad. Sci. USA* 77, 1224-8

Aldridge, W.N. (1953) The differentiation of true and pseudocholinesterase by organophosphorus compounds. *Biochem J. 53* 62-7

Anctil, M. (1981) Luminescence control in isolated notopods of the tubeworm *Chaetopterus variopedatus*: effects of cholinergic and GABAergic drugs. *Comp. Biochem. Physiol. 68C*, 187-94

Anderson, D.C., King, S.C. and Parsons, S.M. (1983) Pharmacological characterization of the acetylcholine transport system in purified *Torpedo* electric organ synaptic vesicles. *Molec. Pharmacol. 24*, 48-54

Anderson, M. and Mrose, H. (1976) Chemical excitation of the proventriculus of the polychaete worm, *Syllis spongiphila. J. exp. Biol. 75*, 113-22

Aoshima, H., Cash, D. and Hess, G.P. (1981) Mechanism of inactivation (desensitization) of acetylcholine receptor. Investigations by fast reaction techniques with membrane vesicles. *Biochemistry 20*, 3467-73

Aracava, Y. and Albuquerque, E.X. (1984) Meproadifen enhances activation and desensitization of the acetylcholine receptor-ionic channel complex (AChR): single channel studies. *FEBS Lett. 174*, 267-74

Aronstam, R.S., Triggle, D.J. and Eldefrawi, M.E. (1979) Structural and stereochemical requirements for muscarinic receptor binding. *Molec. Pharmacol. 15*, 227-34

Arpagaus, M. and Toutant, J.-P. (1985) Polymorphism of acetylcholinesterase in adult *Pieris brassicae* heads. Evidence for detergent-insensitive and Triton X-100-interacting forms. *Neurochem. Int. 1*, 793-804

Artola, A., Callec, J.J., Hue, B., David, J.A. and Sattelle, D.B. (1984) Actions of amantadine at synaptic and extrasynaptic cholinergic receptors in the central nervous system of the cockroach *Periplaneta americana. J. Insect Physiol. 30*, 185-90

Ascher, P., Marty, A. and Neild, T.O. (1978) The mode of action of antagonist of excitatory response to acetylcholine in *Aplysia* neurones. *J. Physiol. (Lond.) 278*, 207-35

Badamchian, M. and Carroll, P.T. (1985) Molecular weight determinations of soluble and membrane-bound fractions of choline o-acetyltransferase in rat brain. *J. Neurosci. 5*, 1955-64

Barker, D.L., Herbert, E., Hildebrand, J.G. and Kravitz, E.A. (1972) Acetylcholine and lobster sensory neurones. *J. Physiol. (Lond.) 226*, 205-29

Barker, L.R., Bueding, E. and Timms, A.R. (1966) The possible role of acetylcholine in *Schistosoma mansoni. Br. J. Pharmacol. 26*, 656-65

Benishin, C.G. and Carroll, P.T. (1981) Differential sensitivity of soluble and membrane-bound forms of choline-O-acetyltransferase to inhibition by coenzyme A. *Biochem. Pharmacol. 30*, 2483-4

Beranek, A.P. (1974) Esterase variation and organophosphate resistance in populations of *Aphis fabae* and *Myzus persicae. Entom. Exp. Appl. 17*, 129-42

Berridge, M.J. (1985) The molecular basis of communication within the cell. *Sci. Amer. 253*, 142-52

Betz, H. and Pfeiffer, F. (1984) Monoclonal antibodies against the α-bungarotoxin-binding protein of chick optic lobe. *J. Neurosci. 4*, 2095-105

Bevan, S. and Steinbach, J.H. (1977) The distribution of α-bungarotoxin binding sites on mammalian skeletal muscle developing *in vivo. J. Physiol. (Lond.) 267*, 195-213

Birdsall, N.J.M. and Hulme, E.C. (1983) Muscarinic receptor subclasses. *Trends Pharmacol. Sci. 4*, 459-63

Birdsall, N.J.M., Hulme, E.C. and Stockton, J.M. (1983) Muscarinic receptor heterogeneity . *Trends Pharmacol. Sci. Suppl.* In: Hirschowitz, B.I., Hammer, R., Giachetti, A., Keirns, J.J. and Levine, R.R. (eds) *Proc. Int. Symp. on Subtypes of Muscarinic Receptors,* pp. 4-8. Elsevier, Amsterdam

Blagburn, J.M., Beadle, D.J. and Sattelle, D.B. (1985) Development of chemosensitivity of an identified insect interneurone. *J. Neurosci. 5,* 1167-75

Bon, S. and Massoulie, J. (1976) Collagen-tailed and hydrophobic components of acetylcholinesterase in *Torpedo marmorata* electric organ. *Proc. Natl. Acad. Sci. USA 77,* 4464-8

Breer, H. (1981a) Characterization of synaptosomes from the central nervous system of insects. *Neurochem. Int. 3,* 155-63

Breer, H. (1981b) Properties of putative nicotinic and muscarinic cholinergic receptors in the central nervous system of *Locusta migratoria. Neurochem. Int. 3,* 43-52

Breer, H. (1982) Uptake of [N-Me-^{3}H] choline by synaptosomes from the central nervous system of *Locusta migratoria. J. Neurobiol. 13,* 107-17

Breer, H. (1983) Choline transport by synaptosomal membrane vesicles isolated from insect nervous tissue. FEBS Lett. *153,* 345-8

Breer, H. and Benke, D. (1985) Synthesis of acetycholine receptors in *Xenopus* oocytes induced by poly (A)$^+$-mRNA from locust nervous tissue. *Naturwiss. 72,* 213-14

Breer, H. and Jeserich, G. (1984) Invertebrate synaptosomes — implications for comparative neurochemistry. In: *Current topics in research on synapses 1,* pp. 165-210. Alan Liss, New York

Breer, H., Kleene, R. and Hinz, G. (1985) Molecular forms and subunit structure of the acetylcholine receptor in the central nervous system of insects. *J. Neurosci. 5,* 3386-92

Breer, H. and Knipper, M. (1985) Effects of neurotoxins on the high affinity translocation of choline in synaptosomal membrane vesicles from insects. *Comp. Biochem. Physiol. 81C,* 219-22

Breer, H. and Lueken, W. (1983) Transport of choline by membrane vesicles prepared from synaptosomes of insect nervous tissue. *Neurochem. Int. 5,* 713-20

Bregestovski, P.D., Bukharaeva, E.A. and Iljin, V.I. (1979) Voltage clamp analysis of acetylcholine receptor desensitization in isolated mollusc neurones. *J. Physiol. 297,* 581-95

Brisson, A. and Unwin, P.N.T. (1985) Quaternary structure of the acetylcholine receptor. *Nature (Lond.) 315,* 474-7

Brown, J.H., Goldstein, D. and Masters, S.B. (1985) The putative M$_1$ muscarinic receptor does not regulate phosphoinositide hydrolysis. Studies with pirenzepine and McN-A343 in chick heart and astrocytoma cells. *Molec. Pharmacol. 27,* 525-31

Brownstein, M., Saavedra, J.M., Axelrod, J., Zeman, G.H. and Carpenter, D.O. (1974) Coexistence of several putative neurotransmitters in single identified neurons of *Aplysia. Proc. Natl. Acad. Sci. USA 71,* 4662-5

Buchner, E. and Rodrigues, V. (1983) Autoradiographic localization of [^{3}H]choline uptake in the brain of *Drosophila melanogaster. Neurosci. Lett. 42,* 25-31

Carbonetto, S.T., Fambrough, D.M. and Muller, K.J. (1978) Non-equivalence of *o*-bungarotoxin receptors and acetylcholine receptors in chick sympathetic neurons. *Proc. Natl. Acad. Sci. USA 75,* 1016-20

Chad, J.E., Kerkut, G.A. and Walker, R.J. (1979) Ramped voltage-clamp study of the action of acetylcholine on three types of neurons in the snail (*Helix aspersa*) brain. *Comp. Biochem. Physiol. 63C,* 269-78

Chadwick, L.E. (1963) Actions on insects and other invertebrates. In: Koelle, G.B. (ed.) *Cholinesterases and anticholinesterase agents*, pp. 741-98. Springer Verlag, Berlin

Chance, M.R.A. and Mansour, T.E. (1953) A contribution to the pharmacology of movement in the liver fluke. *Br. J. Pharmacol. 8*, 134-8

Chang, Y.C. (1975) The end plate graded potentials from the neuromuscular system of the earthworm, *Pheretima hawayana* R. *Comp. Biochem. Physiol. 51A*, 237-40

Changeux, J.-P., Devillers-Thiery, A. and Chemouilli, P. (1984) Acetylcholine receptor: an allosteric protein. *Science 225*, 1335-45

Chaudhary, K.D., Srivastava, V. and Lemonde, A. (1966) Acetylcholinesterase in *Tribolium confusum* Duval. *Arch. Int. Physiol. Biochim. 74.*, 416-28

Clarke, B.S. and Donnellan, J.F. (1982) Concentrations of putative neurotransmitters in the CNS of quick-frozen insects. *Insect Biochem. 12*, 623-38

Conti-Tronconi, B.M., Dunn, S.M.J., Barnard, E.A., Dolly, J.O., Lai, F.A., Ray, N. and Raftery, M.A. (1985) Brain and muscle nicotinic acetycholine receptors are different but homologous proteins. *Proc. Natl. Acad. Sci. USA 82*, 5208-12

Cottrell, G.A. and Powell, B. (1970) Choline acetyltransferase in the snail brain. *Comp. gen. Pharmacol. 1*, 251-3

Crawford, G., Slemmon, J.R. and Salvaterra, P.M. (1982) Monoclonal antibodies selective for *Drosophila melanogaster* choline acetyltransferase. *J. Biol. Chem. 257*, 3853-6

Culotti, J.G. and Klein, W.L. (1983) Occurrence of muscarinic acetylcholine receptors in wild type and cholinergic mutants of *Caenorhabditis elegans. J. Neurosci. 3*, 359-68

Dadi, H.K. and Morris, R.J. (1984) Muscarinic cholingeric receptor of rat brain. Factors influencing migration in electrophoresis and gel filtration in sodium dodecyl sulphate. *Eur. J. Biochem. 144*, 617-28

Dauterman, W.C. and Mehrotra, K.N. (1963) The *N*-alkyl group specificity of cholinesterase from the house-fly, *Musca domestica* L. and the two-spotted spider mite, *Tetraychus telarius* L. *J. Insect Physiol. 9*, 257-63

David, J.A., Crowley, P.J., Hall, S.G., Battersby, M. and Sattelle, D.B. (1984) Action of synthetic piperidine derivatives on an insect acetylcholine receptor/ion channel complex. *J. Insect Physiol. 30*, 191-6

David, J.A. and Sattelle, D.B. (1984) Actions of cholinergic pharmacological agents on the cell body membrane of the fast coxal depressor motoneurone of the cockroach (*Periplaneta americana*). *J. exp. Biol. 108*, 119-36

Denburg, J.L., Eldefrawi, M.E. and O'Brien, R.D. (1972) Macromolecules from lobster axon membranes that bind cholinergic ligands and local anesthetics. *Proc. Natl. Acad. Sci. USA 69*, 177-81

Dowdall, M.J. and Whittaker, V.P. (1973) Comparative studies in synaptosome formation: the preparation of synaptosomes from the head ganglion of the squid, *Loligo pealli. J. Neurochem. 20*, 921

Driskell, W.J., Weber, B.H. and Roberts, E. (1978) Purification of choline acetyltransferase from *Drosophila melanogaster. J. Neurochem. 30*, 1135-41

Ducis, I. and Whittaker, V.P. (1985) High-affinity, sodium-gradient-dependent transport of choline into vesiculated presynaptic plasma membrane fragments from the electric organ of *Torpedo marmorata* and reconstitution of the solubilized transporter into liposomes. *Biochim. Biophys. Acta 815*, 109-27

Dudai, Y. (1978) Properties of an α-bungarotoxin-binding cholinergic nicotinic receptor from *Drosophila melanogaster. Biochim. Biophys. Acta 539*, 505-17

Dudai, Y. and Ben-Barak, J. (1977) Muscarinic receptor in *Drosophila melano-*

gaster demonstrated by binding of [³H]-quinuclidinyl benzilate. *FEBS Lett. 81,* 134-6

Eckenstein, F. and Sofroniew, M.V. (1983) Identification of central cholinergic neurons containing both choline transferase and acetylcholinesterase and of central neurons containing only acetylcholinesterase. *J. Neurosci. 3,* 2286-91

Eldefrawi, A.T. (1984) Acetylcholinesterases and anticholinesterases. In: Kerkut, G.A. and Gilbert, L.I. (eds) *Comprehensive insect physiology, biochemistry and pharmacology,* pp. 115-30. Pergamon Press, Oxford

Eldefrawi, M.E. and Eldefrawi, A.T. (1980) Putative acetylcholine receptors in housefly brain. In: Sattelle, D.B., Hall, L.M. and Hildebrand, J.G. (eds) *Receptors for neurotransmitters, hormones and pheromones in insects,* pp. 59-70. Elsevier North Holland, Biomed. Press, Amsterdam

Eldefrawi, M.E. and Eldefrawi, A.T. (1987) Nervous system based insecticides. In: Hodgson, E. and Kuhr, R.J. (eds) *Safer insecticides: development and use,* Marcel Dekker, New York

Eldefrawi, A.T., Miller, E.R. and Eldefrawi, M.E. (1982) Binding of depolarizing drugs to the ionic channel sites of the nicotinic acetylcholine receptor. *Biochem. Pharmacol. 31,* 1819-22

Eldefrawi, A.T. and O'Brien, R.D. (1970) Binding of muscarone by extracts of housefly brain: relationship to receptors for acetylcholine. *J. Neurochem. 17,* 1287-93

Eldefrawi, M.E., Tripathi, R.K. and O'Brien, R.D. (1970) Acetylcholinesterase isozymes from the house-fly brain. *Biochim. Biophys. Acta 212,* 308-14

Eldefrawi, M.E., Warnick, J.E., Schofield, G.G., Albuquerque, E.X. and Eldefrawi, A.T. (1981) Interaction of imipramine with the ionic channel of the acetylcholine receptor of motor endplate and electric organ. *Biochem. Pharmacol. 30,* 1391-4

El-Fakahany, E.F., Eldefrawi, A.T. and Eldefrawi, M.E. (1982) Nicotinic acetyl-choline receptor densensitization studied by [³H]perhydrohistrionicotoxin binding. *J. Pharmacol. Exp. Ther. 221,* 694-700

Emson, P.C., Malthe-Sorenssen, D. and Fonnum, F. (1974) Purification and properties of choline acetyltransferase from the nervous system of different invertebrates. *J. Neurochem. 22,* 1089-98

Erzen, I. and Brzin, M. (1979) Cholinergic mechanisms in *Planaria torva. Comp. Biochem. Physiol. 64C,* 207-16

Feigenbaum, P. and El-Fakahany, E.E. (1985) Regulation of muscarinic cholinergic receptor density in neuroblastoma cells by brief exposure to agonist: possible involvement in desensitization of receptor function. *J. Pharmacol. Exp. Ther. 233,* 134-40

Finer-Moore, J. and Stroud, R.M. (1984) Amphipathic analysis and possible formation of the ion channel in an acetylcholine receptor. *Proc. Natl. Acad. Sci. USA 81,* 155-9

Fitzpatrick-McElligott, S. and Stent, G.S. (1981) Appearance and localization of acetylcholinesterase in embryo of the leech *Helobdella triserialis. J. Neurosci. 1,* 901-7

Florey, E. (1973) Acetylcholine as sensory transmitter in crustacean. *J. Comp. Physiol. 83,* 1-16

Florey, E. and Florey, E. (1965) Cholinergic neurones in the Onychophora: a comparative study. *Comp. Biochem. Physiol. 15,* 125-36

Florio, V.A. and Sternweis, P.C. (1985) Reconstitution of resolved muscarinic cholinergic receptors with purified GTP-binding proteins. *J. Biol. Chem. 260,* 3477

Fluck, R.A. and Jaffe, M.J. (1975) Cholinesterases from plant tissues. *Biochim.*

31

Biophys. Acta 410, 130-4

Frank, E. and Fischbach, G.D. (1977) ACh receptors accumulate at newly formed nerve–muscle synapses *in vitro*. In: Lash, J.W. and Burger, M.M. (eds) *Cell and tissue interactions*, pp. 285-92. Raven Press, New York

Frontali, N., Piazza, R. and Scopelliti, R. (1971) Localization of acetylcholinesterase in the brain of *Periplaneta americana*. *J. Insect Physiol. 17*, 1833-42

Fulton, B.P. (1982) Presynaptic acetylcholine receptors at the excitatory amino acid synapse in locust muscle. *Neuroscience 7*, 2117-24

Gardner, C.R. and Cashin, C.H. (1975) Some aspects of monoamine function in the earthworm, *Lumbricus terrestris*. *Neuropharmacology 14*, 493-500

Gardner, C.R. and Walker, R.J. (1982) The roles of putative neurotransmitters and neuromodulators in annelids and related invertebrates. *Progr. Neurobiol. 18*, 81-120

Geffard, M., Vieillemaringe, J., Heinrich-Rock, A.-M. and Duris, P. (1985) Anti-acetylcholine antibodies and first immunocytochemical application in insect brain. *Neurosci. Lett. 57*, 1-6

Gepner, J.I., Hall, L.M. and Sattelle, D.B. (1978) Insect acetylcholine receptors as a site of insecticide action. *Nature (Lond.) 276*, 188-90

Giller, E., Jr and Schwartz, J.H. (1971) Choline acetyltransferase in identified neurons of abdominal ganglion of *Aplysia californica*. *J. Neurophysiol. 34*, 93-107

Goodman, C.S. and Spitzer, N.C. (1979) Embryonic development of identified neurones: differentiation from neuroblast to neurone. *Nature (Lond.) 280*, 208-14

Goodman, C.S. and Spitzer, N.C. (1980) Embryonic development of neurotransmitter receptors in grasshoppers. In: Sattelle, D.B., Hall, L.M. and Hildebrand, J.G. (eds) *Receptors for neurotransmitters, hormones and pheromones in insects*, pp. 195-207. Elsevier North-Holland Biomedical Press, Amsterdam

Greenspan, R.J. (1980) Mutation of choline acetyltransferase and associated neural defects in *Drosophila melanogaster*. *J. Comp. Physiol. 137*, 83-92

Gunderson, C.B., Katz, B. and Miledi, R. (1982) The antagonism between botulinum toxin and calcium in motor nerve terminals. *Proc. Roy. Soc. Lond. B216*, 369-76

Haga, K., Haga, T., Ichiyama, A., Katada, T., Kurose, H. and Ui, M. (1985) Functional reconstitution of purified muscarinic receptors and inhibitory guanine nucleotide regulatory protein. *Nature (Lond.) 316*, 731-3

Haim, N., Nahum, S. and Dudai, Y. (1979) Properties of a putative muscarinic cholinergic receptor from *Drosophila melanogaster*. *J. Neurochem. 32*, 543-52

Hall, J.C. and Kankel, D.R. (1976) Genetics of acetylcholinesterase in *Drosophila melanogaster*. *Genetics 83*, 517-35

Hall, L.M. (1980) Biochemical and genetic analysis of an *o*-bungarotoxin-binding receptor from *Drosophila melanogaster*. In: Sattelle, D.B., Hall, L.M. and Hildebrand, J.G. (eds) *Receptors for neurotransmitters, hormones and pheromones in insects*, pp. 111-24. Elsevier North-Holland, Amsterdam

Hall, L.M. and Teng, N.N.H. (1975) Localization of acetylcholine receptors in *Drosophila melanogaster*. In: McMahon, D. and Fox, C.F. (eds) *Developmental biology — pattern formation — gene regulation*, pp. 282-9 Benjamin, Menlo Park, CA

Hall, L.M., Von Borstel, R.W., Osmond, B.C., Hoeltzli, S.D. and Hudson, T.H. (1978) Genetic variants in an acetylcholine receptor from *Drosophila melanogaster*. *FEBS Lett. 95*, 243-6

Hammer, R., Berrie, C.P., Birdsall, N.J.M., Burgen, A.S.V. and Hulme, E.C.

32

(1980) Pirenzepine distinguishes between different subclasses of muscarinic receptors. *Nature (Lond.) 283*, 90-2

Harris, H. (1975) *The principles of human biochemical genetics.* Elsevier North-Holland, Amsterdam

Harris, R., Cattell, K.J. and Donnellan, J.F. (1981) The purification and molecular characterisation of a putative nicotinic muscarinic acetylcholine receptor from housefly heads. *Insect Biochem. 11*, 371-85

Harvey, A.L. and Karlsson, E. (1984) Polypeptide neurotoxins from mamba venoms that facilitate transmitter release. *Trends Pharmacol. Sci.* Feb., 71-2

Herron, G.S. and Schimerlik, M.I. (1983) Glycoprotein properties of the solubilized atrial muscarinic acetylcholine receptor. *J. Neurochem. 41*, 1414-20

Hersh, L.B., Wainer, B.H. and Andrews, L.P. (1984) Multiple isoelectric and molecular weight variants of choline acetyltransferase: artifact or real? *J. Biol. Chem. 259*, 1253-8

Hildebrand, J.G., Hall, L.M. and Osmond, B. (1979) Distribution of binding sites for ^{125}I-labeled α-bungarotoxin in normal and deafferented antennal lobes of *Manduca sexta. Proc. Natl. Acad. Sci. USA 76*, 499-503

Hootman, S.R., Picado-Leonard, T.M. and Burnham, D.B. (1985) Muscarinic acetylcholine receptor structure in acinar cells of mammalian exocrine glands. *J. Biol. Chem. 260*, 4186-94

Houk, E.J., Hardy, J.L. and Cruz, W.J. (1981) Acetylcholinesterases of the mosquito *Culex tarsalis* Coquillett. *Comp. Biochem. Physiol. 69C*, 117-20

Husain, S.S. and Mautner, H.G. (1973) Purification of choline acetyltransferase of squid head ganglia. *Proc. Natl. Acad. Sci. USA 70*, 3749-53

Ikeda, S.R., Aronstam, R.S. and Eldefrawi, M.E. (1980) Nature of regional and chemically-induced differences in the binding properties of muscarinic acetyl-choline receptors from rat brain. *Neuropharmacology 19*, 575-85

Ito, Y., Kuriyama, H. and Tashiro, N. (1970) Effects of catecholamines on the neuromuscular junction of the somatic muscle of the earthworm. *J. exp. Biol. 54*, 167-86

Jaffe, M.J. (1970) Evidence for the regulation of phytochrome-mediated processes in bean roots by the neurohumor, acetylcholine. *Plant Physiol. 46*, 768-77

Johnson, C.D. and Stretton, A.O.W. (1985) Localization of choline acetyltrans-ferase within identified motoneurons of the nematode *Ascaris. J. Neurosci. 5*, 1984-92

Jones, S.W., Galasso, R.T. and O'Brien, R.D. (1977) Nicotine and α-bungarotoxin binding to axonal and non-neural tissue. *J. Neurochem. 29*, 803-9

Jones, S.W., Sudershan, P. and O'Brien, R.D. (1981) α-Bungarotoxin binding in house fly heads and *Torpedo* electroplax. *J. Neurochem. 36*, 447-53

Jope, R.S. and Johnson, V.W. (1986) Quinacrine and 2-(4-phenylpiperidino)cyclo-hexanol (AH5183) inhibit acetylcholine release and synthesis in rat brain slices. *Molec. Pharmacol. 29*, 45-51

Kandel, E.R. (1979) Cellular insights into behavior and learning. *Harvey Lect. Ser. 73*, 19-92

Karlin, A. (1983) The anatomy of a receptor. *Neurosci. Commen. 1*, 111-23

Katz, B. and Miledi, R. (1973) The characteristics of 'end-plate noise' produced by different depolarizing drugs. *J. Physiol. (Lond.) 230*, 707-17

Katz, B. and Miledi, R. (1978) A re-examination of curare action at the motor endplate. *Proc. Roy. Soc. Lond. B 203*, 119-33

Kehoe, J. (1972a) Ionic mechanisms of a two-component cholinergic inhibition in *Aplysia* neurones. *J. Physiol. (Lond,) 225*, 85-114

Kehoe, J. (1972b) Three acetylcholine receptors in *Aplysia* neurones. *J. Physiol.*

(Lond.) 225, 115-46

Kehoe, J., Sealock, R. and Bon, C. (1976) Effects of α-toxins from *Bungarus multicinctus* and *Bungarus caeruleus* on cholinergic responses in *Aplysia* neurons. *Brain Res. 107*, 527-40

Keller, K.J., Martino, A.M., Hall, D.P., Schwartz, R.D. and Taylor, R.L. (1985) High affinity binding of [³H]acetylcholine to muscarinic cholinergic receptors. *J. Neurosci. 5*, 1577-82

Kelly, L.E. (1981) The regulation of protein phosphorylation in synaptosomal fractions from *Drosophila* heads: the role of cyclic adenosine monophosphate and calcium/calmodulin. *Comp. Biochem. Physiol. 69B*, 61-7

Kemp, G., Bentley, L., McNamee, M.G. and Morley, B.J. (1985) Purification and characterization of the α-bungarotoxin binding protein from rat brain. *Brain Res. 347* 274-83

Kerkut, G.A., Pitman, R.M. and Walker, R.J. (1969) Ionophoretic application of acetylcholine and GABA onto insect central neurones. *Comp. Biochem. Physiol. 31*, 611-33

Kiefer, G. (1959) Pharmakologische Untersuchungen über den Automatismus der Lateralherzen des Regenwurmes *Lumbricus terrestris. Z. Wiss. Zool. 162*, 357-66

Kubo, T., Noda, M., Takai, T., Tanabe, T., Kayano, T., Shimizu, S., Tanaka, K.-I., Takahashi, H., Hirose, T., Inayama, S., Kikuno, R., Miyata, T. and Numa, S. (1985) Primary structure of delta subunit precursor of calf muscle acetylcholine receptor deduced from cDNA sequence. *Eur. J. Biochem. 149*, 5-13

Kuffler, D.P. (1978) Neuromuscular transmission in longitudinal muscles of the leech *Hirudo medicinalis. J. Comp. Physiol. 124*, 333-8

Kupfer, C. and Koelle, G.B. (1951) A histochemical study of cholinesterase during formation of the motor endplate of the albino rat. *J. Exp. Zool. 116*, 397-413

Lees, G., Beadler, D.J. and Botham, R.P. (1983) Cholinergic receptors on cultured neurones from the central nervous system of embryonic cockroaches. *Brain Res. 288*, 49-59

Lentz, T.L. (1966) Histochemical localization of neurohumors in a sponge. *J. Exp. Zool. 162*, 171-80

Lester, D.S. and Gilbert, L.I. (1985) Choline acetyltransferase activity in the larval brain of *Manduca sexta. Insect Biochem. 15*, 685-94

Lester, H.A., Changeux, J.-P. and Sheridan, R.E. (1975) Conductance increases produced by bath application of cholinergic agonists to *Electrophorus electricus* electroplaques. *J. gen. Physiol. 65*, 797-816

Levitan, H. and Tauc, L. (1972) Acetylcholine receptors: topographic distribution and pharmacological properties of two receptor types on a single molluscan neurone. *J. Physiol. (Lond.) 222*, 537-58

Livingstone, M.S. (1985) Genetic dissection of *Drosophila* adenylate cyclase. *Proc. Natl. Acad. Sci. USA 82*, 5992-6

Lummis, S.C.R. and Sattelle, D.B. (1985) Binding of N-[propionyl-[³H]propionylated α-bungarotoxin and L-(benzilic-4,4′-[³H]quinuclidinyl) benzilate to CNS extracts of the cockroach *Periplaneta americana. Comp. Biochem. Physiol. 80C*, 75-83

Lummis, S.C.R., Sattelle, D.B. and Ellory, J.C. (1984) Molecular weight estimates of insect cholinergic receptors by radiation inactivation. *Neurosci. Lett. 44*, 7-12

Lundberg, J.M., Hedlund, B. and Bartfai, (1982) Vasoactive intestinal polypeptide enhances muscarinic ligand binding in cat submandibular salivary gland. *Nature (Lond.) 295*, 147-9

Luthin, G.R. and Wolfe, B.B. (1985) Characterization of [³H]pirenzepine binding to muscarinic cholinergic receptors solubilized from rat brain. *J. Pharmacol. Exp. Ther. 234*, 37-44

McCaman, R.E. and Ono, J.K. (1982) *Aplysia* cholinergic synapses: a model for central cholinergic function. In: Hanin, I. and Goldberg, A.M. (eds) *Progress in cholinergic biology: model cholinergic synapses*, pp. 23-43. Raven Press, New York

McCaman, R.E., Weinreich, D. and Borys, H. (1973) Endogenous levels of acetylcholine and choline in individual neurons of *Aplysia*. *J. Neurochem. 21*, 473-6

McMahon, K.K. and Hosey, M.M. (1985) Agonist interactions with cardiac muscarinic receptors. Effects of Mg^{2+}, guanine nucleotides, and monovalent cations. *Molec. Pharmacol. 28*, 400-9

MacPhee-Quigley, K., Taylor, P. and Taylor, S. (1985) Primary structures of the catalytic subunits from two molecular forms of acetylcholinesterase. *J. Biol. Chem. 260*, 12185-9

Malthe-Sorenssen, D. (1976) Choline acetyltransferase. Evidence for acetyl transfer by a histidine residue. *J. Neurochem. 27*, 873-81

Mansour, N.A., Eldefrawi, M.E. and Eldefrawi, A.T. (1977) Isolation of putative acetylcholine receptor proteins from housefly brain. *Biochemistry 16*, 4126-32

Mansour, T.E. (1979) Chemotherapy of parasitic worms: new biochemical strategies. *Science 205*, 462-9

Manulis, S., Ishaaya, I. and Perry, A.S. (1981) Acetylcholinesterase of *Aphis criticola*: properties and significance in determining toxicity of systemic organophosphorus and carbamate compounds. *Pest. Biochem. Physiol. 15*, 267-74

Marder, E. and Paupardin-Tritsch, D. (1978) The pharmacological properties of some crustacean neuronal acetylcholine, α-aminobutyric acid and *L*-glutamate responses. *J. Physiol. (Lond.) 280*, 213-36

Marder, E. and Paupardin-Tritsch, D. (1980) The pharmacological profile of the acetylcholine response of a crustacean muscle. *J. exp. Biol. 88*, 147-59

Marsden, J.R., Bgata, N. and Cain, H. (1981) Evidence for a cerebral cholinergic system and suggested pharmacological patterns of neural organization in the prostomium of the polychaete *Nereis virens* (Sars). *Tiss. Cell 13*, 255-67

Marshall, L.M. (1981) Synaptic localization of α-bungarotoxin binding which blocks nicotinic transmission at frog sympathetic neurons. *Proc. Natl. Acad. Sci. USA 78*, 1948-52

Massoulie, J. and Bon, S. (1982) The molecular forms of cholinesterase and acetylcholinesterase in vertebrates. *Ann. Rev. Neurosci. 5*, 57-106

Mattissan, A., Nilsson, S. and Fange, R. (1974) Light microscopical and ultrastructural organisation of muscles of *Priapulus caudatus* (Priapulida) and their responses to drugs, with phylogenetic remarks. *Zool. Scripta 3*, 209-18

Meedel, T.H. and Whittaker, J.R. (1979) Development of acetylcholinesterase during embryogenesis of the ascidian *Ciona intestinalis. J. exp. Zool. 210*, 1-10

Meyer, M.R. and Reddy, G.R. (1985) Muscarinic and nicotinic cholinergic binding sites in the terminal abdominal ganglion of the cricket (*Acheta domesticus*). *J. Neurochem. 45*, 1101-12

Michaelson, D.M. and Angel, I. (1981) Saturable acetylcholine transport into purified cholinergic synaptic vesicles. *Proc. Natl. Acad. Sci. USA 78*, 2048-52

Miledi, R., Molenaar, P.C. and Polak, R.L. (1983) Electrophysiological and chemical determination of acetylcholine release at the frog neuromuscular junction. *J. Physiol. (Lond.) 334*, 245-54

Mishina, M., Kurosaki, T., Tobimatsu, T., Morimoto, Y., Noda, M., Yamamoto, T., Terao, M., Lindstrom, J., Takahashi, T., Kuno, M. and Numa, S. (1984) Expression of functional acetylcholine receptor from cloned cDNAs. *Nature*

(Lond.) 307, 604-8

Mishina, M., Tobimatsu, T., Imoto, K., Tanaka, K.-I., Fujita, Y., Fukuda, K., Kurasaki, M., Takahashi, H., Morimoto, Y., Hirose, T., Inayama, S., Takahashi, T., Kuno, M. and Numa, S. (1985) Location of functional regions of acetylcholine receptor α-subunit by site-directed mutagenesis. *Nature (Lond.) 313,* 364-9

Mumenthaler, M. and Engel, W.K. (1961) Cytological localization of cholinesterase in developing chick embryo skeletal muscle. *Acta Anat. (Basel) 47,* 274-99

Muneoka, Y., Ichimura, Y., Shiba, Y. and Kanno, Y. (1981) Mechanical responses of the body wall strips of an echiuroid worm *Urechis unicinctus* agents and amino acids. *Comp. Biochem. Physiol. 69C,* 171-7

Murray, T.F., Mpitsos, G.J., Siebenaller, J.F. and Barker, D.L. (1985) Stereoselective L-[³H]quinuclidinyl benzilate-binding sites in nervous tissue of *Aplysia californica*: evidence for muscarinic receptors. *J. Neurosci. 5,* 3184-8

Natoff, I.L. (1969) The pharmacology of the cholino-receptor in the muscle of *Ascaris lumbricoides* var. *suum. Br. J. Pharmacol. 37,* 251-7

Newkirk, R.F., Ballou, E.W., Vickers, G. and Whittaker, V.P. (1976) Comparative studies in synaptosome formation: preparation of synaptosomes from the ventral nerve cord of the lobster (*Homarus americanus*). *Brain Res. 101,* 103-11

Newkirk, R.F., Sukumar, R., Thomas, W.E. and Townsel, J.G. (1981) The preparation and partial characterization of synaptosomes from central nervous tissue of *Limulus. Comp. Biochem. Physiol. 70C,* 177-84

Nicholas, M.T., Moreau, M. and Guerrier, P. (1978) Indirect nervous control of luminescence in the polynoid worm *Harmothoe lumulata. J. exp. Zool. 206,* 427-32

Nistri, A., Cammelli, E, and De Bellis, A.M. (1978) Pharmacological observations on the cholinesterase activity of the leech central nervous system. *Comp. Biochem. Physiol. 61C,* 203-5

Noda, M., Takahashi, H., Tanabe, T., Toyoato, M., Kikyotani, S., Furutani, Y., Hirose, T., Takashima, H., Inayama, S., Miyata, T. and Numa, S. (1983) Structural homology of *Torpedo californica* acetylcholine receptor subunits. *Nature (Lond.) 302,* 528-32

Norman, R.I., Mehraban, F., Barnard, E.A. and Dolly, J.O. (1982) Nicotinic acetylcholine receptor from chick optic lobe. *Proc. Natl. Acad. Sci. USA 79,* 1321-25

O'Donohue, T.L., Millington, W.R., Handelmann, G.E., Contreras, P.C, and Chronwall, B.M. (1985) On the 50th anniversary of Dale's law: mutiple neurotransmitter neurons. *Trends Pharmacol. Sci. 6,* 305-8

Oh, T.H., Chyu, J.Y. and Max, S.R. (1977) Release of acetylcholinesterase by cultured spinal cord cells. *J. Neurobiol. 8,* 469-76

Olianas, M.C., Onali, P., Schwartz, J.P., Neff, N.H. and Costa, E. (1984) The muscarinic receptor cyclase complex of rat striatum: desensitization following chronic inhibition of acetylcholinesterase activity. *J. Neurochem. 42,* 1439-43

Ono, J. and Salvaterra, P.M. (1981) Snake α-toxin effects on cholinergic and noncholinergic responses of *Aplysia californica* neurons. *J. Neurosci. 1,* 259-70

Oswald, R.E. and Freeman, J.A. (1981) Alpha-bungarotoxin binding and central nervous system nicotinic acetylcholine receptors. *Neuroscience 6,* 1-14

Ozaki, H. (1974) Localization and multiple forms of acetylcholinesterase in sea urchin embryos. *Dev. Growth Differ. 16,* 267-79

Parsons, S.M., Carpenter, R.S., Koenigsberger, R. and Rothlein, J.E. (1982) Transport in the cholinergic synaptic vesicle. *Fed. Proc. 41,* 2765-8

Patrick, J. and Stallcup, W.B. (1977) Immunological distinction between acetylcholine receptor and the α-bungarotoxin-binding component on sympathetic neurons. *Proc. Natl. Acad. Sci. USA 74*, 4689-92

Perkins, B.A. and Cottrell, G.A. (1972) Choline acetyl transferase activity in nervous tissue of *Hirudo medicinalis* (leech) and *Nephrops norvegicus* (Norway lobster). *Comp. gen. Pharmacol. 3*, 19-21

Peterson, G.L., Herron, G.S., Yamaki, M., Fullerton, D.S. and Schimerlik, M.I. (1984) Purification of the muscarinic acetylcholine receptor from porcine atria. *Proc. Natl. Acad. Sci. USA 81*, 4993-7

Polsky, R. and Shuster, L. (1976) Preparation and characterization of two iso-enzymes of choline acetyltransferase from squid head ganglia. II. Self-association, molecular weight determinations, and studies with inactivating antisera. *Biochim. Biophys. Acta 455*, 43-6

Potter, L.T., Glover, V.A.S. and Saelens, J.K. (1968) Choline acetyltransferase from rat brain. *J. Biol. Chem. 243*, 3864-70

Prempeh, A.B., Prince, A.K. and Hide, E.G. (1972) The reaction of acetyl-coenzyme A with choline acetyltransferase. *Biochem. J. 129*, 991-4

Prescott, D.J., Hildebrand, J.G., Sanes, J.R. and Jewett, S. (1977) Biochemical and developmental studies of acetylcholine metabolism in the central nervous system of the moth *Manduca sexta. Comp. Physiol. 56*, 77-84

Rauh, J.J., Lambert, M.P., Cho. N.J., Chiu, H. and Klein, W.L. (1986) Glyco-protein properties of muscarinic acetylcholine receptors from bovine cerebral cortex. *J. Neurochem. 46*, 23-32

Reichardt, L.F. and Kelly, R.B. (1983) A molecular description of nerve terminal function. *Ann. Rev. Biochem. 52*, 871-926

Robinson, T., Aguilar, O. and Lunt, G. (1984) Interaction of a nicotinic acetyl-choline receptor from locust (*Schistocerca gregaria*) central nervous system with concanavalin A: comparison with vertebrate receptor. *Biochem. Soc. Trans. 12*, 807-8

Ross, D.H. and Triggle, D.J. (1972) Further differentiation of cholinergic receptors in leech muscle. *Biochem. Pharmacol. 21*, 2533-6

Rozhkova, E.K., Malyutina, T.A. and Shishov, B.A. (1980) Pharmacological characteristics of cholinoreception in somatic muscles of the nematode, *Ascaris suum. Gen. Pharmacol. 11*, 141-6

Rudloff, E. (1978) Acetylcholine receptors in the central nervous system of *Drosophila melanogaster. Exp. Cell. Res. 111*, 185-90

Ruess, K.-P. and Lieflander, M. (1979) Action of detergents on covalently labeled, membrane-bound muscarinic acetylcholine receptor of bovine nucleus caudatus. *Biochim. Biophys. Res. Commun. 88*, 627-33

Said, S. (1984) Isolation, localization, and characterization of gastrointestinal peptides. *Clin. Biochem. 17*, 65-7

Sakai, M. (1967) Hydrolysis of acetylthiocholine and butyrylthiocholine by cholin-esterases of insects and a mite. *Appl. Ent. Zool. 2*, 111-12

Sakmann, B., Methfessel, C., Mishina, M., Takahashi, T., Takai, T., Kurasaki, M., Fukuda, K. and Numa, S. (1985) Role of acetylcholine receptor subunits in gating of the channel. *Nature (Lond.) 318*, 538-43

Sakmann, B., Patlak, J. and Neher, E. (1980) Single acetylcholine-activated channels show burst-kinetics in presence of desensitizing concentrations of agonist. *Nature (Lond.) 286*, 71-3

Sanes, J.R., Prescott, D.J. and Hildebrand, J.G. (1977) Cholinergic neurochemical development of normal and deafferented antennal lobes during metamorphosis of the moth, *Manduca sexta. Brain Res. 119*, 389-402

Sastry, B.V.R. and Sadavongvivad, C. (1979) Cholinergic systems in non-nervous tissues. *Pharmacol. Rev. 30*, 65-132

Sattelle, D.B. (1980) Acetylcholine receptors of insects. In: Berridge, M.J. and Treherne, J.E. (eds) *Advances in insect physiology*, pp. 215-315. Academic Press, London

Sattelle, D.B. and Breer, H. (1985) Purification by affinity chromatography of a nicotinic acetylcholine receptor from the CNS of the cockroach *Periplaneta americana*. *Comp. Biochem. Physiol. 82C*, 349-52

Sattelle, D.B. and David, J.A. (1983) Voltage-dependent block by histrionicotoxin of the acetylcholine-induced current in an insect motoneurone cell body. *Neurosci. Lett. 43*, 37-41

Sattelle, D.B., Harrow, I.D., David, J.A., Pelhate, M., Callec, J.J., Gepner, J.I. and Hall, L.M. (1985a) Nereistoxin: actions on a CNS acetylcholine receptor/ion channel in the cockroach *Periplaneta americana*. *J. exp. Biol. 118*, 37-52

Sattelle, D.B., Harrow, I.D., Hue, B., Pelhate, M., Gepner, J.I. and Hall, L.M. (1983) α-Bungarotoxin blocks excitatory synaptic transmission between cercal sensory neurones and giant interneurone 2 of the cockroach, *Periplaneta americana*. *J. exp. Biol. 107*, 473-89

Sattelle, D.B., Hue, B., Pelhate, M., Sherby, S.M., Eldefrawi, A.T. and Eldefrawi, M.E. (1985b) Actions of phencyclidine and its thienylpyrrolidine analogue on synaptic transmission and axonal conduction in the central nervous system of the cockroach *Periplaneta americana*. *J. Insect. Physiol. 31*, 917-24

Schmidt-Nielsen, B.K., Gepner, J.I., Teng, N.N.H. and Hall, L.H. (1977) Characterization of an α-bungarotoxin binding component from *Drosophila melanogaster*. *J. Neurochem. 29*, 1013-29

Schreiber, G. and Sokolovsky, M. (1985) Muscarinic receptor heterogeneity revealed by interaction with berylium tosylate. Different ligand–receptor conformations versus different receptor subclasses. *Molec. Pharmacol. 27*, 27-31

Seeman, G.R. and Houlihan, R.K. (1951) Enzyme systems in *Tetrahymena geleii* S. II. Acetylcholinesterase activity. Its relation to the motility of the organism and to coordinated ciliary action in general. *J. Cell. Comp. Physiol. 37*, 309-21

Shain, W., Greene, L.A., Carpenter, D.O., Sytkowski, A.J. and Vogel, Z. (1974) *Aplysia* acetylcholine receptors: blockade by and binding of α-bungarotoxin. *Brain Res. 72*, 225-40

Shaker, N. and Eldefrawi, A. (1981) Muscarinic receptor in house fly brain and its interaction with chlorobenzilate. *Pest. Biochem. Physiol. 15*, 14-20

Shaker, N., Eldefrawi, A.T., Aguayo, L.G., Warnick, J.E., Albuquerque, E.X. and Eldefrawi, M.E. (1982) Interactions of *d*-tubocurarine with the nicotinic acetylcholine receptor/channel molecule. *J. Pharmacol. Exp. Ther. 220*, 172-7

Shaw, K.-P., Aracava, Y., Akaike, A., Daly, J.W., Rickett, D.L. and Albuquerque, E.X. (1985) The reversible cholinesterase inhibitor physostigmine has channel-blocking and agonist effects on the acetylcholine receptor–ion channel complex. *Molec. Pharmacol. 28*, 527-38

Sherby, S.M., Eldefrawi, A.T., Albuquerque, E.X. and Eldefrawi, M.E. (1985) Comparison of the actions of carbamate anticholinesterases on the nicotinic acetylcholine receptor. *Molec. Pharmacol. 27*, 343-8

Sherby, S.M., Eldefrawi, A.T., David, J.A., Sattelle, D.B. and Eldefrawi, M.E. (1986) Interactions of charatoxins and nereistoxin with the nicotinic acetylcholine receptors of *Torpedo* electric organ and insect CNS. *Arch. Insect Biochem. Physiol. 3*, 431-45

Shibahara, S., Kubo, T., Perski, H.J., Takahashi, H., Noda, M. and Numa, S. (1985) Cloning and sequence analysis of human genomic DNA encoding γ subunit precursor of muscle acetylcholine receptor. *Eur. J. Biochem. 146*, 15-22

Silinsky, E.M. (1985) The biophysical pharmacology of calcium-dependent acetylcholine secretion. *Pharmacol. Rev. 37*, 81-132

Silver, A. (1974) *The biology of cholinesterases.* Elsevier North-Holland, Amsterdam

Skau, K.A. and Brimijoin, S. (1978) Release of acetylcholinesterase from rat hemidiaphragm preparations stimulated through the phrenic nerve. *Nature (Lond.) 275*, 224-6

Soeda, Y., Eldefrawi, M.E. and O'Brien, R.D. (1975) Lobster axon acetylcholinesterase: a comparison with acetylcholinesterases of bovine erythrocytes, house fly head and *Torpedo* electroplax. *Comp. Biochem. Physiol. 50C*, 163-8

Souccar, C., Varanda, W.A., Aronstam, R.S., Daly, J.W. and Albuquerque, E.X. (1984) Interactions of gephyrotoxin with the acetylcholine receptor–ionic channel complex. II. Enhancement of desensitization. *Molec. Pharmacol. 25*, 395-400

Spielholz, N.I. and van der Kloot, W.G. (1973) Localization and properties of the cholinesterase in crustacean muscle. *J. Cell Biol. 59*, 407-20

Spivak, C.E., Maleque., Oliveira, A.C., Masukawa, L.M., Tokuyama, T., Daly, J.W. and Albuquerque, E.X. (1982) Actions of the histrionicotoxins at the ion channel of the nicotinic acetylcholine and the voltage sensitive ion channels of muscle membranes. *Molec. Pharmacol. 21*, 351-61

Stockton, J.M., Birdsall, N.J.M., Burgen, A.S.V. and Hulme, E.C. (1983) Modification of the binding properties of muscarinic receptors by gallamine. *Molec. Pharmacol. 23*, 551-7

Strauss, W.L. and Nirenberg, M. (1985) Inhibition of choline acetyltransferase by monoclonal antibodies. *J. Neurosci. 1*, 175-80

Suter, C. and Usherwood, P.N.R. (1985) Action of acetylcholine and antagonists on somata isolated from locust central neurons. *Comp. Biochem. Physiol. 80C*, 221-9

Takai, T., Noda, M., Mishina, M., Shimizu, S., Furutani, Y., Kayano, T., Ikeda, T., Kubo, T., Takahashi, H., Takahashi, T., Kuno, M. and Numa, S. (1985) Cloning, sequencing and expression of cDNA for a novel subunit of acetylcholine receptor from calf muscle. *Nature (Lond.) 315*, 761-4

Tauc, L. and Baux, G. (1982) Are there intracellular acetylcholine receptors in the cholinergic synaptic nerve terminals? *J. Physiol. (Paris) 78*, 366-72

Thomas, W.E., Brady, R.N. and Townsel, J.G. (1978) A characterization of α-bungarotoxin-binding in the brain of the horseshoe crab, *Limulus polyphemus*. *Arch. Biochem. Biophys. 187*, 53-60

Tiedt, T., Albuquerque, E.X., Bakry, N.M., Eldefrawi, M.E. and Eldefrawi, A.T. (1979) Voltage- and time-dependent actions of piperocaine on the ionic channel of the acetylcholine receptor. *Molec. Pharmacol. 16*, 909-21

Trimmer, B.A. and Berridge, M.J. (1985) Inositol phosphates in the insect nervous system. *Insect Biochem. 15*, 811-15

Tsai, M.-C., Mansour, N.A., Eldefrawi, A.T., Eldefrawi, M.E. and Albuquerque, E.X. (1978) Mechanism of action of amantadine on neuromuscular transmission. *Molec. Pharmacol. 14*, 787-803

Usherwood, P.N.R. and Cull-Candy, S.G. (1975) Pharmacology of somatic nerve–muscle synapses. In: Usherwood, P.N.R. (ed.) *Insect muscle*, pp. 207-80. Academic Press, London

Varanda, W.A., Aracava, Y., Sherby, M., Van Meter, W.G., Eldefrawi, M.E. and Albuquerque, E.X. (1985) The acetylcholine receptor of the neuromuscular junction recognizes mecamylamine as a noncompetitive antagonist. *Molec. Pharmacol. 28*, 128-37

Venter, J.C., Eddy, B., Hall., L.M. and Fraser, C.M. (1984) Monoclonal antibodies detect the conservation of muscarinic cholinergic receptor structure from *Drosophila* to human brain and detect possible structural homology with α-adrenergic receptor. *Proc. Natl. Acad. Sci. USA 81*, 272-6

Vigny, M., Gisiger, V. and Massoulie, J. (1978) 'Nonspecific' cholinesterase and acetylcholinesterase in rat tissues: molecular forms, structural and catalytic properties, and significance of the two enzyme systems. *Proc. Natl. Acad. Sci. USA 75*, 2588-92

Voss, G. and Matsumura, F. (1965) Biochemical studies on a modified and normal cholinesterase found in the Leverkusen strains of the two spotted spider mite, *Tetranychus urticae. Can. J. Biochem. 43*, 63-72

Vyas, S. and O'Regan, S. (1985) Reconstitution of carrier-mediated choline transport in proteoliposomes prepared from presynaptic membranes of *Torpedo* electric organ, and its internal and external ionic requirements. *J. Membr. Biol. 85*, 111-19

Wagner, J.A., Carlson, S.S. and Kelly, R.B. (1978) Chemical and physical characterization of cholinergic synaptic vesicles. *Biochemistry 17*, 1199-1206

Walker, R.J., Woodruff, G.N. and Kerkut, G.A. (1968) The effect of ACh and 5-HT on electrophysiological recordings from muscle fibres of the leech *Hirudo medicinalis. Comp. Biochem. Physiol. 24*, 987-90

Watson, M., Roeske, W.R. Vickroy, T.W., Smith, T.L., Akiyama, K., Gulya, K., Duckles, S.P., Serra, M., Adem, A., Nordberg, A., Gehlert, D.R., Wamsley, J.K. and Yamamura, H.I. (1986) Biochemical and functional basis of putative muscarinic receptor subtypes and its implications. *Trends Pharm. Sci. Suppl. Proc. 2nd Int. Symp. Subtypes of Muscarinic Receptors II*, 46-55

Watson, M., Vickroy, T.W., Roeske, W.R. and Yamamura, H.I. (1984) Subclassification of muscarinic receptors based upon the selective antagonist pirenzepine. *Trends Pharmacol. Sci. Suppl.* 9-11

Watson, M., Yamamura, H.I. and Roeske, W.R. (1983) A unique regulatory profile and regional distribution of [^{3}H]pirenzepine binding in the rat provides evidence for distinct M_1 and M_2 muscarinic receptor subtypes. *Life Sci. 32*, 3001-11

Wells, G.P. (1937) Studies on the physiology of *Arenicola marina* I. The pacemaker role of the oesophagus and the action of adrenaline and acetylcholine. *J. exp. Biol. 14*, 117-57

White, H.L. and Wu, J.C. (1973) Kinetics of choline acetyltransferases (EC 2.3.1.6) from human and other mammalian central and peripheral nervous tissues. *J. Neurochem 20*, 297-307

Winkler, H., Schmidt, W., Fischer-Colbrie, R. and Weber, A. (1983) Molecular mechanisms of neurotransmitter storage and release: a comparison of the adrenergic and cholinergic system. In: Changeux, J.-P., Glowinski, J., Imbert, M. and Bloom, F.E. (eds) *Molecular and cellular interactions underlying higher brain functions, progress in brain research, 58*, pp. 11-20 Elsevier, Amsterdam

Wolfe, L.S. and Smallman, B.N. (1956) The properties of cholinesterase from insects. *J. Cell. Comp. Physiol. 48*, 215-35

Woodruff, G.N., Walker, R.J. and Newton, L.C. (1971) The actions of some muscarinic and nicotinic agonists on the Retzius cells of the leech. *Comp. gen. Pharmacol. 2*, 106-17

Wu, C.-F. Berneking, J.M. and Barker, D.L. (1983) Acetylcholine synthesis and accumulation in the CNS of *Drosophila* larvae: analysis of *shibire*^{ts}, a mutant with a temperature-sensitive block in synaptic transmission. *J. Neurochem. 40*, 1386-95

Yamamura, H.I. and Snyder, S.H. (1974) Muscarinic cholinergic binding in rat brain. *Proc. Natl. Acad, Sci. USA 71*, 1725-9

Young, E.F., Ralston, E., Blake, J., Ramachandran, J., Hall, Z.W. and Stroud, R.M. (1985) Topological mapping of acetylcholine receptor: evidence for a model with five transmembrane segments and a cytoplasmic COOH-terminal peptide. *Proc. Natl. Acad. Sci. USA 82*, 626-30

Yu, C.-C. and Booth, G.M. (1971) Inhibition of choline acetylase from the house fly (*Musca domestica* L.) and mouse. *Life Sci. 10*, 337-47

Zeimal, E.V. and Vulfius, E.A. (1968) The action of cholinomimetics and cholinolytics on gastropod neurons. In: Salanki, J. (ed.) *Neurobiology of Invertebrates*, pp. 255-65. Acad. Kiado, Budapest

Ziskind, L. and Werman, R. (1975) Sodium ions are necessary for cholinergic desensitization in molluscan neurones. *Brain Res. 88*, 171-6

2

Glutamate

Ian R. Duce

The acidic amino acid L-glutamic acid subserves a number of important functions in the metabolism of nervous and other tissues. However, it is the putative neurotransmitter function of this dicarboxylic acid in a wide range of animal groups (Table 2.1) which has focused attention on its neurochemistry.

The evidence that L-glutamate was a likely neurotransmitter emerged from evidence that ionophoretic application caused excitation of spinal neurones (Curtis *et al.*, 1959; Curtis and Watkins, 1960). These and other elecrophysiological studies on vertebrate species were concurrent with studies demonstrating that L-glutamate depolarised crustacean muscle and induced contractions (Robbins, 1958, 1959; van Harreveld and Mendelson, 1959). Since these early studies a healthy debate has ensued in the literature as to whether or not L-glutamate is indeed a neurotransmitter. These arguments have revolved around the thesis that a compound with a wide distribution and such an important place in metabolism would not be ideally suited for a specific function such as neurotransmission. The controversy has proved difficult to resolve because of the absence of a potent specific glutamate antagonist. However, the bulk of evidence now appears to suggest that glutamate is an excitatory neurotransmitter in vertebrates and many invertebrates (see reviews by Usherwood, 1978, and Fonnum, 1984). In addition to these excitatory responses involving depolarisation of the postsynaptic cell by L-glutamate, application of L-glutamate has also been shown to produce hyperpolarisation of insect muscle (Cull-Candy and Usherwood, 1973; Lea and Usherwood, 1973; Usherwood and Cull-Candy, 1974); molluscan central neurones (Cottrell *et al.*, 1972; Oomura *et al.*, 1974; Judge *et al.*, 1977). Invertebrate tissues sensitive to L-glutamate are shown in Table 2.1.

Table 2.1: Glutamate-sensitive sites on invertebrate nerves and muscles

Site	Effect	Reference
Cnidaria		
Actinia equina, sea anemone: sphincter muscle	Inhibition of electrically induced contraction	Carlyle (1974)
Annelida		
Hirudo medicinalis, leech: Retzius cell	Excitation, depolarisation, sodium and potassium permeability	Walker *et al.* (1980)
Pheretima hawayana, earthworm: body-wall muscle	Depolarisation, contraction	Chang (1975)
Lumbricus terrestris, earthworm: body-wall muscle	Muscle contraction, potentiates evoked contraction	Gardner (1981)
Mollusca		
Helix aspersa, snail: CNS giant 5HT cell	Inhibition, hyperpolarisation	Cottrell *et al.* (1972)
H. aspersa: visceral ganglion neurone	Hyperpolarisation, biphasic synaptic response	Judge *et al.* (1977)
H. aspersa: central neurones		Piggott *et al.* (1975)
E4	Biphasic	
F30	Depolarisation	
F1	Hyperpolarisation	
Onchidium verruculatum: central neurones	Inhibition hyperpolarisation, increased potassium permeability	Oomura *et al.* (1974)
H. aspersa: pharyngeal retractor muscle	Contraction	Kerkut *et al.* (1965)
Aplysia, Helix: central neurones	Excitatory and inhibitory responses; after Con A all responses excitatory	Kehoe (1978)
Aplysia: gill muscles	Contraction	Carew *et al.* (1974)
Aplysia: buccal mass protractor muscle	Contraction	Taraskevich *et al.* (1977)
Busycon canaliculatum: radula protractor muscle	Contraction	Hill (1970)
Loligo vulgaris, squid: stellate ganglion giant synapse	Depolarisation of giant axon	Miledi (1967)
Arthropoda		
Limulus polyphemus, horseshoe crab: central neurones	Excitation, inhibition	Walker *et al.* (1980)

Table 2.1 continued

Arthropoda — Crustacea		
Astacus leptodactylus, crayfish: vas deferens	Contraction	Murdock (1971)
Astacus astacus, crayfish: hindgut	Contraction	Jones (1962)
Crustacean skeletal muscle	Contraction, excitation, increased sodium permeability	Numerous authors. See reviews by: Usherwood (1978), Nistri and Constanti (1979), Leake and Walker (1980), Atwood (1976. 1982)
Arthropoda — Insecta		
Schistocerca gregaria, locust: isolated neuronal somata	Inhibition, hyperpolarisation, few excitatory depolarising responses, reduced input resistance	Usherwood *et al.* (1980); Giles and Usherwood (1985)
Periplaneta americana, cockroach: fast coxal depressor motor neurone	Hyperpolarisation	Wafford and Sattelle (1986)
Insect skeletal muscle	Contraction, depolarisation, increased permeability to cations, also extrajunctional hyperpolarising glutamate receptors gate chloride channels	Numerous authors. See reviews by: Usherwood and Cull-Candy (1975), Usherwood (1978), Nistri and Constanti (1979), Leake and Walker (1980), Piek (1985a)
Leucophaea maderae, cockroach: hindgut	Contraction	Holman and Cook (1970)

SYNTHESIS OF *L*-GLUTAMATE

One of the criteria often cited as essential for a compound to be considered as a neurotransmitter is the localisation of its synthetic machinery in the presynaptic cell and the storage of the putative neurotransmitter in the presynaptic terminal (Werman, 1966). In the case of neurones releasing acetylcholine, noradrenaline or GABA, the restricted distribution of these chemicals has made it possible to demonstrate the location and dynamics of their synthesis. The wide distribution and complex involvement of *L*-glutamate in metabolism make it difficult to distinguish between pools of glutamate involved in neurotransmission and those concerned with other cellular functions.

Many studies have been carried out on vertebrate preparations to determine the source(s) of neurotransmitter glutamate. The usual strategy employed has been to use radiolabelled precursors and to determine the specific radioactivity of released glutamate or glutamate present in particular cellular compartments. It can be seen from Figure 2.1 that in nervous tissue L-glutamate can be produced from glutamine by the action of glutaminase, from the TCA cycle intermediate α-keto-glutarate, by transamination or the action of glutamate dehydrogenase, or possibly from ornithine (Yoneda *et al.*, 1982) or proline (Roberts, 1981) via glutamate semialdehyde. The latter pathway may be involved in the formation of glutamate as a precursor for GABA (Roberts, 1981).

In vertebrate nervous tissue it has been shown that radiolabelled glucose is metabolised differently from radiolabelled acetate, resulting in metabolic pools of amino acids containing different specific activities. Glucose appears to be metabolised in a compartment high in glutamate with little glutamine synthesis whereas acetate is metabolised in a different compartment which is high in glutamine synthesis (van den Berg, 1973). These compartments have been interpreted as representing neuronal and glial cells, respectively (see Fonnum, 1984, for review).

Figure 2.1: Possible routes for the synthesis of neurotransmitter L-glutamate. The most important pathways in invertebrates are unknown, but interconversion of glutamate and alanine (Huggins *et al.*, 1967; Bradford *et al.*, 1969) and glutamate and glutamine (Irving *et al.*, 1979b) have been identified in excitable tissues of invertebrates

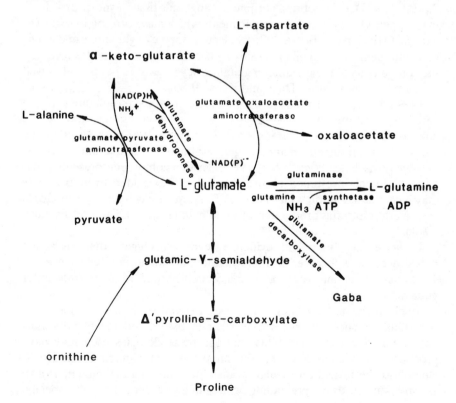

The formation of released ('neurotransmitter') glutamate could utilise as immediate precursors either glutamine or α-keto-glutarate, and several groups of investigators have attempted to resolve which of these two sources is exploited in the formation of transmitter glutamate.

There is good evidence from a number of laboratories that glutamine can be taken up by nerve endings and used to form releasable glutamate (Bradford and Ward, 1976; Bradford et al., 1978; Hamberger et al., 1979; Shank and Aprison, 1979; Reubi, 1980). Glutaminase, the enzyme necessary to mediate this conversion, has been shown to be enriched in nerve endings (Bradford and Ward, 1976), and glutamine is present in high concentrations in cerebrospinal fluid. However, glutamine could also be produced in glial cells from glutamate by the action of glutamine synthetase, which has been shown to be localised in glial cells (Norenberg and Martinez-Hernandez, 1979). The compartmentation of these enzymes has led several authors to suggest that there may be a glutamate/glutamine cycle present in nervous tissue, where glutamate released from neurones could be taken up by glial cells and converted to glutamine; this glutamine is then transported back to neurones for conversion back to glutamate and subsequent reuse (Hamberger, 1979; Cotman et al., 1981; Shank and Aprison, 1981). In vertebrate peripheral ganglionic tissue, synthesis of glutamate appears to involve two compartments which may correspond to nerves and glia (Minchin and Beart, 1975). Selective uptake of glutamine and glutamate into neurones and glia respectively has been demonstrated, suggesting the components of the glutamate/glutamine cycle exist in peripheral as well as central nervous tissue (Duce and Keen, 1983).

Production of neurotransmitter glutamate from α-keto-glutarate in the vertebrate nervous system is thought to involve transamination by the enzyme aspartate amino transferase (Fonnum, 1968), which may be localised in presumed glutamatergic terminals in retina and cochlear nucleus (Altschuler et al., 1981, 1982). Evidence has been presented that nerve terminals presumed to use excitant amino acid neurotransmitters may have mitochondria with 'blocked' TCA cycles, such that intermediates are diverted into the production of glutamate (Hajos and Kerpel-Fronius, 1971).

In vertebrate brain the evidence tends to suggest that glutamate dehydrogenase is localised in non-synaptic mitochondria (Reijnerse et al., 1975) and thus may not be a major contributor to neurotransmitter glutamate.

Much of the above evidence has been derived from in vitro studies on subcellular fractions or slices of vertebrate tissue, and recent appraisals have pointed out the possibility that the methodologies used may have produced an overemphasis on the importance of glutamine relative to glucose in the formation of neurotransmitter glutamate (Fonnum, 1984). In vivo studies using preloading with glutamine generated low specific activities for the released glutamate, which might reflect dilution of the administered glutamine or a low conversion rate of glutamine to glutamate (Thanki *et al.*, 1983). However, release of tritiated glutamate derived from labelled glutamine was shown to be potentiated during the initiation of long-term potentiation in rat hippocampus, suggesting a close relationship between glutamine and transmitter glutamate in this particular system (Dolphin *et al.*, 1982; Feasey *et al.*, 1986).

Synthesis of neurotransmitter *L*-glutamate in invertebrates

In the vertebrate nervous system, which has been extensively studied, the precise precursors and route of metabolism of transmitter glutamate are still unresolved. It is therefore not surprising that neurotransmitter glutamate synthesis is not clearly defined in invertebrate species. Several studies have been carried out involving incubation of excitable tissues in radiolabelled precursors followed by chromatographic analysis of tissue and body fluids. ^{14}C originating as trehyalose or glucose in the haemolymph of the cockroach *Periplaneta americana* was found to be rapidly incorporated into aspartate, glutamate and glutamine in the nerve cord (Treherne, 1960). When experiments were carried out *in vivo*, over 50 per cent of the label was incorporated into glutamine and glutamate. *In vitro* incubations resulted in reduced incorporation into glutamate, glutamine and aspartate but a large incorporation into alanine.

The importance of the Krebs TCA cycle in glutamate metabolism of invertebrate nervous tissue was further suggested by comparative studies on molluscs (*Mytilus edulis*, mussel, and *Helix aspersa*, snail), and arthropods (*Carcinus maenas*, crab, and *P. americana*, cockroach) (Huggins *et al.*, 1967). In these experiments TCA intermediates and aspartate, alanine and glutamine were labelled following incubation of nerves and muscles with [^{14}C]glutamate. The distribution of the label varied between the tissues and the species, and the authors concluded that the differences were related more to phylogeny than to physiology of the tissues. Interspecific variation in neuronal glutamate metabolism was also described (Bradford *et al.*, 1969) in a comparative study of *H. aspersa*, *Schistocerca gregaria* (locust), *Eledone cirrhosa* (octopus) and rat. The major labelled metabolites formed from glucose or glutamate were again aspartate, glutamate, glutamine and alanine. The production of alanine was much greater in the invertebrates than in the rat (cf. Treherne, 1960).

Although these experiments show that glutamate in the brains of various species can originate from exogenous glucose or glutamate, such experiments cast little light on the existence or origins of specific pools of glutamate involved in neurotransmission. The possible role of axonal flow in the provision of releasable glutamate was indicated by Kerkut *et al.*

47

(1967) who demonstrated the release of radiolabelled glutamate from a snail muscle preparation following incubation of the attached brain in [^{14}C]glutamate or glucose. However, these authors obtained similar results using a frog sciatic nerve/gastrocnemius muscle preparation. The significance of the latter data in a preparation where glutamate is unlikely to be the transmitter is unclear.

UPTAKE OF GLUTAMATE

Despite the fact that comparative information on the synthesis of neurotransmitter glutamate is sparse, the process by which this amino acid is taken up by components of the nervous and muscular systems has been quite widely investigated in the context of neurotransmission. Glutamate uptake systems have been examined both in terms of their kinetics using radiotracer techniques and in terms of their localisation by means of light and electron microscope autoradiography.

It has been shown by many authors that neural tissue can accumulate glutamate even against a large concentration gradient (e.g. Stern et. al., 1949; Lathja, 1968). However, a resurgence of interest in glutamate uptake was stimulated by the discovery that uptake systems with high affinity for L-glutamate existed. A high-affinity uptake system for L-glutamate was recognised in vertebrate CNS preparations (Balcar and Johnston, 1972, 1973; Hammerschlag and Weinreich, 1972; Logan and Snyder, 1972; Bennett et al., 1973). These uptake processes were shown to be highly sodium dependent (Bennett et al., 1973). Sodium-dependent high-affinity uptake systems are thought to be of particular importance in the functioning of synapses involving the release of amine neurotransmitters such as noradrenaline, dopamine and 5HT or amino acids such as GABA and glycine (Fonnum et al., 1980), where inactivation of the neurally released compound is thought to be mediated by active uptake. In vertebrate nervous systems the bulk of evidence suggests that the synaptic action of glutamate is terminated by uptake into nerve terminals and/or glial cells (Cox and Bradford, 1979; Fonnum, 1984). The high-affinity carrier system appears to be specific for L-glutamate as well as D-aspartate and L-aspartate, to the exclusion of other amino acids (Logan and Snyder, 1972; Balcar and Johnson, 1972, 1973). Several investigators have taken advantage of this property by using the uptake of the metabolically inactive isomer D-aspartate to label cells with high-affinity glutamate carriers (Streit, 1980).

Regional variation in glutamate uptake, and changes in this uptake following lesions, are often presented as evidence for glutamatergic transmission (see review by Cotman et al., 1981). However, as these authors point out, glutamate uptake is not likely to be a specific marker for glutamatergic transmission, as aspartate shares the same uptake process. Nevertheless, uptake processes have been extensively surveyed in vertebrate species, and recent reviews include: Cox and Bradford (1979), Fagg and Lane (1979), Shank and Aprison (1979), Vincent and McGeer (1980), Cotman *et al.* (1981) and Fonnum (1984).

Uptake of *L*-glutamate by invertebrate excitable cells

Uptake of *L*-glutamate has also been investigated in several groups of invertebrates, although the bulk of information concerns arthropod species. An understanding of glutamate uptake mechanisms in arthropods is particularly relevant in view of the numerous references to large haemolymph concentrations of glutamate (see review by Florkin and Jeuniaux, 1974). This observation has frequently been used to question the candidacy of glutamate as a neuromuscular transmitter in arthropods, where the bathing glutamate would be expected to interfere with neuromuscular transmission. Usherwood and Cull-Candy (1975) proposed three possible explanations: (1) *L*-glutamate is predominantly in blood cells rather than plasma; (2) *L*-glutamate is in the plasma but is in an inactive state, e.g. bound to blood proteins; (3) *L*-glutamate is in a free state but access to synapses is prevented by a blood/synapse barrier. A number of laboratories have directed their efforts to resolving this problem.

The haemocytes have been shown to be enriched in glutamate and aspartate in insects (Evans and Crossley, 1974) and in crustaceans (Evans, 1972; Miller *et al.*, 1973; Murdock and Chapman, 1974). In contrast, Miller *et al.* (1973) suggested that in insects the total haemocyte concentration was only about 4 per cent of the blood concentration. However, Irving *et al.* (1979) pointed out disparities in centrifugation parameters and other preparation techniques between studies, which may explain these differences, and they found in their own studies higher glutamate levels in the haemocytes than in the haemoplasm. Miller *et al.* (1973) also showed that freshly obtained haemolymph had lower activity in pharmacological assays on insect muscles than 'aged' blood, suggesting that plasma glutamate levels rose during storage. This phenomenon was reduced if haemocytes were removed from the blood on extraction. Irving *et al.* (1979c) concluded that the haemolymph glutamate levels were very low ($<10^{-5}$M in several insect and crustacean species), and that fresh *Lucilia sericata* (sheep blowfly) larval haemolymph had no pharmacological activity, but activity developed with storage at room temperature over a time course of minutes. They attributed this change to an increase in haemolymph glutamate concentration, largely due to conversion of glutamine which was present at high concentration in their haemolymph samples. It was also pointed out by these authors that the high levels of glutamate found by other workers may result from the formation of glutamate from glutamine during preparation of haemolymph samples.

Little evidence has been produced for non-cellular binding of L-glutamate in a pharmacologically inactive form (Clements and May, 1974). However, these authors have suggested the existence of diffusion barriers which could protect the synapse from haemolymph glutamate. Clements and May (1974) described experiments in which perfusion of glutamate into intact locust legs was ineffective but when the retractor unguis muscle was dissected from the leg it contracted when L-glutamate or haemolymph was applied. They concluded that in the intact system diffusion barriers exist which are disrupted during dissection. Clements and May regarded the basal lamina as a more likely candidate for this barrier than the glia or tracheoblasts as the latter structures were apparently intact following dissection. Observations that synapses more deeply embedded between muscle fibres are less susceptible to exogenous glutamate (Usherwood and Cull-Candy, 1975) lend weight to this idea.

Perhaps the clearest data concerning the actions of haemolymph L-glutamate come from examining the effects of injecting L-glutamate into the haemolymph of larval *Lucilia sericata* and adult *Locusta migratoria* (Irving *et al.*, 1979a). Raising the haemolymph concentration of L-glutamate above about 10^{-3}M produced dose-dependent motor impairment. As *L. sericata* has motor terminals free of glial investment, and in view of the similar data obtained from locusts, the implication is that there is unlikely to be a haemolymph/neuromuscular barrier. There is of course no requirement for such a diffusion barrier if glutamate is indeed absent from the haemolymph as seems likely from the work of Irving *et al.* (1979c). If the haemolymph glutamate concentration is $<10^{-5}$M, the question arises as to how the presence of this ubiquitous amino acid is maintained at such low levels.

Uptake systems for L-glutamate have been discovered in the peripheral nerves of crabs *C. maenas* (Evans, 1973) and *Maia squinado* (Baker and Potashner, 1971, 1973) and squid (Baker and Potashner, 1973), as well as cockroach CNS (Evans, 1975). These systems had similar properties in that they were partially sodium dependent with relatively low affinity (Table 2.2), specific for glutamate and similar acidic amino acids, stereospecific and dependent to a degree on metabolic energy. In the crab peripheral nerve preparation, the uptake system appears to be preferentially located in the glial cells (Evans, 1974). However, similar uptake systems for L-glutamate have been detected in frog (Wheeler and Boyarsky, 1968) and rat (Yamaguchi *et al.*, 1970) peripheral nerve, where glutamate is not thought to be involved in neurotransmission. Thus they probably represent a general feature of neural biochemistry rather than having particular relevance to glutamatergic neurotransmission.

In larval *L. sericata* and adult locusts (Irving *et al.*, 1979b) and adult *L.*

Table 2.2: Some properties of invertebrate glutamate uptake systems

Preparation	K_m (µM)	V_{max}	Sodium dependence (%)	Reference
Carcinus maenas, crab: nerve	280	65.4 pmol/ml water/min	Yes (also sodium-insensitive system)	Evans (1973)
Maia squinado, crab: nerve	—	—	Yes	Baker and Potashner (1971)
Homarus americanus, lobster: strips of leg muscle	—	—	50	Freeman *et al.* (1981)
Periplaneta americana, cockroach: CNS	330	15.9 pmol/mg net weight/min	Yes (also sodium-independent system)	Evans (1975)
Lucilia serivata, blowfly: larvae	(Uptake and metabolism in haemolymph, muscle and other tissues examined)			Irving *et al.* (1979b)
Locusta migratoria, locust: various tissues				
Lucilia sericata: larval muscle	1.7	2.1 pmol/g wet wt/min	Yes	Boden (1983)
Schistocerca gregaria, locust: thoracic muscle membranes	21.6	175 pmol/mg protein	Yes	James *et al.* (1977a)
Schistocerca gregaria: thoracic muscle membranes	19.4	200 pmol/min/ mg protein	71	Briley *et al.* (1982)
	664	950 pmol/min/ mg protein	46	
Schistocerca gregaria: femoral muscle membranes	8.82	8.74 µmol/mol ACh/min	Partial	Cleworth (1981)

sericata (Hart *et al.*, 1977), injected ^{14}C-labelled *L*-glutamate was rapidly accumulated by tissues in these insects such as the muscle and fat body using a transport process for which *L*-aspartate could compete. [14C]Amino acid uptake by the body wall muscle of *L. sericata in vitro* (Boden, 1983) had a characteristic time course; following an initial rapid rise, intramuscular labelled free amino acid levels stabilised between 10 and 30 min. The values obtained for *L*-glutamate, *L*-glutamine, *L*-aspartate and *L*-leucine at 30 min are shown in Table 2.3. The glutamate uptake process was found to be highly sodium dependent, being reduced from 1.01 ± 0.18 nmol/g wet weight to 0.181 ± 0.028 nmol/g wet weight in saline with lithium substituted for sodium. The approximate K_m for the glutamate uptake process was 1.7 µM. Incorporation of [14C]glutamate into the soluble fraction of intact extensor tibiae nerve/muscle preparations of the locust *Locusta migratoria in vitro* (Botham, 1977) showed similar characteristics. In both of the previous investigations and in an earlier study

Table 2.3: Amino acid uptake by body-wall muscles of *Lucilia sericata*. Amount of [^{14}C]amino acid in body-wall muscles following 30 min incubation in 500 nM amino acid in physiological saline at room temperature. Results expressed in nmol/g wet weight; each value is the mean of five experiments ± standard error of mean

	Non-protein bound	Protein bound
L-Glutamate	0.918 ± 0.084	0.0093 ± 0.005
L-Aspartate	0.902 ± 0.090	0.0034 ± 0.005
L-Glutamine	0.508 ± 0.064	0.0004 ± 0.002

involving the cockroach *Gromphadorhina portentosa* (Faeder and Salpeter, 1970), stimulation of the neuromuscular preparations prior to incubation in radiolabelled glutamate resulted in enhanced uptake of *L*-glutamate. In *L. sericata* (Boden, 1983) and *L. migratoria* (Botham *et al.*, 1979b), high-frequency stimulation at 100 Hz produced maximal incorporation of glutamate, two or three times higher than that achieved by unstimulated preparations.

Slices of lobster neuromuscular tissue also accumulated [^{14}C]glutamate by a mechanism that was partially sodium dependent. The majority of the sodium-dependent uptake was inhibited by *L*-aspartate, *D*-aspartate, *L*-cysteate or *D,L*-β-hydroxythreoaspartate at a concentration of 0.2 mM (Freeman *et al.*, 1981). It is of interest that the same authors found that *L*-aspartate potentiated the effects of *L*-glutamate in electrophysiological studies. However, their data do not support the idea that this potentiation is a result of inhibition of uptake, as the other compounds that inhibited uptake had no effect on the postsynaptic response.

In an electrophysiological study on another crustacean, *Maia squinado*, Crawford and McBurney (1977) found that the decay of the excitatory postsynaptic current (e.p.s.c.) was prolonged in the presence of *L*-glutamate or *L*-aspartate. They interpreted these data as being a consequence of *L*-aspartate competing for a high-affinity uptake site, thus prolonging the time during which the transmitter (presumably *L*-glutamate) remained in the synaptic cleft. Although these results are compelling, particularly since inactivation by a high-affinity uptake system appears to be widely accepted as the inactivation mechanism for excitant amino acid neurotransmitters in vertebrates, some caution must be exercised as it has been demonstrated that at the locust neuromuscular junction bath application of *L*-glutamate and *L*-aspartate have negligible effects on the time course of e.p.s.c. (Clark *et al.*, 1980). The time courses of e.p.s.c. recorded from the body-wall muscles of *L. sericata* larvae were extended by application of *L*-glutamate but not by *L*-aspartate (Boden, 1983). This phenomenon was interpreted in terms of an effect on the gating kinetics of the glutamate receptor ion channel rather than inhibition of an uptake system. Other agents such as organic mercury compounds also increased the time constant of the e.p.s.c. decay. However, these chemicals also affected the amplitude of the response and might have been mediating their effects directly on receptor ion channel function.

Overall, then, it is clear that sodium-dependent glutamate uptake sites are present in neural and neuromuscular tissue from arthropods and in some cases operate in the high-affinity range. However, it is not yet clear that such systems are directly responsible for the inactivation of neurally released glutamate in arthropods (see Freeman et al., 1981). It has been argued that diffusion from the synaptic cleft may be sufficiently rapid to account for the termination of transmitter action at the locust neuromuscular synapse (Clark et al., 1980). Of course it is also clear that uptake systems have an important role in maintenance of haemolymph concentrations at low levels, protection of the synapse, and presumably other general metabolic functions.

The evidence reviewed so far concerning L-glutamate uptake has concerned intact in vivo or in vitro tissues. The anatomical complexity of these experimental systems impedes a precise understanding of the cellular compartments involved. Two alternative approaches have proved profitable in furthering our understanding of L-glutamate uptake by invertebrate neural and neuromuscular tissues, namely autoradiography and subcellular fractionation.

Autoradiographic localisation of L-glutamate uptake systems

In conjunction with work on the uptake of L-glutamate by crab peripheral nerve (Evans, 1973), the distribution of the accumulated amino acid was examined by light and electron microsope autoradiography (Evans, 1974). The principal finding was that glia were more heavily labelled than axons. An autoradiographic examination of the excitatory neuromuscular junction of G. portentosa also provided evidence that glial cells play a major part in the uptake of L-glutamate, and further that this uptake is enhanced following nerve stimulation (Faeder and Salpeter, 1970). This work and later experiments demonstrating inhibition of the synaptic but not the non-synaptic uptake process by chlorpromazine (Faeder et al., 1974) are often cited as evidence that removal of L-glutamate from the synaptic cleft of insects depends on high-affinity uptake.

This line of work was further extended on locust leg muscle by Botham et al. (1979b), who showed that if preparations were stimulated to fatigue at 100 Hz prior to incubation with L-[³H]glutamate, not only the glial cytoplasm but also the nerve terminal axoplasm had significantly more label than any other structure. They also monitored the nerve terminal morphology over the time course of the experiment and found that the uptake of L-[³H]glutamate was coincident with recovery of the normal vesicle population following stimulus-induced depletion, the implication being that nerve terminal vesicles contain glutamate (Botham *et al.*, 1979a).

Recently attempts have been made to relate autoradiographic data from locust muscle to some of the properties of the uptake systems determined using other radiotracer studies (van Marle *et al.*, 1983, 1984, 1985). The results from this work are encouraging in that it appears that uptake sites in nerve terminals and glia operate by a high-affinity mechanism which is sodium dependent and is inhibited by 0.5 mM aspartate. A further development from this work has been the finding that a toxin from the sphecid wasp *Philanthus triangulum*, δ-philanthotoxin, inhibits the high-affinity uptake of *L*-glutamate (van Marle *et al.*, 1984), but not GABA (van Marle, *et al.*, 1985). This may prove to be an important finding, as there is a particular shortage of tools for neurochemical and pharmacological studies of glutamate binding sites. Progress in this area would be enhanced by agents which could distinguish between transport sites and neuroreceptors, and it will be particularly interesting to see how these compounds perform in binding and uptake assays.

A common finding from *L*-glutamate autoradiography has been the observation that nerve terminal glial cells take up this amino acid avidly. This has led to conjecture that glial uptake might be important in neurotransmission. However, some glutamatergic synapses, e.g. neuromuscular junctions of segmental muscles of larval blowflies (Osborne, 1975), do not have a terminal glial sheath. Examination of the uptake of amino acids by blowfly larvae (Boden, 1983), using light microscopy of whole muscles, demonstrated that [³H]glutamate and [³H]glutamine were accumulated along the length of the segmental motor nerves, but it was not clear whether this accumulation was in the axon or preterminal glia. Using high-resolution electron microscope autoradiography of the nerve terminal and adjacent tissues, Boden (1983) found that glutamate and glutamine were present in higher concentration in the nerve terminal than in the post-synaptic or non-synaptic areas, whereas aspartate was more concentrated in the muscle fibres and sarcoplasmic reticulum. However, following a thorough statistical analysis he concluded that no compartment was significantly more heavily labelled than any other.

Autoradiography has been valuable in elucidating structures with glutamate uptake systems and has also provided some data on the properties of these processes. The technique is, however, time-consuming and has many inherent characteristics which limit the amount of data that are forthcoming. It was pointed out by Briley *et al.* (1982) that one of the reasons for our meagre knowledge of the neurochemistry of insect neuromuscular systems was the lack of well characterised tissue preparations in which to study the cellular and subcellular components of the neurotransmitter systems. This comment can equally well be applied to the neurochemistry of L-glutamate throughout the invertebrates, and is in stark contrast to the number of preparations which have been developed to examine the neurochemistry of amino acids in the vertebrate CNS. This deficiency has nevertheless been addressed by several people, and useful information has been published on both glutamate receptors (considered later) and transport systems in invertebrates.

L-Glutamate uptake by subcellular fractions from invertebrates

Arthropod muscle has always been attractive to neurochemists as a potential source of subcellular fractions for the examination of the biochemistry of neurotransmitter L-glutamate, because of its multi-terminal innervation. It is disappointing that as yet a highly purified nerve terminal fraction has proved elusive. However, recent work (P.J. Richardson, personal communication) on *Nephrops norvegicus* (Dublin Bay prawn) muscle appears to suggest that extensive purification of glutamate decarboxylase (a nerve terminal enzyme) in membrane-bound structures has been achieved using conventional subcellular fractionation techniques.

An extensive fractionation of *Sarcophaga barbata* flight muscle combined with a comprehensive examination of the enzymes associated with L-glutamate metabolism and marker enzymes did not yield fractions apparently enriched in nerve terminals (Donellan *et al.*, 1974). The thoracic muscle of the locust *Schistocera gregaria* was used as the starting material for the production of a subcellular fraction which accumulated L-glutamate (Briley *et al.*, 1982). These authors initially homogenised muscles in locust saline containing 0.25 M sucrose. The homogenate was centrifuged via a two-step procedure to produce a pellet containing mitochondria and membrane vesicles. The pellet (P2) was further separated on a discontinuous Ficoll gradient to produce a pellet (P3) and three bands (B1, B2, B3). B3 contained large membrane-bound bodies (1–3 μm in diameter) and was used to examine the uptake of amino acids (Briley *et al.*, 1982). An examination of marker enzymes suggested that B3 contained a significant enrichment of plasma membrane. L-Glutamate was taken up by a high-affinity (K_m 19.4 μM) and a low-affinity (K_m 664 μM) system into this fraction. The high-affinity site was highly sodium dependent but the low-affinity site was less so. Uptake was also reduced by freeze-thawing or osmotic disruption of the membrane fraction. The kinetics of the transport system for L-glutamate were closely mimicked by L-aspartate and D-aspartate, and both aspartate isomers were potent inhibitors of uptake, although the process was insensitive to D-glutamate. A number of compounds which had been reported to affect L-glutamate uptake in vertebrate preparations (Roberts, 1981) were tested and it was found that none was particularly potent in this system at 100 μM. However, the results obtained were similar to those obtained from vertebrate preparations, and it would be interesting to see how other groups of compounds such as the wasp toxins described by Piek and colleagues (Piek, 1985a) performed in this assay.

In a different study on subcellular fractions of locust leg muscle Cleworth *et al.*, (1980) (see also Cleworth, 1981) obtained a fraction enriched in plasma membrane using a different centrifugation regime from that of Briley *et al.* (1982). A ligand binding assay on this material revealed L-glutamate sites with two affinities. The lower affinity site appeared to have many of the properties to be expected of a transport system. L-Glutamate binding was slow, non-saturable and reduced both in the absence of sodium and following freeze–thaw treatment, and was inhibited by chlorpromazine. In the same study the properties of the putative uptake system in subcellular fractions were compared with those of muscles incubated *in situ*, and a large measure of comparability was found. Uptake of radiolabelled L-glutamate (1 µM) into whole muscles was reduced by L-glutamate (100 µM) to 59 per cent of control values, by L-cysteine sulphinate (100 µM) to 67 per cent and by L-aspartate (100 µM) to 20 per cent. Other glutamate agonists such as quisqualate, ibotenate and kainate were ineffective at 100 µM. However, chlorpromazine (10 µM), which has been shown to reduce glutamate uptake in several systems, including insect muscle (Faeder *et al.*, 1974), reduced uptake in these experiments to 69 per cent.

An earlier examination of glutamate binding sites in locust thoracic muscle (James *et al.*, 1977a) resolved a sodium-sensitive binding site with a lower affinity (K_d 21.6 µM) than the proposed glutamate receptor (K_d 0.53 µM). The lower affinity site was postulated to be an uptake site in part on the basis of its sodium dependency and inhibition of binding by chlorpromazine.

L-Glutamate uptake sites, at least in arthropods, appear to be important in the functioning of L-glutamate as a neurotransmitter. Low- and high-affinity uptake systems in various tissues appear to be involved in maintaining haemolymph concentrations at low levels. High-affinity systems in the nerve terminal and surrounding glia may be involved in maintaining a pool of releasable neurotransmitter. It is not yet clear that high-affinity uptake is the primary mode of neurotransmitter inactivation, but even if a diffusional model is proposed, high-affinity uptake could be important in maintaining a concentration gradient from the vicinity of the receptors. The central position of L-glutamate uptake in glutamatergic neurotransmission implies that a deeper understanding of this process and a more comprehensive range of chemicals to use as probes would facilitate research in comparative neurobiology and in methodology for the control of arthropod pests.

RELEASE OF *L*-GLUTAMATE

Neurotransmitter release from invertebrate nerve endings, where *L*-glutamate is the putative neurotransmitter, has been examined in detail using electrophysiological methods, and excellent reviews have been published (Usherwood and Cull-Candy, 1975; Usherwood, 1978; Atwood, 1976, 1982; Piek, 1985a). It is also worthy of note that the squid giant synapse, which has been used for many studies on basic mechanisms of neurotransmitter release (Miledi, 1967; Llinas and Nicholson, 1975), may well utilise *L*-glutamate as its chemical mediator (Miledi, 1969). In general, neurotransmitter release at these sites has been found to be a quantal process involving calcium ions in excitation–secretion coupling. The presence of electron-lucent vesicles in electron micrographs of glutamatergic nerve terminals has inevitably led to the equation of quantal release with vesicular release. Some evidence for this contention has been derived from examination of synaptic vesicle populations and their response to neural activity (Atwood *et al.*, 1972; Rees and Usherwood, 1972; Mckinlay and Usherwood, 1973; Reinecke and Walther, 1978, 1981; Botham *et al.*, 1979a). A problem with interpreting these data has always been the use of glutaraldehyde fixatives, which in themselves cause depolarisation and might be expected to affect the vesicle population. Recent attempts to visualise the fusion of vesicles during the release process by means of freeze-fracture of unfixed rapidly frozen locust retractor unguis muscle (Newman and Duce, 1984; Newman, 1986) did not result in the clear evidence for vesicular release of acetylcholine produced using this technique (Heuser *et al.*, 1979). However, there may be several reasons for this disparity other than a difference in the release process, such as differences in the amount of transmitter required to elicit a postsynaptic response in the two systems. Non-vesicular release of neurotransmitters has of course been proposed by a number of neuroscientists (see review by Dunant and Israel, 1985) and although a full summary of the evidence cannot be included here, it has been shown, using venom from a polychaete worm, *Glycera convoluta*, that quantal neurotransmitter release can be massively enhanced without affecting vesicle populations (Manaranche *et al.*, 1980) at crayfish excitatory neuromuscular junctions.

The physiology of neurotransmitter release from vertebrate synapses where excitant amino acids are reported to be the chemical transmitter has been examined in less detail, because of the inaccessibility of these central neurones to intracellular recording and presynaptic manipulation, as well as their multiple inputs. Paradoxically a wealth of neurochemical evidence for the release of *L*-glutamate via a calcium-dependent process has been obtained from several areas of the brain *in vivo*, as well as brain slices, hemisected spinal cords, synaptosomes, and glial preparations *in vitro*. Glutamate release has been initiated following depolarising stimuli of

57

various sorts including: elevated extracellular potassium concentration, electrical field stimulation, veratridine, activation of particular pathways; or alternatively by the use of calcium ionophores, venoms or toxins (see reviews by Fagg and Lane, 1979; Cox and Bradford, 1979; Fonnum, 1984).

Release of L-glutamate from intact invertebrate preparations

In vertebrates, release studies provide some of the most compelling evidence for the neurotransmitter status of L-glutamate. This is unfortunately not the case in invertebrates. Publications showing evoked release of L-glutamate are reviewed below but again the lack of a preparation enriched in nerve terminals has impeded progress in this area. More than two decades ago Kerkut et al. (1965) showed that endogenous glutamate was released from perfused legs of shore crabs C. maenas and cockroaches P. americana in sufficient amounts to be measured using thin layer chromatography. In addition, only glutamate was detectable and the amount released was proportional to the number of stimuli applied. A certain amount of criticism has been directed at these findings concerning the methods used and the large amount of glutamate released (Kravitz et al., 1970; Usherwood and Cull-Candy, 1975). An amino acid analyser was used by Usherwood et al. (1968) to measure the evoked release of amino acids from an isolated locust retractor unguis muscle. The release of glutamate, as well as of alanine, glycine and aspartate, was detected, and the amount of glutamate released was modified by changing the stimulus frequency. Glutamate release was increased in elevated calcium concentration and reduced in the presence of magnesium. Application of similar methodology to an isolated preparation from lobster muscle (Kravitz et al., 1970) resulted in the stimulus-evoked release of a small amount of glutamate superimposed on a large background rate of release. In an attempt to resolve some of these differences Daoud and Miller (1976) returned to studying perfused legs of crabs and locusts. An amino acid analyser was used to quantify the amino acids in perfusates before and during stimulation. Stimulation caused an increase in the release not only of glutamate but also of many of the amino acids in the perfusate. A confusing feature of some of this work which was alluded to by Usherwood and Cull-Candy (1975) was the variability of results from preparation to preparation. This problem may in part result from trying to recognise the release of a small pool of neurotransmitter from nerve terminals, superimposed on a system containing considerable amounts of L-glutamate.

In order to overcome this difficulty it would obviously be desirable to label the neuronal glutamate specifically. The location of a high-affinity uptake system at the neuromuscular junction of arthropods should allow preferential loading of structures at this site if labelled glutamate is provided at low concentration. The release of [^{14}C]glutamate from the cheliped opener muscle of crayfish which had been incubated in the labelled amino acid was enhanced by 87 ± 21 per cent above the resting rate during stimulation of the excitatory nerve at 50 Hz (Wang and Boyarsky, 1979). Evoked release was calcium dependent and was only slightly reduced in the presence of [5×10^{-4}M] glutamate, which prevented contraction and eliminated the possibility that muscle movement had led to the release of extracellular labelled glutamate. Release was inhibited by stimulation of inhibitory fibres or bath application of GABA.

In the previous section it was pointed out that exogenous glutamate was shown by several authors to be predominantly taken up by glial cells. It is therefore possible that depolarisation of the nerve terminals in Wang and Boyarsky's experiments led to release of [^{14}C]glutamate from glia and/or nerve terminals.

The problem of glial release may be minimised if a preparation is chosen which is devoid of nerve terminal glia. Boden (1983) examined the release of ^{3}H-labelled L-glutamate from larval $L.$ $sericata$ body-wall muscles. He also prestimulated the segmental nerves supplying the body-wall muscles at 100 Hz for 2 min prior to incubation in L-[^{3}H]glutamate (0.5 μM) for 30 min, conditions which have been shown to increase the labelling of nerve terminal axoplasm in locust motor nerves (Botham et $al.$, 1979b). Preparations were then perfused with saline. After 50 min the nerves were stimulated for 20 min at 50 Hz, and perfusion was continued for a further 50 min following stimulation. Perfusate was collected in 10-min fractions throughout the experiment and the amount of ^{3}H in each fraction was counted (Figure 2.2). During stimulation, release of ^{3}H material was increased, and this increase was abolished in the absence of calcium. When contractions of the muscle were prevented by the addition of L-glutamate (500 μM), a reduction in release of about 32 per cent was observed. The mean release of glutamate in these experiments was 1.05 ± 2 pmol per 10 min per preparation. These results do appear to demonstrate the evoked release of glutamate, but the released tritiated material was not analysed and it might have contained metabolites other than glutamate. Irving et $al.$ (1979b) showed that 80 min after injection of labelled glutamate into the haemolymph 24 per cent of original material remained as glutamate. More recent studies on the release of material from insect neuromuscular preparations (Bates et $al.$, 1985, using methods similar to those of Boden described above) revealed that elevation of extracellular potassium was an effective way of inducing release of preloaded L-[^{3}H]glutamate from $L.$ $sericata$ body-wall muscles (Figure 2.3) and $S.$ $gregaria$ isolated extensor tibiae muscles.

A major advance was made in studies of L-glutamate release by the use of the gas chromatograph–mass spectrometer (GCMS) (Takeuchi et $al.$,

Figure 2.2: The efflux of ^3H from a larval preparation of the blowfly *Lucilia sericata*. The preparation was initially stimulated via the segmental nerves at 100 Hz for 2 min and then incubated for 30 min in saline containing L-[^3H]glutamate, 0.5 μM. The preparation was then perfused with saline at a rate of 0.2 ml/min (bath volume 5 ml). After 50 min the segmental nerves were stimulated for 20 min at 50 Hz (black bar on figure). The level of ^3H monitored in the efflux during this time rose (dotted line). The continuous line was obtained by fitting the curve from the exchange constants for unstimulated preparations.

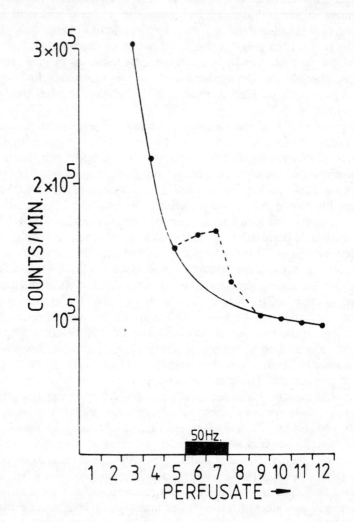

Source: from Boden (1983).

1980). The use of this sensitive instrument has enabled these authors to clearly identify neurally stimulated release of endogenous glutamate (Takeuchi *et al.*, 1980; Kawagoe *et al.*, 1981, 1982). Furthermore the ability of GCMS to distinguish between glutamate and deuterated glutamate (glutamate D_5) allowed experiments to be carried out which monitored the simultaneous release of endogenous glutamate and exogenous glutamate D_5 (Kawagoe *et al.*, 1984). Following incubation in 0.5 mM glutamate D_5 the release of glutamate and glutamate D_5 from crayfish abdominal slow flexor muscle was monitored. Resting release of glutamate D_5 and glutamate was similar. However, neural stimulation produced a large increase in release of endogenous glutamate but no change in the release of glutamate D_5. In contrast, application of 50 mM potassium solution induced a slight but significant increase in glutamate D_5 release in addition to release of endogenous glutamate. The intriguing conclusion from these data is that in this system exogenous glutamate is taken up into a different compartment from neurally released glutamate and that high potassium produces effects on neurotransmitter release different from or in addition to nerve stimulation.

There is of course a discrepancy between these data and the release of exogenous glutamate described by Wang and Boyarsky (1979). One possibility is that in the latter study, in which postsynaptic potentials were not recorded, the stimuli applied were sufficient to depolarise other cells such as glia, or the higher stimulation frequency may have led to elevated potassium levels in the vicinity of the glia, resulting in the release of glial glutamate. It would be interesting to see how electrical field stimulation affected release of endogenous and exogenous glutamate. Differences may of course be due to variability between muscle preparations.

Release of *L*-glutamate from subcellular fractions

As an alternative to examining release from intact neuromuscular preparations, Briley *et al.* (1982) incubated the B3 fraction from locust thoracic muscle (see above) with radiolabelled *L*-glutamate or *L*-aspartate. The material was washed and resuspended in normal saline or media containing 55 mM potassium with or without calcium. Elevated potassium produced a 10- to 20-fold increase in acidic amino acid release, but this effect was independent of calcium. This is in contrast to experiments on mammalian brain synaptosomes, where the release of endogenous glutamate (de Belleroche and Bradford, 1972; Osborne *et al.*, 1973) and preloaded amino acids (Levy *et al.*, 1973) was found to be calcium dependent. However, there have been a significant number of experiments where the potassium-induced release of radiolabelled amino acids from mammalian brain has been shown to have a calcium-independent component (Fagg

Figure 2.3: Efflux of L-[³H]glutamate from larvae of *Lucilia sericata* and the effect of muscle depolarisation using elevated potassium (a) and electrical stimulation (b) at 50 Hz (S.E. Bates, unpublished data). The preparations were stimulated at 100 Hz for 5 min, and immediately incubated in 0.5 µM L-[³H]glutamate for 30 min. After incubation the larvae were perfused with saline at a rate of 1 ml/min (bath volume 1 ml). Two-millilitre fractions were collected and the levels of L-glutamate in the effluent were measured. After 30 min the preparations were depolarised for 10 min (shaded area of histogram) and perfusion was continued for a further 30 min. In both cases depolarisation resulted in an increase in ³H in the perfusate but the electrical stimulation always resulted in a smaller increase than elevated extracellular potassium.

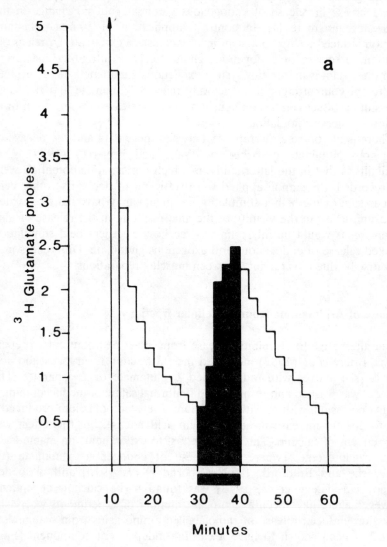

Figure 2.3: (continued)

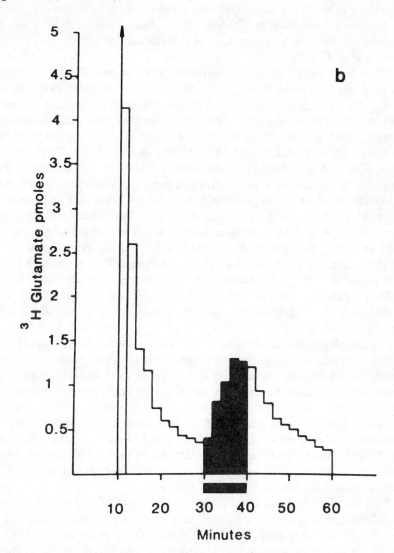

and Lane, 1979). In view of the non-neuronal plasma membrane and sarcoplasmic reticulum which are probably present in B3 (Briley *et al.*, 1982), non-neuronal uptake and release is likely to predominate in these studies. Cotman *et al.* (1976) proposed that neurotransmitter release would be a calcium-dependent process, and they carried out experiments in which GABA-loaded synaptosomes were perfused with high-potassium zero-calcium medium to evoke calcium-independent release. Subsequently

63

'neurotransmitter GABA' was released by addition of calcium. It would be interesting to see if calcium-dependent release like that described in insect whole muscle preparations (Boden, 1983; Bates *et al.*, 1985) could be detected in subcellular fractions using methodology similar to that of Cotman *et al.* (1976).

It has been pointed out that there are inherent problems in interpretation of data obtained from studies on the release of preloaded labelled amino acids (Fagg and Lane, 1979). The amounts of material released from invertebrate preparations may be estimated from recent studies. Using radiolabelled glutamate the amounts range from 0.1 to 100 pmol per 10-min period of stimulation at 50 Hz (Wang and Boyarsky, 1979; Boden, 1983; S.E. Bates, personal communication). The specific activity of the releasable material is of course unknown, and total amounts released may be higher. However, the implication from these studies is that very sensitive analytical methods will be required to monitor this process. The experiments of Takeuchi and colleagues using GCMS measured release in the range of 1–10 pmol per 10-min period. These experiments appear to be the most significant yet, but GCMS is a technique that is unlikely to be routinely available to the majority of invertebrate neurochemists. Other modern techniques such as HPLC and derivatisation of amino acids with agents such as *o*-phthaldehyde would appear to be potentially useful methodologies.

Calcium-dependent evoked release of glutamate has been described in several systems, and may now be regarded as tangible evidence that *L*-glutamate is a neurotransmitter at arthropod neuromuscular junctions. It would be interesting to examine other invertebrate systems where *L*-glutamate may be a transmitter, to see if the release process can be identified. Insect central neurones have been shown to be sensitive to *L*-glutamate (Usherwood *et al.*, 1980; Giles and Usherwood, 1985; Wafford and Sattelle, 1986), and the recently developed insect CNS synaptosome preparation (Breer, 1981, 1986) should be an ideal system in which to study *L*-glutamate release.

Reproducible methods for studying *L*-glutamate release would help to provide insight into phenomena such as long-term facilitation (Atwood *et al.*, 1975). Questions such as whether releasable *L*-glutamate is found in vesicles or in a cytoplasmic pool (de Belleroche and Bradford, 1977) might also be answered in well characterised invertebrate preparations.

A variety of tools which can manipulate the release process have been discerned from electrophysiological work, and it would be of interest to see their effect on the neurochemistry of *L*-glutamate release. For example, *Habrobracon hebetor* venom blocks synaptic vesicle release, but does not prevent calcium influx (Walther and Reinecke, 1983), whereas black widow spider venom produces a large increase in quantal release (Cull-Candy *et al.*, 1973). The effects of these and other venoms have recently been reviewed (Piek, 1985b; Zlotkin, 1984) and should prove useful in our attempts to understand the synaptic release of L-glutamate.

RECEPTORS FOR L-GLUTAMATE

Once L-glutamate has been released, it is thought to function as a neuro-transmitter by binding to specific receptors on the postsynaptic cell. Freeze-fracture electron microscopy of the postsynaptic membrane has revealed the presence of arrays of intramembranous particles (IMPs) in several putative glutamatergic synapses. Postsynaptic IMPs at cholinergic neuromuscular junctions and electric organ synapses have been positively identified as nicotinic acetylcholine receptors. An interesting difference between the ultrastructure of these two types of chemical synapse relates to the fracturing characteristics of these IMPs. In cholinergic synapses they fracture with the P-face of the postsynaptic membrane (Rash and Ellisman, 1974; Heuser and Salpeter, 1979) whereas in insect muscle (Rheuben and Reese, 1978; Newman and Duce, 1983, 1985; Figure 2.4), crustacean muscle (Franzini-Armstrong, 1976), mammalian cerebellum (Gulley and Reese, 1977) and cochlear nucleus (Gulley *et al.*, 1977), where glutamate may be the neurotransmitter, postsynaptic IMPs fracture with the E-face. The functional significance of the different fracturing properties of these postsynaptic proteins is unknown.

Binding of glutamate to postsynaptic receptors results in a change in membrane permeability which affects the excitability of the target cell. The properties of glutamate receptors and their associated ion channels have been examined physiologically, pharmacologically and biochemically, and the synthesis of information from these fields has resulted in our current understanding of these molecules. The information which is available is naturally determined by the techniques used to acquire it.

Electrophysiological studies on vertebrate brain have always been hindered by the inaccessibility of deep-lying structures, so that much of the earlier work involved extracellular recording coupled with ionophoretic administration of L-glutamate and potential agonists and antagonists. This work generated much valuable information on the pharmacology of glutamate receptors but less about the mechanisms underlying changes in excitability (see reviews by Puil, 1981; Watkins and Evans, 1981). However, recently the use of brain slices and cultured cells has enabled methods such as patch clamp and voltage clamp to resolve more details of glutamate receptor function (e.g. Collingridge *et al.*, 1983; Nowak *et al.*, 1984; Crunelli *et al.*, 1985).

As a result of a close collaboration between organic chemists, who have synthesised or isolated novel acidic amino acids and analogues, and pharmacologists and neurochemists, it has been possible to define four

65

Figure 2.4: Freeze-fracture electron micrograph of a locust retractor unguis muscle. A large area of the E-face of the muscle (EM) is exposed with the characteristic particle array found at the postsynapse (PSS). This array may represent neurotransmitter receptors. The P-face of a motor nerve is exposed at the synapse (PN). Magnification × 30 000. Courtesy Dr T.M. Newman, unpublished micrograph

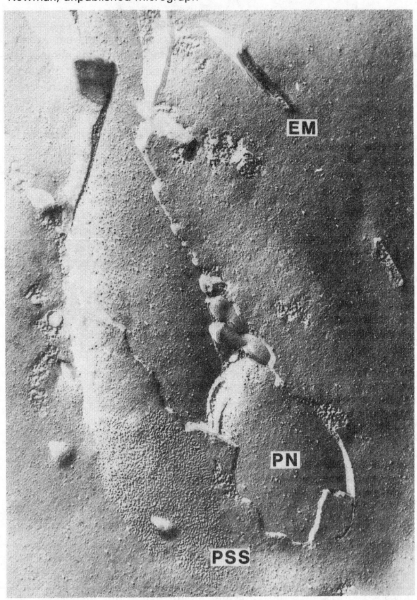

classes of receptor for acidic amino acids (Table 2.4) (Foster and Fagg, 1984; Fagg, 1985). Neurochemical examination of vertebrate brain L-glutamate receptors was initiated in the early 1970s using ligand binding assays to measure the specific binding of L-glutamate to CNS membranes (Michaelis et al., 1974; Roberts, 1974). The ligand binding technique involves incubation of a suitable preparation containing receptors with a radiolabelled ligand during which time binding takes place. It is then necessary to separate the receptor fraction with its bound ligand from the unbound ligand, usually by centrifugation or filtration. The amount of bound radioactive ligand can then be counted, and with a suitable control for non-specific binding the kinetics and characteristics of the process can be determined. This technique has been considerably refined in its application to acidic amino acid binding to vertebrate nervous tissue, and the methodology and the relevance of current data to the scheme in Table 2.4 were thoroughly reviewed by Foster and Fagg (1984).

Ligand binding studies of high-affinity sites (most authors have found K_d values in the range 100–1000 nM for L-glutamate) require ligands with high specific radioactivity. Ligands with these characteristics have only recently become available for acidic amino acid binding sites: indeed high-specific-activity forms of some diagnostic ligands, e.g. quisqualic acid, are not yet available. Although this deficiency can be overcome to some extent, undoubtedly the major impediment to studies of glutamate receptors by pharmacological, anatomical or biochemical techniques is the continuing absence of a specific ligand for glutamate receptors with irreversible or slowly reversible binding properties.

Table 2.4: Classification of acidic amino acid receptors in vertebrate brain

Nomenclature:	A1	A2	A3	A4
Trivial name:	NMDA	Quisqualate	Kainate	L-APB
Most potent and selective agonists	NMDA, ADCP, ibotenate	Quisqualate, AMPA	Kainate, Domoate	L-APB (unclear whether an agonist or antagonist)
Most potent and selective antagonists	D-AP5, D-AP7, Asp-AMP	Glu-tau, GAMS, GDEE		
Radioactive ligands available	L-glutamate, D-AP5	L-glutamate, AMPA	L-glutamate, kainate	L-glutamate, D,L-APB

Source: redrawn from Foster and Fagg, 1984; Foster, 1985

Abbreviations: NMDA, N-Methyl-D-aspartate; ADCP, 1-amino-1,3-dicarboxycyclopentane; AMPA,α-amino-3-hydroxy-5-methyl-isoxazole-4-propionic acid; L-APB, L-2-amino-4-phosphonobutyric acid; D-AP5, D-2-amino-5-phosphonovalerate; Glu-tau, γ-D-glutamyltaurine; D-AP7, D-2-amino-7-phosphonoheptanoate; GAMS, γ-D-glutamylaminomethylsulphonate; Asp-AMP,β-D-aspartylaminomethylsulphonate; GDEE, glutamate diethylester

Results from binding studies confirm observations from work on L-glutamate release and electrophysiology that L-glutamate is probably a widespread neurotransmitter in the vertebrate CNS. This fact, coupled with the high density of synaptic connections in the brain, ensures that vertebrate brain is a good starting material for biochemical investigations of glutamate receptors. Most authors find the density of binding sites in mammalian brain to be in the range 10–100 pmol/mg of protein.

L-Glutamate receptors in the central nervous system of invertebrates

Central glutamate receptors in invertebrates are less thoroughly studied than those in the periphery, and even in annelids (Walker *et al.*, 1980), molluscs (Gerschenfeld, 1973; Oomura *et al.*, 1974; Kehoe, 1978; Usherwood, 1978; Walker *et al.*, 1980) and arthropods (Usherwood *et al.*, 1980; Walker *et al.*, 1980; Wafford and Sattelle, 1986), where they have been examined electrophysiologically, their density is unknown. Invertebrate CNS has not frequently been used for binding studies on L-glutamate receptors. However, recent demonstrations of specific binding to receptor sites in snail *H. aspersa*, CNS (Pin *et al.*, 1986), honeybee and housefly CNS (Sherby *et al.*, 1986) and in locusts CNS in our own laboratory (G. Usoh, personal communication), suggest that such studies may prove profitable.

L-Glutamate binding to *H. aspersa* brain was to a single saturable site with a K_d of 0.12 μM and B_{max} 30 pmol/mg protein (Table 2.5) (Pin *et al.*, 1986). Kainate was the most potent inhibitor of glutamate binding, and the properties of [^{3}H]kainate binding suggested a single kainate/glutamate binding site. The pharmacology of this glutamate binding does not unfortunately correspond to the pharmacology of L-glutamate responses described in gastropod CNS (Oomura *et al.*, 1974; Kehoe, 1978; Leake and Walker, 1980; Walker *et al.*, 1980). One possible explanation for this may be that the membrane preparation used in these studies was prepared from whole CNS which has an extensive epineural sheath containing muscles and other cells. Pin *et al.* (1986) were able to show that the majority of binding sites were present in the sheath material rather than in the neural tissue, and they were unable to characterise the binding sites on the neurones. It is therefore possible that this study is predominantly concerned with muscle glutamate receptors. This point is an important one, and the heterogeneity of tissues associated with invertebrate central ganglia, including those of insects, should perhaps be considered in future investigations, particularly where 'whole head' preparations are involved.

An extensive study has recently been carried out on insect glutamate receptors, comparing the properties of binding sites on housefly and honeybee central nervous system and thoracic muscle and also including

Table 2.5: Glutamate binding sites in invertebrate nerve and muscle tissues

Preparation	K_d (μM)	B_{max} pmol/mg protein	Reference
Schistocerca gregaria, locust: thoracic muscle membrane proteolipids	0.5	—	Lunt (1973)
Artemisia longinaris, brine shrimp: muscle	13		Fiszer de Plazas and De Robertis (1974)
Musca domestica, housefly: leg muscle proteolipids	2	12 000	Fiszer de Plazas *et al.* (1977)
Schistocerca gregaria: thoracic muscle proteolipids	8 50 (small amount of low-affinity binding in separate proteolipid peak)	7200	James *et al.* (1977)
thoracic muscle membranes	0.53	25	James *et al.* (1977a)
	30	52	Filbin *et al.* (1980)
	0.012	5.7	Briley *et al.* (1982)
	1.3	11	
femoral muscle membranes	0.3	0.75 μmol/mol ACh/min	Cleworth (1981)
Musca domestica, housefly: thoracic muscle membranes	0.5	5–60 (mean 37)	Filbin *et al.* (1985)
	0.9	23	Sherby *et al.* (1986)
Apis mellifera, honeybee: thoracic muscle membranes	0.87	7.6	Sherby *et al.* (in press)
Musca domestica: brain membranes	0.63	5.7	Sherby *et al.* (in press)
Apis mellifera: brain membranes	1.36	3.8	Sherby *et al.* (in press)
Periplaneta americana, cockroach: nerve cord	0.16	5	Sherby *et al.* (in press)
Schistocerca gregaria: nerve cord	0.02	0.3	G. Usoh (personal communication)
Helix aspersa, snail: perioesophageal ganglia; CNS	0.12	30	Pin *et al.* (1986)

observations on cockroach central ganglia and rat brain (Sherby *et al.*, 1986). This work suggests that high-affinity sites for L-glutamate are present in housefly, honeybee and cockroach CNS with a similar abundance (Table 2.5). The density of binding sites appears to be lower on CNS membranes than on those of thoracic muscles, particularly in housefly where B_{max} for thoracic muscle is three to four times higher. However, B_{max} values are all lower than for rat brain. The levels of binding in these experiments were initially found to be very low, but addition of 2.5 mM calcium chloride enhanced binding by 50 to 100 per cent, most of this

69

enhancement being attributable to chloride ions. Displacement of L-glutamate binding by various analogues revealed that L-aspartate was the most potent inhibitor followed by L-glutamate, L-cysteate and ibotenate. The sites are, however, relatively insensitive to L-2-amino-phosphono-butyric acid (L-APB) and thus differ from the chloride-sensitive sites in the mammalian brain (Foster and Fagg, 1984).

Both the experiments of Pin *et al.* (1986) and Sherby *et al.* (1986) used filtration to separate bound and free ligands. In our laboratory we are currently using a centrifugation assay to examine the binding of glutamate to membranes produced from dissected nerve cords of locusts *S. gregaria.* We have found that this method has a number of advantages over the filtration assay for small amounts of CNS tissue. As pointed out by Filbin *et al.* (1980), the dissociation of L-glutamate from its receptor is rapid and the filtration assay which involves washing of the ligand receptor complex is likely to produce dissociation, which may be a particular disadvantage where a low density of binding sites is suspected. Using this method we have found high-affinity binding of L-glutamate to locust CNS neurones (Table 2.5). This material has yet to be fully characterised.

L-Glutamate receptors in arthropod muscle

Of all types of L-glutamate receptors, those on arthropod muscle are arguably the most thoroughly characterised electrophysiologically. Their accessibility to microelectrode recording has enabled detailed examination of such features as: the structural requirements for ligand potency; ionic currents gated by glutamate receptors (Usherwood and Cull-Candy, 1975; Nistri and Constanti, 1979; Leake and Walker, 1980); desensitisation (Takeuchi and Takeuchi, 1964; Usherwood and Machili, 1968; Dudel, 1975; Clark *et al.*, 1979; Shinozaki and Ishida, 1979; Stettmeir *et al.*, 1983a). Analysis of single-receptor ion channel complexes using current noise analysis (Anderson *et al.*, 1976; Cull-Candy and Miledi, 1980; Stettmeir *et al.*, 1983b) or patch clamp (Patlak *et al.*, 1979; Gration *et al.*, 1981a, 1982; Cull-Candy and Parker, 1982) has now provided precise information concerning the behaviour of these macromolecules, and a model has been produced to describe the gating of the locust glutamate receptor channel complex (Kerry *et al.*, 1986).

Arthropod muscle is therefore an obvious source of material for binding studies, and experiments to characterise glutamate binding sites from insect muscle were carried out in the early 1970s by Lunt (1973). These experiments involved production of a crude membrane fraction from locust thoracic muscle which was then solubilised in the detergent deoxycholate, or in chloroform/methanol. Binding to the detergent extract was measured by equilibrium dialysis and revealed a high-affinity binding site for gluta-

70

mate which did not bind aspartate or glutamine. The dissociation constant is shown in Table 2.5. The proteolipid extract solubilised in chloroform/methanol was labelled with [^{14}C]glutamate and separated on a Sephadex LH20 column to produce two proteolipid peaks with bound radioactivity.

A fuller characterisation of L-glutamate binding to locust muscle proteolipids produced a number of interesting observations (James *et al.*, 1977b). Two fractions with K_d 50 μM and K_d 8 μM were separated. The latter high-affinity binding proteolipid peak accounted for the vast majority of the glutamate binding and was further characterised in terms of its binding and protein content. Glutamate binding was relatively insensitive to N-methyl-D, L-aspartic acid, L-leucine or D-glutamate, but 10 μM glutamate diethyl ester, 10 μM N-methyl-D, L-glutamic acid or 5 μM L-aspartic acid reduced binding of 5 μM L-glutamate by 9.8, 10.6 and 47 per cent respectively. SDS polyacrylamide gel electrophoresis detected two prominent protein bands in this fraction with molecular weights of 77000 and 95000 daltons. In a separate publication combining this technique with electrophysiology, D,L-2-amino-4-phosphonobutyric acid was suggested to behave as an antagonist at locust muscle glutamate receptors (Cull-Candy *et al.*, 1976).

When hydrophobic proteins were extracted from housefly legs and examined using a glutamate binding assay based on LH20 Sephadex (Fiszer de Plazas *et al.*, 1977), binding appeared to be to a single site with an apparent dissociation constant of 3 μM and absolute stereospecificity for L-glutamate. Binding of L-glutamate (5 μM) was inhibited by 500 μM glutamic acid diethyl ester and 200 μM L-nuciferine by 28 and 59 per cent, respectively. Crustacean muscle has also been examined using chloroform/methanol extraction and Sephadex chromatography (Fiszer de Plazas and De Robertis, 1974). These brine shrimp, *Artemisia longinaris*, proteolipids bound glutamate with a K_d of 13 μM.

The early studies by Lunt and colleagues described above were particularly ambitious in seeking to isolate the receptor complexes from their membranes as well as examining their properties. Some criticism has been levelled at receptor isolation based on chloroform/methanol extraction (Levinson and Keynes, 1972). However, as pointed out by Filbin *et al.* (1980), there is also evidence that proteolipid extraction may be a valid method for isolating receptors (Taylor, 1978). Isolation of receptor complexes may ultimately be necessary for the full characterisation of receptor properties, but this approach necessarily involves complex and time-consuming methodology. It may therefore be more appropriate for studies on the pharmacology, kinetics and ion dependency of glutamate binding to be carried out on membrane fractions where the receptor is presumably operating in its normal phospholipid environment in an aqueous medium.

In tandem with their work on proteolipids James *et al.* (1977a) carried

71

out a series of experiments using a microsomal fraction from locust thoracic muscle to provide additional information on glutamate binding sites. Differential centrifugation yielded a microsomal fraction (P3) which was low in glutamate-metabolising enzymes and was enriched in glutamate binding sites. The presence or absence of sodium ions in the incubation medium revealed two types of glutamate binding. In the absence of sodium a high-affinity binding site (K_d 0.53 μm, B_{max} 25 pmol/mg protein) predominated, whereas in the presence of sodium a low-affinity binding site was detected which was proposed as a glutamate transport site and has already been discussed under 'uptake of glutamate'.

Displacement of [^{14}C]glutamate from the high- and low-affinity sites was compared using agents which have been suggested either to affect glutamate uptake or alternatively to act as agonists or antagonists at neurotransmitter receptors. James *et al.* (1977a) found that uptake blockers such as chlorpromazine or L-glutamyl-γ-hydroxamate reduced binding to low-affinity sites, whereas agonists such as quisqualate or the antagonist D,L-2-amino-4-phosphonobutryic acid (which reduced glutamate binding to proteolipid extracts) (Cull-Candy *et al.*, 1976) were potent competitors for high-affinity binding sites. L-Aspartate blocked binding to both sites. Competing ligands were all used at high concentrations (100 μM), and it is therefore difficult to compare their relative potencies.

The techniques used in the work described above, particularly details of the filtration assay, were reappraised by Filbin *et al.* (1980). In sodium-free medium they detected a single binding site for L-[^{3}H]glutamate with a dissociation constant K_d of 30 μM and B_{max} of 52 pmol/mg. A separate [^{3}H]kainate binding site was also found, but no interaction between glutamate and kainate at these sites was detected. This result is in concordance with electrophysiological evidence that kainate is a poor agonist at locust L-glutamate receptors (Daoud and Usherwood, 1975) and the observation that separate kainate receptors where glutamate is only slightly effective are found in vertebrates (Foster and Fagg, 1984). In a later paper Lunt and colleagues (Briley *et al.*, 1982) had further modified their membrane preparation and developed a centrifugation assay to examine [^{3}H]glutamate binding. They reported the presence of two sodium-independent sites with K_d values of 12.5 nM and 1.3 μM and B_{max} values of 0.57 pmol/mg and 10.9 pmol/mg, respectively. However, they emphasised that binding studies on insect muscle membranes remained capricious and results inconsistent, a point of view which has been expressed by several other groups working in this field!

Despite such problems glutamate binding studies have continued in other laboratories. Recently specific binding to thoracic muscles of housefly and honeybee was described by Eldefrawi and colleagues (Eldefrawi *et al.*, 1985; Filbin *et al.*, 1985; Sherby *et al.*, 1986).

In the second of these studies (Filbin *et al.*, 1985) a centrifugation assay was used to examine binding of 100 nM [³H]glutamate in sodium-free conditions. Apparently binding was only seen in the presence of protease inhibitors, and the level of specific binding was small relative to non-specific binding. However, high-affinity stereospecific binding with a K_d of 0.5 μM was detected. The effect of inhibitors was interesting. The order of potency for displacing L-glutamate (100 nM) by various putative ligands (1 mM) (expressed as per cent of control binding remaining) was: L-glutamate (0), L-aspartate (0), L-glutamate diethylester (34), D-glutamate (39), N-methyl-D-aspartate (NMDA) (40), ibotenate (52), N-methyl-DL-aspartate (78), quisqualate (84), N-methyl-L-aspartate (86), kainate (97), D-aspartate (100). Overall these data contain features expected of a glutamate receptor; however, the relatively low displacement of glutamate by quisqualate, which is a more potent agonist than glutamate at depolarising receptors on locust muscle (Gration et al., 1981b), appears anomalous. This may reflect the high concentration of ligands used in these competition experiments, and IC_{50} values would give a much clearer impression of their effectiveness. Alternatively species or tissue variability might underlie the difference. This observation does, however, cast some doubts on whether the population of binding sites examined in this work contains more than a low density of synaptic receptors.

Recent work in Eldefrawi's laboratory (Sherby et al., in press) employed a filtration assay to examine L-glutamate binding to insect brain membranes (see above) and thoracic muscle. In this system protease inhibitors were apparently not required, but assays took place in the presence of 2.5 mM calcium chloride, as binding of 10 nM L-[³H]glutamate was enhanced 50–100 per cent in the presence of chloride ions. L-glutamate binding was saturable with a K_d of 0.9 μM for housefly and 0.87 μM for honeybee, with B_{max} values of 23.1 and 7.6 pmol/mg protein, respectively. The binding was about 75 per cent dissociable within 1 h of addition of L-glutamate 100 μM, and calculating the equilibrium dissociation constant from dissociation and association rates a K_d of 0.55 μM was determined for housefly thoracic muscle. IC_{50} values were derived for several putative ligands, and the order of potency shows some interesting differences from the previous work of Filbin et al. (1985). The binding was sterospecific and the L-isomers of glutamate and aspartate were the most potent ligands in inhibiting [³H]glutamate binding. Other effective amino acids in inhibiting binding were quisqualate> ibotenate> L-cysteate. Compounds such as dihydrokainate and NMDA were ineffective, suggesting that the binding sites in these experiments did not conform to the properties of either the kainate or NMDA receptors of mammalian brain (Foster and Fagg, 1984).

At least three populations of L-glutamate receptors have been detected on insect muscles (Cull-Candy and Usherwood, 1973). Junctional receptors and extrajunctional D receptors which cause depolarisation at normal membrane potential by gating cationic channels are particularly sensitive to quisqualate, whereas extrajunctional H receptors which gate channels for chloride ions and usually produce hyperpolarisation are activated by ibotenate (Lea and Usherwood, 1973). Junctional receptor populations were suggested to be heterogeneous on the basis of structure activity studies (Gration, 1980). The majority of receptors appear to have a preference for L-glutamate in a partially folded configuration. However, using L-aspartate (a fully folded glutamate analogue) and ibotenate (a fully extended glutamate analogue), subpopulations of junctional glutamate receptors were discovered with high affinity for these alternative configurations. It would therefore be of great interest to see whether ibotenate, quisqualate and aspartate inhibited binding to the same population of receptors and whether the enhancement of binding produced by chloride ions was ibotenate sensitive. It does not appear from the work of Sherby *et al.* (1986) that insect muscle L-glutamate binding sites conform precisely to any of the excitant amino acid receptor subtypes proposed for mammalian brain (Foster and Fagg, 1984). However, until more electrophysiological and neurochemical data have been gathered from invertebrate preparations with regard to the effects of some of the ligands shown to have specificity and high potency on mammalian receptors (Fagg, 1985), similarities and differences will remain obscure.

Thoracic muscle of locust is a plentiful source of muscle membrane, but it has some disadvantages in that it is likely to be less densely innervated and certainly its pharmacology is less well characterised than the muscles in the femur of the locust. With this in mind, L-glutamate binding to femoral muscle membranes was examined (Cleworth *et al.*, 1980; Cleworth, 1981). A filtration assay was used to look at L-[³H]glutamate binding to a microsomal fraction from locust femoral muscle. Acetylcholinesterase activity, which was used in these studies as a marker for plasma membrane, was enriched in this fraction. A sodium-independent binding site was detected with a K_d of 0.3 μM. Binding to this high-affinity site was reduced by aspartate (100 μM) or cysteine sulphinate (100 μM) by 30 to 50 per cent, but quisqualate (100 μM) or ibotenate (10 μM) produced only slight reductions in binding. One peculiar feature of the high-affinity binding site described by Cleworth (1981) was increased binding of L-[³H]glutamate in the presence of low concentrations ($<$ 1 μM) of unlabelled ligands including glutamate, cysteine sulphinate, kainate, quisqualate, dihydroxyphenylalanine and concanavalin A. This was ascribed to positive cooperativity between the [³H]glutamate binding site and ligand binding elsewhere to the receptor complex. There was also some evidence that denervation of muscles two weeks prior to preparation of the membrane fraction led to an increase in the number of high-affinity binding sites. The main problem encountered in this work was the consistent presence of glutamate transport systems and other contaminants in the membrane fraction used for assay.

74

We have recently attempted to obtain material from locust femoral muscle with more desirable properties for ligand binding studies (Quicke *et al.*, 1986). Some progress has been made in this respect. An initial homogenisation in 5 mM Tris acetate followed by a low-speed centrifugation step produced a supernatant which is almost a pure suspension of small membrane vesicles. This material is enriched in plasma membrane, on the basis of binding of the lectin concanavalin A which was used in these studies as a plasma membrane marker. The membrane was pelleted and resuspended in 50 mM Tris acetate and, following several washes, L-[^{3}H] glutamate binding was examined using a centrifugation assay. Specific L-glutamate binding was detected and could be displaced by quisqualate, ibotenate and aspartate. A fuller characterisation of this material is in progress.

Future directions for neurochemical studies on invertebrate L-glutamate receptors

The work reviewed above demonstrates that, despite technical problems imposed by invertebrate preparations, significant progress has been made in developing an understanding of glutamate binding, and it is beginning to be possible to make comparisons with binding data on vertebrate acidic amino acid receptors. It is clear that populations of glutamate receptors examined in these experiments are heterogeneous, and pharmacological differences will probably be the only way of distinguishing them. So far most of the work carried out on invertebrates has relied on the use of ligands to displace radiolabelled glutamate. This method is not ideal in view of the many functions fulfilled by glutamate which involve binding to a receptor site. Binding studies on other neurotransmitters in both invertebrates and vertebrates have benefited from the use of ligands other than the transmitter itself (e.g. bungarotoxin for the nicotinic acetylcholine receptor, QNB for the muscarinic acetylcholine receptor, muscimol for the GABA receptor). Recently a number of alternative radiolabelled glutamate receptor ligands have become available, but most of these are directed towards the NMDA receptor, probably the least valuable group from evidence so far obtained in invertebrates. In particular, labelled quisqualate and ibotenate, whose effects on insect muscle are well known, would be very useful tools.

An alternative approach which has been suggested by Eldefrawi and colleagues (Eldefrawi *et al.*, 1985) might be to examine the allosteric effects which channel blockers at glutamate receptor linked ion channels might have on glutamate binding. Examples of compounds in this category include curare and gallamine (Cull-Candy and Miledi, 1983), streptomycin and ketamine (Boden *et al.* in preparation), diltiazem (Ishida and Shino-

GLUTAMATE

zaki, 1980), chlorisondamine. (Lingle *et al.*, 1981), and MLV 5860 (Shinozaki and Ishida, 1986). Of course radiolabelled derivatives of such compounds could be used to detect the ion channel component of the receptor ion channel complex. Some of the compounds mentioned above have fairly slow unblocking rates, but more promising in this respect are the properties of some arthropod venoms. δ-Philanthotoxin is a potent channel blocker at insect glutamate receptors and its effects are well characterised (Piek, 1985a, b). The channel block which it causes appears to be very slowly reversible in the absence of glutamate, but activation of the receptor complexes results in rapid unblocking. Toxin from the spider *Nephila clavata* was reported to act as a glutamate antagonist at crustacean, squid and mammalian receptors (Kawai *et al.*, 1982a, b, 1983a; Abe *et al.*, 1983). Venom components from other orb-web spiders, *Araneus ventricosus* (Kawai *et al.*, 1983b) and *Argiope lobata* (Tashmukhamedov *et al.*, 1983), also antagonised glutamate receptors on crustacean and insect muscle respectively. The mode of action of spider venom from orb-web spiders on insect glutamate receptors appears to involve channel block which is activation induced and slowly reversible; recovery is slowed by activation of the receptor (Usherwood *et al.*, 1984; Bateman *et al.*, 1985; Usherwood and Duce, 1985). The active constituents include compounds with molecular weight of less than 1000 daltons which are effective at concentrations in the nanomolar range. These molecules are in the process of further characterisation and are potentially useful tools for identification of glutamate receptor ion channel components and their purification using affinity chromatography. A preliminary report of isolation of glutamate receptors from insects using spider venom and their reconstitution in a functional form in an artificial membrane has been published (Usmanov *et al.*, 1985).

Ultimately the isolation and purification of glutamate receptors is a goal aspired to by vertebrate and invertebrate neurobiologists. A glutamate binding protein from rat brain has been purified and characterised (Michaelis *et al.*, 1983), and immunological and binding studies have been carried out on this material by Michaelis and colleagues. The potential of slowly reversible toxins or the development of photoaffinity labels might enhance the likely yield of glutamate binding protein from invertebrates and enable invertebrate neurochemists to progress in this direction.

If it does prove possible to isolate glutamate binding material from an aqueous preparation of invertebrate tissue, its properties can be compared with preparations such as the proteolipid glutamate binding fractions described by Lunt and co-workers. Reconstitution of a functional receptor into artificial membranes would be of great value in understanding the kinetics of glutamate receptors.

One aspect of receptor function which is well known but less well understood is desensitisation. Desensitisation of *L*-glutamate receptors has

76

been described for insect muscle (Usherwood and Machili, 1968; Daoud and Usherwood, 1978), crustacean muscle (Takeuchi and Takeuchi, 1964; Dudel, 1975) and rat olfactory cortex (Braitman, 1986). In insect muscle, onset and recovery from desensitisation are agonist-dependent phenomena (Anis *et al.*, 1981). Several mechanisms have been suggested to account for desensitisation (Triggle, 1981), including a change in receptor conformation or an increase in affinity for the agonist. Invertebrate peripheral glutamate receptors may be particularly helpful in examining these possibilities because desensitisation can be blocked by lectins such as concanavalin A in insects (Mathers and Usherwood, 1976; Evans and Usherwood, 1985) or crustaceans (Shinozaki and Ishida, 1979). In *Aplysia* or *Helix* central ganglia, concanavalin A had a different effect on glutamate responses, altering inhibitory glutamate responses into excitatory responses, but it had no effect on ibotenate or quisqualate responses (Kehoe, 1978). It is not known whether glutamate receptors in membrane fractions used for binding assays exhibit the same desensitisation characteristics, but the effect of lectins on the glutamate binding parameters would be interesting to examine. It is also probable that lectins will be useful tools in affinity purification of glutamate binding proteins.

CONCLUSIONS AND FUTURE DEVELOPMENTS

Many of the developments in the field of characterisation, using conventional isolation and purification techniques, are likely to be overtaken by the application of molecular neurobiology. Isolation of the messenger RNA for invertebrate glutamate receptors would allow many exciting developments.

Expression of the receptor in a novel situation such as the *Xenopus* oocyte, which has already been used to study several receptor types including mammalian CNS glutamate receptors (Gunderson *et al.*, 1984), would be valuable in two respects. First, the well characterised pharmacology would demonstrate that functional messenger RNA had been isolated. Secondly, single channel studies in a novel cellular environment would be of interest in themselves. Acquisition of messenger RNA for glutamate receptors may ultimately allow the application of cloning techniques and subsequent sequencing of the constituent proteins.

The value of molecular biology in studies of receptors has been clearly demonstrated by the rapid advances made in the last few years in understanding the vertebrate nicotinic acetylcholine receptor. It is now becoming possible to compare the structures of nicotinic acetylcholine receptors purified from insect CNS with those from vertebrates. Speculation has arisen that insect acetylcholine receptors might represent the ancestral receptor protein (Hanke and Breer, 1986). In discussing this point Lunt

(1986) speculated that molecular biology should be able to resolve questions of this nature. This seems to be a good example of the way in which comparative neurochemistry might be important in the future.

However, the neurochemical analysis of glutamatergic systems in invertebrates has great potential in more immediately accessible areas such as the development of pesticides. The motor innervation of insect skeletal muscle is a particularly accessible site, and so far agents have not been specifically designed to perturb normal function at this site. There is evidence that some existing insecticides, e.g. pyrethroids, affect neuromuscular transmission in insects (Salgado *et al.*, 1983), but the biochemistry of the toxicity has not been examined. The widespread use of existing pesticides and closely related compounds will lead inevitably to resistance and the necessity to develop novel compounds with different modes of action. If chemicals directed against glutamatergic synapses are to be developed, it is obviously important that comparative data on both the potential target systems and the mammalian systems become available.

There may also be instances in which studies on anatomically simple invertebrate preparations can provide relevant information for the understanding of processes taking place in the mammalian brain. Some aspects of glutamatergic neurotransmission may fall into this category. For instance, a model to explain this cytotoxic effects of prolonged exposure to acidic amino acids was produced by examining the effects of glutamate on insect muscle (Duce *et al.*, 1983). When receptor desensitisation was prevented, glutamate produced degenerative effects which appear to result from entry of calcium through the activated ion channels (Donaldson *et al.*, 1983).

In conclusion, information on the comparative neurochemistry of glutamate in invertebrate species is patchy both in terms of quality and quantity. However, in areas such as synaptic release, uptake and receptor characterisation there are signs that a clearer understanding of the way invertebrate excitable cells utilise this amino acid as a neurotransmitter may be close at hand.

REFERENCES

Abe, T., Kawai, N. and Niwa, A. (1983) Effect of a spider toxin on the glutamatergic synapse of a lobster muscle. *J. Physiol. (Lond.) 339*, 243-52

Altschuler, R.A., Neises, G.R., Harmison, G.G., Wenthold, R.J. and Fex, J. (1981) Immunocytochemical localization of aspartate aminotransferase immunoreactivity in cochlear nucleus of the guinea pig. *Proc. Natl. Acad. Sci. USA 78*, 6553-7

Altschuler, R.A., Mosinger, J.L., Harmison, G.G., Parakkal, M.H. and Wenthold, R.J. (1982) Aspartate aminotransferase-like immunoreactivity as a marker for aspartate/glutamate in guinea pig photoreceptors. *Nature (Lond.) 298*, 657-9

Anderson, C.R., Cull-Candy, S.G. and Miledi, R. (1976) Glutamate and quisqualate noise in voltage clamped locust muscle fibres. *Nature (Lond.) 261*, 151-3

Anis, N.N., Clark, R.B., Gration,˙K.A.F. and Usherwood, P.N.R. (1981) Influence of agonists on desensitization of glutamate receptors on locust muscle. *J. Physiol. (Lond.) 312*, 345-64

Atwood, H.L. (1976) Organization and synaptic physiology of crustacean neuromuscular systems. *Progr. Neurobiol. 7*, 291-391

Atwood, H.L. (1982) Synapses and neurotransmitters. In: Atwood, H.L. and Sandeman, D.C. (eds) *The biology of Crustacea, Vol. 3. Neurobiology: structure and function*, pp. 105-50. Academic Press, New York and London

Atwood, H.L., Lang, F. and Morin, W.A. (1972) Synaptic vesicles: selective depletion in crayfish excitatory and inhibitory axons. *Science 176*, 1353-5

Atwood, H.L., Swenarchuk, L.E. and Gruenwald, C.R. (1975) Long-term synaptic facilitation during sodium accumulation in nerve terminals. *Brain Res. 100*, 198-204

Baker, P.F. and Potashner, S.J. (1971) The dependence of glutamate uptake by crab nerve on external Na^+ and K^+. *Biochim. Biophys. Acta 249*, 616-22

Baker, P.F. and Potashner, S.J. (1973) The role of metabolic energy in the transport of glutamate by invertebrate nerve. *Biochim. Biophys. Acta 318*, 123-39

Balcar, V.J. and Johnston, G.A.R. (1972) The structural specificity of the high affinity uptake of *L*-glutamate and *L*-aspartate by rat brain slices. *J. Neurochem. 19*, 2657-66

Balcar, V.J. and Johnston, G.A.R. (1973) High affinity uptake of transmitters: studies on the uptake of *L*-aspartate, GABA, *L*-glutamate and glycine in cat spinal cord. *J. Neurochem. 20*, 529-39

Bateman, A., Boden, P., Dell, A., Duce, I.R., Quicke, D.L.J. and Usherwood, P.N.R. (1985) Postsynaptic block of a glutamatergic synapse of low molecular weight fractions of spider venom. *Brain Res. 339*, 237-44

Bates, S.E., Duce, I.R., Usherwood, P.N.R. and Wilson, R.G. (1985) Uptake and release of glutamate in *Lucilia sericata* larvae. *Pest. Sci. 16*, 530

Bennett, J.P. Jr, Logan, W.J. and Snyder, S.W. (1973) Amino acids as central nervous transmitters. The influence of ions, amino acid analogues and ontogeny on transport systems for *L*-glutamic and *L*-aspartic acids and glycine into central nervous synaptosomes of the rat. *J. Neurochem. 21*, 1533-50

Boden, P. (1983) The fate of neurotransmitter released from insect motor nerve endings. Ph.D. Thesis, University of Nottingham

Botham, R.P. (1977) An examination of the ultrastructure of the insect neuromuscular junction and investigation of the neurotransmitters. Ph.D. Thesis, CNAA

Botham, R.P., Beadle, D.J., Hart, R.J., Potter, C. and Wilson, R.G. (1979a) Changes in the distribution and size of synaptic vesicles in neuromuscular junctions of *Locusta migratoria* after stimulation and rest. *Cell Tissue Res. 203*, 373-8

Botham, R.P., Beadle, D.J., Hart, R.J., Potter, C. and Wilson, R.G. (1979b) Glutamate uptake after stimulation induced depletion of vesicle numbers in neuromuscular junctions of *Locusta migratoria*. *Cell Tissue Res. 203*, 379-86

Bradford, H.F., Chain, E.B., Cory, H.T. and Rose, S.P.R. (1969) Glucose and amino acid metabolism in some invertebrate nervous systems. *J. Neurochem. 16*, 969-79

Bradford, H.F. and Ward, H.K. (1976) On glutaminase activity in mammalian synaptosomes. *Brain Res. 110*, 115-25

Bradford, H.F., Ward, H.K. and Thomas, A.J. (1978) Glutamine: a major substrate for nerve endings. *J. Neurochem. 30*, 1453-9

Braitman, D.J. (1986), Desensitisation to glutamate and aspartate in rat olfactory

79

cortex slice. *Brain Res. 364*, 199-203

Breer, H. (1981) Characterisation of synaptosomes from the central nervous system of insects. *Neurochem. Int. 3*, 155-63

Breer, H. (1986) Synaptosomes — systems for studying insect neurochemistry. In: Ford, M.G., Lunt, G.G., Reay, R.C. and Usherwood, P.N.R. (eds) *Neuropharmacology and pesticide action*, pp. 384-413. Ellis Horwood, Chichester

Briley, P.A., Filbin, M.T., Lunt, G.G. and Donellan, J.F. (1982) Binding and uptake of glutamate and γ-aminobutyric acid in membrane fractions from locust muscle. In: *Ciba Foundation Symposium 88, Neuropharmacology of insects*, pp. 158-75. Pitman, London

Carew, J.J., Pinsker, H., Rubinson, K. and Kandel, E.R. (1974) Physiological and biochemical properties of neuromuscular transmission between identified motoneurones and gill muscle in *Aplysia. J. Neurophysiol. 37*, 1020-40

Carlyle, R.F. (1974), The occurrence in and actions of amino acids on isolated supra oral sphincter preparations of the sea anemone *Actinia equina. J. Physiol. (Lond.) 236*, 635-52

Chang, Y.C. (1975), The endplate and graded potentials from the neuromuscular system of the earthworm, *Pheretima hawayana. Comp. Biochem. Physiol. 51A*, 237-40

Clark, R.B., Gration, K.A.F. and Usherwood, P.N.R. (1979) Desensitization of glutamate receptors on innervated and denervated locust muscle fibres. *J. Physiol. (Lond.) 290*, 551-68

Clark, R.B., Gration, K.A.F. and Usherwood, P.N.R. (1980) Influence of glutamate and aspartate on the time course of decay of excitatory synaptic currents at locust neuromuscular junctions. *Brain Res. 192*, 205-6

Clements, A.N. and May, T.E. (1974) Studies on locust neuromuscular physiology in relation to glutamate. *J. exp. Biol. 60*, 673-705

Cleworth, J.F. (1981) Biochemical examination of glutamate receptors on locust muscle. Ph.D. Thesis, University of Nottingham

Cleworth, J.F., Robinson, N.L. and Usherwood, P.N.R. (1980) Biochemical studies on glutamate receptors in insect muscle. In: *Insect neurobiology and pesticide action (Neurotox '79)*, pp. 279-80. Society of Chemical Industry, London

Collingridge, G.L., Kehl, S.J. and McLennan, H. (1983) Excitatory amino acids in synaptic transmission in the Schaffer collateral–commissural pathway of the rat hippocampus. *J. Physiol. (Lond.) 334*, 33-46

Cotman, C.W., Foster, A.C. and Lanthorn, T. (1981) An overview of glutamate as a neurotransmitter. In: Di Chiara, G. and Gessa, G.L. (eds) *Glutamate as a neurotransmitter*, pp. 1-27. Raven Press, New York

Cotman, C.W., Haycock, J.W. and White, W.F. (1976) Stimulus–secretion coupling processes in brain: analysis of noradrenaline and gamma-aminobutyric acid release. *J. Physiol. (Lond.) 254*, 475-505

Cottrell, G.A., Macon, J.B. and Szczepaniak, A.C. (1972) Glutamic acid mimicking of synaptic inhibition on the giant serotonin neurone of the snail. *Br. J. Pharmacol. 45*, 684-7

Cox, D.W.G. and Bradford, H.F. (1979) Uptake and release of excitatory amino acid neurotransmitters. In: McGeer, E.G., Olney, J.W. and McGeer, P.L. (eds) *Kainic acid as a tool in neurobiology*, pp. 71-93. Raven Press, New York

Crawford, A.C. and McBurney, R.N. (1977) The termination of transmitter action at the crustacean neuromuscular junction. *J. Physiol. (Lond.) 268*, 711-29

Crunelli, V., Forda, S. and Kelly, J.S. (1985) Excitatory amino acids in the hippocampus: synaptic physiology and pharmacology. *Trends Neurosci. 8*, 26-30

Cull-Candy, S.G., Donellan, J.F., James, R.W. and Lunt, G.G. (1976) 2-Amino-4-phosphonobutyric acid as a glutamate antagonist on locust muscle. *Nature*

(Lond.) 262, 408-9

Cull-Candy, S.G. and Miledi, R. (1980) Factors affecting the channel kinetics of glutamate receptors in locust muscle fibres. In: Sattelle, D.B., Hall, L.M. and Hildebrand, J.G. (eds) *Receptors for neurotransmitters, hormones and pheromones in insects,* pp. 161-73. Elsevier North-Holland, Amsterdam

Cull-Candy, S.G. and Miledi, R. (1983) Block of glutamate-activated synaptic channels by curare and gallamine. *Proc. Roy. Soc. (Lond.) Ser. B, 218,* 111-18

Cull-Candy, S.G., Neal, H. and Usherwood, P.N.R. (1973) Action of black widow spider venom on an aminergic synapse. *Nature (Lond.) 241,* 353-4

Cull-Candy, S.G. and Parker, I. (1982) Rapid kinetics of single glutamate receptor channels. *Nature (Lond.) 295,* 410-12

Cull-Candy, S.G. and Usherwood, P.N.R. (1973), Two populations of *L*-glutamate receptors on locust muscle fibres. *Nature New Biol. 246,* 62-4

Curtis, D.R., Phillis, J.W. and Watkins, J.C. (1959) Chemical excitation of spinal neurons. *Nature (Lond.) 183,* 611-12

Curtis, D.R. and Watkins, J.C, (1960) The excitation and depression of spinal neurons by structurally related amino acids. *J. Neurochem. 6,* 117-41

Daoud, M.A.R. and Miller, R. (1976) Release of glutamate and other amino acids from arthropod nerve–muscle preparations. *J. Neurochem. 26,* 119-23

Daoud, M.A.R. and Usherwood, P.N.R. (1975) Action of kainic acid on a glutamatergic synapse. *Comp. Biochem. Physiol. C52,* 51-3

Daoud, M.A.R. and Usherwood, P.N.R. (1978) Desensitization and potentiation during glutamate application to locust skeletal muscle. *Comp. Biochem. Physiol. 59,* 105-10

de Belleroche, J.S. and Bradford, H.F. (1972) Metabolism of beds of mammalian cortical synaptosomes: response to depolarizing influences. *J. Neurochem. 19,* 585-602

de Belleroche, J.S. and Bradford, H.F. (1977) On the site of origin of transmitter amino acids released by depolarization of nerve terminals *in vitro. J. Neurochem. 29,* 335-43

Dolphin, A.C., Errington, M.L. and Bliss, T.V.P. (1982) Long-term potentiation of the perforant path *in vivo* is associated with increased glutamate release. *Nature (Lond.) 297,* 496-8

Donaldson, P.L., Duce, I.R. and Usherwood, P.N.R. (1983) Calcium accumulation precedes the degenerative effects of *L*-glutamate on locust muscle fibres. *Brain Res. 274,* 261-5

Donellan, J.F., Jenner, D.W. and Ramsey, A. (1974) Subcellular fractionation of fleshfly flight muscle in attempts to isolate synaptosomes and to establish the localization of glutamate enzymes. *Insect Biochem. 4,* 243-65

Duce, I.R., Donaldson, P.L. and Usherwood, P.N.R. (1983) Investigations into the mechanism of excitant amino acid cytotoxicity using a well characterised glutamatergic system. *Brain Res. 263,* 77-87

Duce, I.R. and Keen, P. (1983) Selective uptake of [^{3}H]glutamine and [^{3}H]glutamate into neurons and satellite cells of dorsal root ganglia *in vitro. Neuroscience 8,* 861-6

Dudel, J. (1975) Potentiation and desensitisation after glutamate induced postsynaptic currents at the crayfish neuromuscular junction. *Pflügers Arch. 356,* 317-27

Dunant, Y. and Israel, M. (1985) The release of acetylcholine. *Sci. Amer. 252,* 40-8

Eldefrawi, M.E., Abalis, I.M., Filbin, M.T. and Eldefrawi, A.T. (1985) Glutamate and GABA receptors of insect muscles: biochemical identification and interactions with insecticides. In: von Keyserlingk, H.C., Jager, A. and von Szczepanski, Ch. (eds) *Approaches to new leads for insecticides,* pp. 101-6.

Springer-Verlag, Berlin and Heidelberg

Evans, M.L. and Usherwood, P.N.R. (1985) The effects of lectins on desensitisation of locust muscle glutamate receptors. *Brain Res. 358*, 34-9

Evans, P.D. (1972) The free amino acid pool of the haemocytes of *Carcinus maenas. J. exp. Biol. 56*, 501-7

Evans, P.D. (1973) The uptake of L-glutamate by the peripheral nerves of the crab *Carcinus maenas. Biochim. Biophys. Acta 311*, 302-13

Evans, P.D. (1974) An autoradiographical study of the localisation of the uptake of glutamate by the peripheral nerves of the crab, *Carcinus maenas. J. Cell Sci. 14*, 351-67

Evans, P.D. (1975) The uptake of L-glutamate by the central nervous system of the cockroach *Periplaneta americana. J. exp. Biol. 62*, 55-67

Evans, P.D. and Crossley, A.C. (1974) Free amino acids in the haemocytes and plasma of the larvae of *Calliphora vicina. J. exp. Biol. 61*, 463-72

Faeder, I.R., Mathews, J.R. and Salpeter, M.M. (1974) [^{3}H]Glutamate uptake at insect neuromuscular junctions: effect of chlorpromazine. *Brain Res. 80*, 53-70

Faeder, I.R, and Salpeter, M.M. (1970) Glutamate uptake by a stimulated insect nerve–muscle preparation. *J. Cell. Biol. 46*, 300-7

Fagg, G.E. (1985) L-Glutamate, excitatory amino acid receptors and brain function. *Trends Neurosci. 8*, 207-10

Fagg, G.E. and Lane, J.D. (1979) The uptake and release of putative amino acid neurotransmitters. *Neurosci. 4*, 1015-36

Feasey, K.J., Lynch, M.A. and Bliss, T.V.P. (1986) Long-term potentiation is associated with an increase in calcium dependent, potassium stimulated release of [^{14}C]glutamate from hippocampal slices; an *ex vivo* study in the rat. *Brain Res. 364*, 39-44

Filbin, M.T., Eldefrawi, M.E. and Eldefrawi, A.T. (1985) Biochemical identification of a putative glutamate receptor in housefly thoracic membranes. *Life Sci. 36*, 1531-9

Filbin, M.T., Lunt, G.G. and Donellan, J.F. (1980) Glutamate receptor biochemistry. In: Sattelle, D.B., Hall, L.M. and Hildebrand, J.G. (eds) *Receptors for neurotransmitters, hormones and pheromones in insects*, pp. 153-60. Elsevier North-Holland, Amsterdam

Fiszer de Plazas, S. and De Robertis, E. (1974) Isolation of hydrophobic proteins binding neurotransmitter amino acids. Glutamate receptor of the shrimp muscle. *J. Neurochem. 23*, 1115-20

Fiszer de Plazas, S., De Robertis, E. and Lunt, G.G. (1977), L-Glutamate and γ-aminobutyrate binding to hydrophobic protein fractions from leg muscle of fly (*Musca domestica*). *Gen. Pharmacol. 8*, 133-7

Florkin, M. and Jeuniaux, C. (1974) Haemolymph: composition. In: Rockstein, M. (ed.) *The physiology of the Insecta, vol. 5*. Academic Press, New York

Fonnum, F. (1968) The distribution of glutamate decarboxylase and aspartate transaminase in subcellular fractions of rat and guinea-pig brain. *Biochem. J. 106*, 401-12

Fonnum, F. (1984) Glutamate: a neurotransmitter in mammalian brain. *J. Neurochem. 42*, 1-11

Fonnum, F., Lund-Karlsen, R., Malthe-Sorenssen, D., Sterri, S. and Walaas, I. (1980) High affinity transport systems and their role in transmitter action. In: Cotman, C.W., Poste, G. and Nicholson, G.L. (eds) *The cell surface and neuronal function*, pp. 455-504. Elsevier, Amsterdam

Foster, A.C. and Fagg, G.E. (1984) Acidic amino acid binding sites in mammalian neuronal membranes: their characteristics and relationship to synaptic receptors. *Brain Res. Rev. 7* 103-64

Franzini-Armstrong, C. (1976) Freeze-fracture of excitatory and inhibitory synapses in crayfish neuromuscular junctions. *J. Microsc. Biol. Cell. 25*, 217-22

Freeman, A.R., Shank, R.P., Kephart, J., Dekin, M. and Wang, M. (1981) A model for excitatory transmission at a glutamate synapse. In: Di Chiara, G. and Gessa, G.L. (eds) *Glutamate as a neurotransmitter*, pp. 227-43. Raven Press, New York

Gardner, C.R. (1981) Effects of neurally active amino acids and monoamines on the neuromuscular transmission of *Lumbricus terrestris. Comp. Biochem. Physiol. 68C*, 85-90

Gerschenfeld, H.M. (1973) Chemical transmission in invertebrate central nervous systems and neuromuscular junctions. *Physiol. Rev. 53*, 1-119

Giles, D. and Usherwood, P.N.R. (1985) The effects of putative amino acid neurotransmitters on somata isolated from neurones of the locust central nervous system. *Comp. Biochem. Physiol. 80C*, 231-6

Gration, K.A.F. (1980) Glutamate receptors at excitatory synapses on locust skeletal muscle. In: *Insect neurobiology and pesticide action (Neurotox 79)*, pp. 169-76. Society of Chemical Industry, London

Gration, K.A.F., Lambert, J.J., Ramsey, R.L., Rand, R.P. and Usherwood, P.N.R. (1981b) Agonist potency determination by patch clamp analysis of single glutamate receptors. *Brain Res. 230*, 400-5

Gration, K.A.F., Lambert, J.J., Ramsey, R.L., Rand, R.P. and Usherwood, P.N.R. (1982) Closure of membrane channels gated by glutamate receptors may be a two-step process. *Nature (Lond.) 295*, 599-601

Gration, K.A.F., Lambert, J.J., Ramsey, R.L. and Usherwood, P.N.R. (1981a) Non random openings and concentration dependent lifetimes of glutamate-gated channels in muscle membrane. *Nature (Lond.) 291*, 423-5

Gulley, R.L. and Reese, T.S. (1977) Freeze fracture studies of the synapses in the organ of Corti. *J. Comp. Neurol. 171*, 517-44

Gulley, R.L., Wenthold, R.J. and Neises, G.R. (1977) Remodelling of neuronal membranes as an early response to deafferentation. A freeze-fracture study. *J. Cell. Biol. 75*, 837-50

Gunderson, C.B., Miledi, R, and Parker, I. (1984) Glutamate and kainate receptors induced by rat brain messenger RNA in *Xenopus* oocytes. *Proc. Roy. Soc. (Lond.) B, 221*, 127-43

Hajos, F. and Kerpel-Fronius, S. (1971) Electron microscope histochemical evidence for a partial or total block of the tricarboxylic acid cycle in the mitochondria of presynaptic axon terminals. *J. Cell. Biol. 51*, 216-22

Hamberger, A.C., Chiang, G.H., Nylen, E.S., Scheff, S.W. and Cotman, C.W. (1979) Glutamate as a CNS transmitter. I. Evaluation of glucose and glutamine as precursors for the synthesis of preferentially released glutamate. *Brain Res. 168*. 513-30

Hammerschlag, R. and Weinreich, D. (1972) Glutamic acid and primary afferent transmission. In: Costa, E., Iversen, L.L. and Paoletti, R. (eds) *Advances in biochemical psychopharmacology, Vol. 6: Studies of neurotransmitters at the synaptic level*, pp. 165-80. Raven Press, New York

Hanke, W. and Breer, H, (1986) Channel properties of an insect neuronal acetylcholine receptor protein reconstituted in planar lipid bilayers. *Nature (Lond.) 321*, 171-4

Hart, R.J., Potter, C. and Wilson, R.G. (1977) Factors governing the toxicity of putative synaptic transmitters and their analogs when injected into the haemocoel of adult male *Lucilia sericata. Pest. Sci. 8*, 722-34

Heuser, J.E., Reese, T.S., Dennis, J.M., Jan, Y., Jan, L. and Evans, L. (1979) Synaptic vesicle exocytosis captured by quick freezing and correlated with quantal transmitter release. *J. Cell. Biol. 81*, 275-300

Heuser, J.E. and Salpeter, S.R. (1979) Organisation of acetylcholine receptors in quick-frozen, deep etched and rotary replicated *Torpedo* post-synaptic membrane. *J. Cell. Biol. 82*, 150-73

Hill, R.B. (1970) Effects of postulated neurohumoral transmitters on the isolated radula protractor of *Busycon canaliculatum*. *Comp. Biochem. Physiol. 33*, 249-58

Holman, G.M. and Cook, B.J. (1970) Pharmacological properties of excitatory neuromuscular transmission in the hindgut of the cockroach *Leucophaea maderae*. *J. Insect Physiol. 16*, 1891-1907

Huggins, A.K., Rick, J.T. and Kerkut, G.A. (1967) A comparative study of the intermediary metabolism of *L*-glutamate in muscle and nervous tissue. *Comp. Biochem. Physiol. 21*, 23-30

Irving, S.N., Osborne, M.P. and Wilson, R.G. (1979a) Studies on *L*-glutamate in insect haemolymph. I. Effect of injected *L*-glutamate. *Physiol. Entomol. 4*, 139-46

Irving, S.N., Wilson, R.G. and Osborne, M.P. (1979b) Studies on *L*-glutamate on insect haemolymph. II. Distribution and metabolism of radiolabelled amino acids injected into the haemolymph. *Physiol. Entomol. 4*, 223-30

Irving, S.N., Wilson, R.G. and Osborne, M.P. (1979c) Studies on *L*-glutamate in insect haemolymph. III. Amino acid analyses of the haemolymph of various arthropods. *Physiol. Entomol. 4*, 231-40

Ishida, M. and Shinozaki, H. (1980) Differential effects of diltiazem on glutamate potentials and excitatory junctional potentials at the crayfish neuromuscular junction. *J. Physiol. (Lond.) 298*, 301-19

James, R.W., Lunt, G.G. and Donnellan, J.F. (1977a) Isolation of a glutamatergic receptor fraction from locust muscle. *Biochem. Soc. Trans. 5*, 170-2

James, R.W., Lunt, G.G. and Donnellan, J.F. (1977b) Characterization of *L*-glutamate binding proteolipids present in locust muscle extracts. *Insect Biochem. 7*, 247-55

Jones, H.C. (1962) The action of *L*-glutamic acid and of structurally related compounds on the hindgut of the crayfish. *J. Physiol. (Lond.) 164*, 295-300

Judge, S.E., Kerkut, G.A. and Walker, R.J. (1977), Properties of an identified synaptic pathway in the visceral ganglion of *Helix aspersa*. *Comp. Biochem. Physiol. 57C*, 101-6

Kawagoe, R., Onodera, K. and Takeuchi, A. (1981) Release of glutamate from the crayfish neuromuscular junction. *J. Physiol. (Lond.) 312*, 225-36

Kawagoe, R., Onodera, K. and Takeuchi, A. (1982) On the quantal release of endogenous glutamate from the crayfish neuromuscular junction. *J. Physiol. (Lond.) 322*, 529-39

Kawagoe, R., Onodera, K. and Takeuchi, A. (1984) The uptake and release of glutamate at the crayfish neuromuscular junction. *J. Physiol. (Lond.) 354*, 69-78

Kawai, N., Niwa, A. and Abe, T. (1982a) Spider venom contains specific receptor blocker of glutaminergic synapses. *Brain Res. 247*, 169-71

Kawai, N., Niwa, A. and Abe, T. (1982b) Effects of spider venom on glutaminergic synapses in the mammalian brain. *Biomed. Res. 3*, 353-5

Kawai, N., Niwa, A. and Abe, T. (1983b) Specific antagonism of the glutamate receptor by an extract from the spider *Araneus ventricosus*. *Toxicon 21*, 438-40

Kawai, N., Yamagishi, S., Saito, M. and Furnya, K. (1983a) Blockade of synaptic transmission in the squid giant synapses by a spider toxin (JSTX). *Brain Res. 278*, 346-9

Kehoe, J.S. (1978) Transformation by concanavalin A of the response of molluscan neurones to *L*-glutamate. *Nature (Lond.) 274*, 866-9

Kerkut, G.A., Leake, L.D., Shapiro, A., Cowan, S. and Walker, R.J. (1965) The presence of glutamate in nerve–muscle perfusates of *Helix, Carcinus* and *Periplaneta. Comp. Biochem. Physiol. 15,* 485-502

Kerkut, G.A., Shapiro, A. and Walker, R.J. (1967) The transport of ^{14}C-labelled material from CNS ⇌muscle along a nerve trunk. *Comp. Biochem. Physiol. 23,* 729-48

Kerry, C.J., Kits, K.S., Ramsey, R.L., Sansom, M.S.P. and Usherwood, P.N.R. (1986) Single channel kinetics of a glutamate receptor. *Biophys. J. 50,* 367-74

Kravitz, G.A., Slater, C.R., Takahashi, K., Bounds, M.D. and Grossfield, R.M. (1970) Excitatory transmission in invertebrates — glutamate as a potential neuromuscular transmitter compound. In: Andersen, P. and Jansen, J.K.S. (eds) *Excitatory synaptic mechanisms,* pp. 85-93. Universitetsforlaget, Oslo

Lathja, A. (1968) Transport as a control mechanism of cerebral metabolite levels. In: Lathja, A. and Ford, D. (eds) *Brain barrier systems: progress in brain research,* vol. 29, pp. 201-18. Elsevier, Amsterdam

Lea, T.J. and Usherwood, P.N.R. (1973) The site of action of ibotenic acid and the identification of two populations of glutamate receptors on insect muscle fibres. *Comp. Gen. Pharmacol. 4,* 333-50

Leake, L.D. and Walker, R.J. (1980) *Invertebrate neuropharmacology.* Blackie, Glasgow and London

Levinson, S.R. and Keynes, R.D. (1972) Isolation of acetylcholine receptors by chloroform methanol extraction: artifacts arising in use of Sephadex LH 20 columns. *Biochim. Biophys. Acta 288,* 241-7

Levy, W.B., Redburn, D.A. and Cotman, C.W. (1973) Stimulus-coupled secretion of γ-aminobutyric acid from rat brain synaptosomes. *Science NY 181,* 676-8

Lingle, C., Eisen, J.S. and Marder, E. (1981) Block of glutamatergic excitatory synaptic channels by chlorisondamine. *Molec. Pharmacol. 19,* 349-53

Llinas, R. and Nicholson, C. (1975) Calcium mediates depolarisation secretion coupling: an aequorin study in squid giant synapse. *Proc. Natl. Acad, Sci. 72,* 187-90

Logan, W.J. and Snyder, S.H. (1972) High affinity systems for glycine, glutamic and aspartic acids in synaptosomes of rat central nervous system. *Brain Res. 42,* 413-31

Lunt, G.G. (1973) Hydrophobic proteins from locust (*Schistocerca gregaria*) muscle with glutamate receptor properties. *Comp. Gen. Pharmacol. 4,* 75-9

Lunt, G.G. (1986) Is the insect neuronal nAChR the ancestral ACh receptor protein? *Trends Neurosci. 9,* 341-2

McKinlay, R.G. and Usherwood, P.N.R. (1973) The role of synaptic vesicles in transmission at the insect nerve–muscle junction. *Life Sci. 13,* 1051-6

Manaranche, R., Thieffry, M. and Israel, M. (1980) Effect of the venom of *Glycera convoluta* on the spontaneous quantal release of transmitter. *J. Cell Biol. 85,* 446-58

Mathers, D.A. and Usherwood, P.N.R. (1976) Concanavalin A blocks desensitization of glutamate receptors on insect muscle fibres. *Nature (Lond.) 259,* 409-11

Michaelis, E.K., Michaelis, M.L. and Boyarski, L.L. (1974) High-affinity glutamate binding to brain synaptic membranes. *Biochim. Biophys. Acta 367,* 338-48

Michaelis, E.K., Michaelis, M.L., Stormann, T.M., Chittenden, W.L. and Grubbs, R.D. (1983) Purification and molecular characterization of the brain synaptic membrane glutamate-binding protein. *J. Neurochem. 40,* 1742-53

Miledi, R. (1967) Spontaneous synaptic potentials and quantal release of transmitter in the stellate ganglion of the squid. *J. Physiol. (Lond.) 192,* 379-406

Miledi, R. (1969) Transmitter action in the giant synapses of the squid. *Nature (Lond.) 223,* 1284-6

Miller, R., Leaf, G. and Usherwood, P.N.R. (1973) Blood glutamate in arthropods.

Comp. Biochem. Physiol. 44A, 991-6

Minchin, M.C.W. and Beart, P.M. (1975) Compartmentation of amino acid metabolism in the rat dorsal root ganglion; a metabolic and autoradiographic study. *Brain Res. 83*, 437-49

Murdock, L.L. (1971) Crayfish vas deferens: contractions in response to *L*-glutamate and gamma aminobutyrate. *Comp. Gen. Pharmacol. 2*, 93-8

Murdock, L.L. and Chapman, G.Y. (1974) *L*-Glutamate in arthropod blood plasma: physiological implications. *J. exp. Biol. 60*, 783-94

Newman, T.M. (1986) Membrane specialisations of the locust neuromuscular system. Ph.D. Thesis, University of Nottingham

Newman, T.M. and Duce, I.R. (1983) Membrane specialisations of the neuro-muscular junction of the locust, *Schistocerca gregaria*. *Cell. Tiss. Res. 234*, 691-706

Newman, T.M. and Duce, I.R. (1985) Preparation dependent distribution of intramembranous particles in freeze fracture replicas of the neuromuscular junction of the locust (*Schistocerca gregaria*). *Cell. Tiss. Res. 243*, 323-7

Nistri, A. and Constanti, A. (1979) Pharmacological characterisation of different types of GABA and glutamate receptors in vertebrates and invertebrates. *Progr. Neurobiol. 13*, 117-236

Norenberg, M.D. and Martinez-Hernandez, A. (1979) Fine structural localization of glutamine synthetase in astrocytes of rat brain. *Brain Res. 161*, 303-10

Nowak, L., Bregostovoki, P., Ascher, P., Herbert, A. and Prochiaritz, A. (1984) Magnesium gates glutamate-activated channels in mouse central neurones. *Nature (Lond.) 307*, 462-5

Oomura, Y., Ooyama, H. and Sawada, M. (1974) Analysis of hyperpolarizations induced by glutamate and acetylcholine on *Onchidium* neurons. *J. Physiol. 243*, 321-41

Osborne, M.P. (1975) The ultrastructure of nerve muscle synapses. In: Usherwood, P.N.R. (ed.) *Insect Muscle*, pp. 181-205. Academic Press, London

Osborne, R.H., Bradford, H.F. and Jones, D.G. (1973) Patterns of amino acid release from nerve-endings isolated from spinal cord and medulla. *J. Neurochem. 21*, 407-19

Patlak, J.B., Gration, K.A.F. and Usherwood, P.N.R. (1979) Single glutamate-activated channels in locust muscle. *Nature (Lond.) 278*, 643-5

Piek, T. (1985) Neurotransmission and neuromodulation in skeletal muscles. In: Kerkut, G.A. and Gilbert, L.I. (eds) *Comprehensive insect physiology bio-chemistry and pharmacology*, Vol. 11, pp. 55-118. Pergamon Press, Oxford

Piek, T. (1985), Insect venoms and toxins. In: Kerkut, G.A. and Gilbert, L.I. (eds) *Comprehensive insect physiology, biochemistry and pharmacology*, Vol. 11, pp. 595-634. Pergamon Press, Oxford

Piggott, S.M., Kerkut, G.A. and Walker, R.J. (1975) Structure activity studies on glutamate receptor sites of three identifiable neurones in the suboesophageal ganglia of *Helix aspersa*. *Comp. Biochem. Physiol. 51C*, 91-100

Pin, J.-P. Bockaert, J. and Recasens, M. (1986) The binding of acidic amino acids to snail, *Helix aspersa*, periesophagic ring membranes reveals a single high affinity glutamate/kainate site. *Brain Res. 366*, 290-9

Puil, E. (1981), S-glutamate: its interactions with spinal neurons. *Brain Res. Rev. 3*, 229-322

Quicke, D.L.J., Usoh, G., Newman, T.M., Duce, I.R. and Usherwood, P.N.R. (1986) Purification of locust muscle plasma membrane with glutamate binding properties. *Neurosci. Lett. 5*, 1986

Rash, J.E. and Ellisman, M.H. (1974) Studies on excitable membranes. I. Macro-molecular specialisations of the neuromuscular junction and the non-junctional sarcolemma. *J. Cell. Biol. 63*, 567-86

Rees, D. and Usherwood, P.N.R. (1972) Fine structure of normal and degenerating motor axons and nerve–muscle synapses in the locust, *Schistocerca gregaria. Comp. Biochem. Physiol. 43A*, 83-101

Reijnerse, G.L.A., Vedstra, H. and van den Berg, C.J. (1975) Subcellular localisation of γ-aminobutyrate transaminase and glutamate dehydrogenase in adult rat brain. *Biochem. J. 152* 469-75

Reinecke, M. and Walther, C. (1978) Aspects of turnover and biogenesis of synaptic vesicles at locust neuromuscular junctions as revealed by zinc iodide–osmium tetroxide (ZIO) reacting with intravesicular SH-groups. *J. Cell. Biol. 78*, 839-55

Reinecke, M. and Walther, C. (1981) Ultrastructural changes with high activity and subsequent recovery at locust motor nerve terminals. A stereological analysis. *Neuroscience 6*, 489-503

Reubi, J.C. (1980) Comparative study of the release of glutamate and GABA newly synthesised from glutamine in various regions of the central nervous system. *Neuroscience 5*, 2145-50

Rheuben, M. and Reese, T.S. (1978) 3-D structure and membrane speicalizations of the moth excitatory neuromuscular synapse. *J. Ultrastruct. Res. 65*, 95-111

Robbins, J. (1958) The effect of amino acids on the crustacean neuromuscular system. *Anat. Rec. 132*, 492-3

Robbins, J. (1959) The excitation and inhibition of crustacean muscle by amino acids. *J. Physiol. (Lond.) 148*, 39-50

Roberts, E. (1981) Strategies for identifying sources and sites of formation of GABA-precursor or transmitter glutamate in brain. In: Di Chiara, G. and Gessa, G.L. (eds) *Glutamate as a neurotransmitter*, pp. 91-102. Raven Press, New York

Roberts, P.J. (1974) Glutamate receptors in rat CNS. *Nature (Lond.) 252*, 399-401

Roberts, P.J. (1981) Binding studies for the investigation of receptors for *L*-glutamate and other excitatory amino acids. In: Roberts, P.J., Storm-Mathisen, J. and Johnston, G.A.R. (eds) *Glutamate: transmitter in the central nervous system*, pp. 35-54. John Wiley, Chichester

Salgado, V.L., Irving, S.N. and Miller, T.A. (1983) Depolarisation of motor nerve terminals by pyrethroids in susceptible and kdr-resistant house flies. *Pest. Biochem. Physiol. 20*, 100-14

Shank, R.P. and Aprison, M.H. (1979) Biochemical aspects of the neurotransmitter function of glutamate. In: Filer, L.J.Jr, Garattini, S., Kane, M.R., Reynolds, W.A. and Wurtman, R.J. (eds) *Glutamic acid: advances in biochemistry and physiology*, pp. 139-149. Raven Press, New York

Shank, R.P. and Aprison, M.H. (1981) Minireview: present status and significance of the glutamine cycle in neural tissues. *Life Sci. 28*, 837-42

Sherby, M.S., Eldefrawi, M.E., Wafford, K.E., Sattelle, D.B. and Eldefrawi, A.T. (1986) Pharmacology of putative glutamate receptors from insect skeletal muscles, insect central nervous system and rat brain. *Comp. Biochem. Physiol.* (in press)

Shinozaki, H. and Ishida, M. (1979) Pharmacological distinction between the excitatory junction potential and the glutamate potential revealed by concanavalin A at the crayfish neuromuscular junction. *Brain Res. 161*, 493-501

Shinozaki, H. and Ishida, M. (1986) A new potent channel blocker: effects on glutamate synapses at the crayfish neuromuscular junction. *Brain Res. 372*, 260-8

Stern, J.R., Eggleston, L.V., Hems, R. and Krebs, H.A. (1949) Accumulation of glutamic acid in isolated brain tissue. *Biochem. J. 44*, 410-18

Stettmeir, H., Finger, W. and Dudel, J. (1983a) Effects of concanavalin A on glutamate operated postsynaptic channels in crayfish muscle. *Pflügers Arch. 397*, 20-4

Stettmeir, H., Finger, W. and Dudel, J. (1983b) Glutamate activated synaptic channels in crayfish muscle investigated by noise analysis. *Pflügers Arch. 397*, 13-19

Streit, P. (1980) Selective retrograde labelling indicating the transmitter of neuronal pathways. *J. Comp. Neurol. 191*, 429-63

Takeuchi, A. and Takeuchi, N. (1964) The effect on crayfish muscle of ionophoretically applied glutamate. *J. Physiol. (Lond.) 170*, 296-317

Takeuchi, A., Onodera, K. and Kawagoe, R. (1980) Release of endogenous glutamate from the neuromuscular junction of the crayfish. *Proc. Japan. Acad. B. 56*, 246-9

Taraskevich, P.S., Gibbs, D., Schmued, L. and Orkand, R.K. (1977) Excitatory effects of cholinergic, adrenergic and glutamatergic agonists on a buccal muscle of *Aplysia. J. Neurobiol. 8*, 325-35

Tashmukhamedov, B.A., Usmamanov, P.B., Kasakov, I., Kalikulov, D., Ya Yukelson, L. and Atakuziev, B.U. (1983) Effects of different spider venoms on artificial and biological membranes. In: *Toxins as tools in neurochemistry*, pp. 312-23. Walter de Gruyther, Berlin

Taylor, R.F. (1978) Isolation and purification of cholinergic receptor proteolipids from rat gastrocnemius tissue. *J. Neurochem. 31*, 1183-98

Thanki, C.M., Sugden, D., Thomas, N.J. and Bradford, H.F. (1983) *In vivo* release from cerebral cortex of [^{14}C]glutamate synthesised from [U-^{14}C]glutamine. *J. Neurochem. 41*, 611-17

Treherne, J.E. (1960) The nutrition of the central nervous system in the cockroach, *Periplaneta americana L.* The exchange and metabolism of sugars. *J. exp. Biol. 37*, 513-33

Triggle, D.J. (1981), Desensitisation. In: Lamble, J.W. (ed.) *Towards understanding receptors*, pp. 28-33. Elsevier, Amsterdam

Usherwood, P.N.R. (1978) Amino acids as neurotransmitters. *Adv. Comp. Physiol. Biochem. 7*, 227-309

Usherwood, P.N.R. and Cull-Candy, S.G. (1974) Distribution of glutamate sensitivity on insect muscle fibres. *Neuropharmacology 13*, 455-62

Usherwood, P.N.R. and Cull-Candy, S.G. (1975) Pharmacology of somatic nerve–muscle synapses. In: Usherwood, P.N.R. (ed.) *Insect muscle*, pp. 207-80. Academic Press, New York

Usherwood, P.N.R. and Duce, I.R. (1985) Antagonism of glutamate receptor channel complexes by spider venom polypeptides. *Neurotoxicology 6*, 239-50

Usherwood, P.N.R., Duce, I.R. and Boden, P. (1984) Slowly-reversible block of glutamate receptor-channels by venoms of the spiders, *Argiope trifasciata* and *Araneus gemma. J. Physiol. (Paris) 79*, 241-5

Usherwood, P.N.R., Giles, D. and Suter, C. (1980) Studies of the pharmacology of insect neurones *in vitro.* In: *Insect neurobiology and pesticide action (Neurotox '79)*, pp. 115-28. Society of Chemical Industry, London

Usherwood, P.N.R. and Machili, P. (1968) Pharmacological properties of excitatory neuromuscular synapses in the locust. *J. exp. Biol. 49*, 341-61

Usherwood, P.N.R., Machili, P. and Leaf, G. (1968) *L*-Glutamate at insect excitatory nerve–muscle synapses. *Nature (Lond.) 219*, 1169-72

Usmanov, P.B., Kalikulov, D., Shadyeva, N.G., Shadyeva, N.G., Nenilin, A.B. and Tashmukhamedov, B.A. (1985) Postsynaptic blocking of glutamatergic and cholinergic synapses as a common property of Araneidae spider venoms. *Toxicon 23*, 528-31

van den Berg, C.J. (1973) A model of compartmentation in mouse brain based on glucose and acetate metabolism. In: Balazs, R. and Cremer, J.E. (eds) *Metabolic compartmentation of the brain*, pp. 137-66. Macmillan, New York

van Harraveld, A. and Mendelson, M. (1959) Glutamate-induced contractions in crustacean muscle. *J. Cell. Comp. Physiol. 54*, 85-94

van Marle, J., Piek, T., Lind, A. and van Weeren-Kramer, J. (1983) Localization of an Na^+-dependent uptake system for glutamate in excitatory neuromuscular junctions of the locust *Schistocerca gregaria. Comp. Biochem. Physiol. 74C*, 191-4

van Marle, J., Piek, T., Lind, A. and van Weeren-Kramer, J. (1984) Inhibition of glutamate uptake in the excitatory neuromuscular synapse of the locust by delta-philanthotoxin: a component of the venom of the solitary wasp *Philanthus triangulum* F. A high resolution autoradiographic study. *Comp. Biochem. Physiol. 79C*, 213-15

van Marle, J., Piek, T., Lind, A. and van Weeren-Kramer, J. (1985) Specificity of two insect toxins as inhibitors of high affinity transmitter uptake. *Comp. Biochem. Physiol. 82C*, 435-7

Vincent, S.R. and McGeer, E.G. (1980) A comparison of sodium dependent glutamate binding with high affinity glutamate uptake in rat striatum. *Brain Res. 184*, 99-105

Wafford, K.A. and Sattelle, D.B. (1986) Effects of amino acid neurotransmitter candidates on an identified insect motoneurone. *Neurosci. Lett. 63*, 135-40

Walker, R.J., James, V.A., Roberts, C.J. and Kerkut, G.A. (1980) Studies on amino acid receptors of *Hirudo, Helix, Limulus* and *Periplaneta*. In: S-Rozsa, K. (ed.) *Neurotransmitters in invertebrates*, pp. 161-90. *Adv. Physiol. Sci. 22*, Akademiai Kiado/Pergamon Press, Oxford

Walther, C. and Reinecke, M. (1983) Block of synaptic vesicle exocytosis without block of Ca^{2+}-influx: an ultrastructural analysis of the paralysing action of *Hobrotracon* venom on locust motor nerve terminals. *Neuroscience 9*, 213-24

Wang, L.D.L. and Boyarsky, L.L. (1979) Release of L-glutamate from the excitatory nerve terminals of the crayfish neuromuscular junction. *Life Sci. 24*, 1011-14

Watkins, J.C. and Evans, R.H. (1981) Excitatory amino acid transmitters. *Ann. Rev. Pharmacol. Toxicol. 21*, 165-204

Werman, R. (1966) A review — criteria for identification of a central nervous system transmitter. *Comp. Biochem. Physiol. 18*, 745-66

Wheeler, D.D. and Boyarsky, L.L. (1968) Influx of glutamic acid in peripheral nerve – characteristics of influx. *J. Neurochem. 15*, 419-33

Yamaguchi, M., Yanos, T., Yamaguchi, T. and Lathja, A. (1970) Amino acid uptake in the peripheral nerve of the rat. *J. Neurobiol. 1*, 419-33

Yoneda, Y., Roberts, E. and Dietz, G.W. (1982) A new synaptosomal biosynthetic pathway of glutamate and GABA from ornithine and its negative feedback inhibition by GABA. *J. Neurochem. 38*, 1686-94

Zlotkin, E. (1984) Toxins derived from arthropod venoms specifically affecting insects. In: Kerkut, G.A. and Gilbert, L.I. (eds) *Comprehensive insect physiology, biochemistry and pharmacology*, pp. 499-546. Pergamon Press, Oxford

3

GABA

Timothy N. Robinson and Richard W. Olsen

INTRODUCTION

γ-Aminobutyric acid (GABA) was discovered at the beginning of the century (Ackerman and Kutscher, 1910), and, shortly after it was first detected in mammalian brain in 1950 (Awapara *et al.*, 1950; Roberts and Frankel, 1950; Udenfriend, 1950), an extract of mammalian brain and spinal cord was shown to contain a factor, 'Factor I', which inhibited the stretch receptor neurone (a single sensory cell with its dendrites entwined in a fine muscle bundle) of the crayfish (Florey, 1954). Through fractional crystallisation of bovine brain Factor I, GABA was shown to be the most active component in terms of its inhibitory activity on the stretch receptor neurone (Bazemore *et al.*, 1957). Following some controversy over the role of GABA as a neurotransmitter, Kravitz and colleagues demonstrated GABA to be present in lobster inhibitory neurones, where it is synthesised, accumulated and released (Kravitz, 1967), and this work will be described later in this chapter. By the early 1970s it was thought that GABA might be the universal transmitter of junctional neuromuscular inhibition in all invertebrate phyla from nematodes to arthropods (though not molluscs), and it was also thought to be involved in the CNS of crustaceans, insects and possibly molluscs (see Gerschenfeld, 1973; Pichon, 1974; Callec, 1974). However, the evidence for this was virtually all electrophysiological, and despite the early studies on GABA involving invertebrate tissues, the bulk of the biochemistry of GABAergic neurotransmission has been elucidated in mammalian tissues. There have been several detailed reviews of the now well-characterised mammalian GABA system (e.g. Bradford, 1986), but it is depicted in a highly simplified form in Figure 3.1.

Essentially GABA is synthesised enzymatically in the nerve terminal from glutamate by glutamate decarboxylase (GAD) and is released into the synaptic cleft in response to stimulation of the neurone. Binding of GABA to receptor complexes in the postsynaptic membrane causes the opening of associated chloride ion channels and hence decreased resistance of the

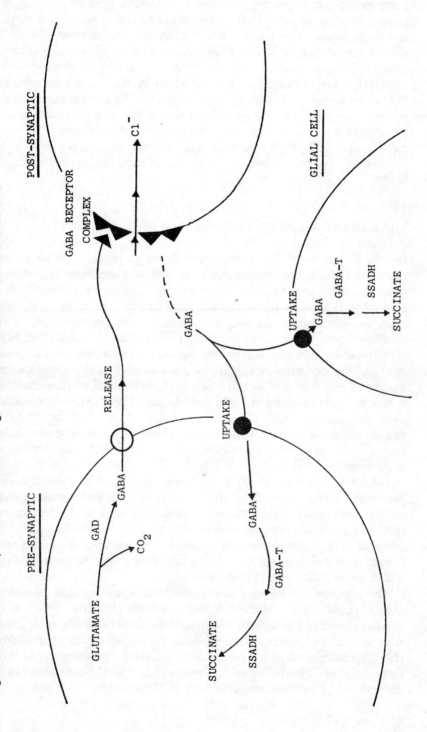

Figure 3.1: Schematic representation of GABAergic neurotransmission in mammalian CNS

postsynaptic cell. Following dissociation of GABA from its receptor, it is cleared from the synaptic cleft by diffusion and uptake by specific carriers into both neurones and glial cells. GABA can then be metabolised by GABA transaminase (GABA-T) to succinic semialdehyde, which in turn is converted to succinate by succinate semialdehyde dehydrogenase (SSADH). Apart from the early work on the synthesis and metabolism of GABA by Kravitz and colleagues, investigations of a comparable system in invertebrates have, until recently, been sparse. However, in recent years a picture has begun to emerge of the invertebrate GABA system. This chapter will deal with the biochemical evidence for the components of this invertebrate system by analogy with the mammalian system described above.

THE GABA CONTENT OF NERVOUS TISSUES

Invertebrate nervous tissues have been shown to possess GABA in concentrations which are comparable with, and in some cases higher than, those measured in mammalian nervous tissues. However, the methods of measuring GABA in tissue extracts have changed over the years from using two-dimensional chromatography and ninhydrin treatment of the resulting spots (Ray, 1964), through detection of 2,4-dinitrophenol (DNP)–amino acid methyl esters by gas–liquid chromatography (Aprison et al., 1973), detection of GABA by gas chromatography–mass spectroscopy (GC–MS) (McAdoo and Coggeshall, 1976), and more recently the microchemical procedure of Witte and Matthaei (1980) whereby GABA is assayed by digestion with GABAse (a mixture of GABA-T and SSADH) and the resulting $NADP^+$ is measured. These variations in methodology should be borne in mind when comparing data for GABA concentrations in different tissues.

GABA was first shown by Kravitz et al. (1962) to be present in crab peripheral nerves (10–12 μg/100 mg dry weight) and lobster ganglia (60 μg/100 mg dry weight). These workers then went on to demonstrate that in lobster inhibitory axons the concentration of GABA (0.1 M) is 100-fold greater than that of glutamate (1 mM), while the GABA concentration of excitatory axons is only 0.6 mM, in accordance with the postulated role of GABA as an inhibitory transmitter.

The situation in the molluscs is less clear since significant levels of GABA could not be detected in octopus brain (Roberts, 1964) or the nervous system of the cephalopod, *Sepula esculenta* (Tsukada et al., 1964). Also in snail and octopus nervous tissue the incorporation of label from [U-14C]glutamate could not be detected in GABA (Bradford et al., 1969; Cory and Rose, 1969). However, subsequent to that, Osborne et al. (1972) demonstrated small amounts of GABA in *Helix* ganglia.

In the annelids endogenous GABA levels were reported to be too low for detection by GC–MS (McAdoo and Coggeshall, 1976), though more recently Sargent (1977) has demonstrated that leech ganglia incubated with labelled glutamate do accumulate labelled GABA.

Insect (honeybee) brain was shown by Frontali (1964) to possess GABA in two- to three-fold higher concentrations than mammalian brain, and Baxter and Torralba (1975) have measured the concentration of GABA in the metathoracic ganglia (12.8 μmol/g dry weight) and abdominal ganglia (11.2 μmol/g dry weight) of the cockroach. Further data for the GABA content of insect and crustacean nervous tissues are compared with parallel data for mammalian tissues in Table 3.1. Clearly there is a lot of evidence — in arthropods at least — that GABA is present in nervous tissues in significant quantities, comparable with mammalian nervous tissues, concordant with its role as a neurotransmitter.

SYNTHESIS AND METABOLISM OF GABA

In the 1960s, Kravitz and colleagues, using extracts from the lobster, found that the metabolic pathway for the synthesis and metabolism of GABA in

Table 3.1: Concentrations of GABA in the nervous tissues of different species. The data for GABA concentrations are expressed as μmol/g wet weight.

Species	Tissue	GABA concentration	Reference[a]
Honeybee	Brain	10.9	Frontali, 1964
Cockroach	Thoracic and abdominal ganglia	2.5	Ray, 1964
Rat	Brain	2–6.1	
Rabbit	Brain	2.7	Tallan, 1962
Hen	Brain	2.7–6	
Lobster	Ganglia	0.95–3.6	Aprison et al., 1973
	Axonal connectives	0.91-2.6	
Locust	Cerebral ganglia	21.5	
	Optic lobes	13.1	Breer and Heilgenberg, 1985
	Thoracic ganglia	10.7	
	Abdominal ganglia	12.1	
Fish	Telencephalon	3.2	
	Tectum opticum	3.8	Breer and Heilgenberg, 1985
	Cerebellum	2.4	
Mouse	Cortex	2.4	Breer and Heilgenberg, 1985
	Cerebellum	2.2	

[a]The different methods employed for the measurement of GABA concentration are as follows: Frontali (1964): Spinco amino acid analyser. Ray (1964); Tallan (1962): ninhyrin treatment of two-dimensional chromatograms. Aprison et al. (1973): measurement of DNP–amino acid esters by GLC. Breer and Heilgenberg (1985): GABAse digestion and measurement of NADP[1].

the crustacean nervous system was very similar to that in the mammalian nervous system (Kravitz, 1967). GABA is produced by the decarboxylation of glutamate, with the concomitant production of CO_2, by the enzyme GAD (L-glutamate 1-carboxy-lyase, EC 4.1.1.15).

Metabolism of GABA is achieved by its transamination in the presence of 2-oxoglutarate to give succinic semialdehyde (SSA) and glutamate, by the enzyme GABA-T (4-aminobutyrate-2-oxoglutarate transaminase, EC 2.6.1.19). The resulting SSA is then oxidised in the presence of NAD^+ by SSA dehydrogenase to give succinate.

This metabolic pathway of GABA in nervous tissues is summarised in Figure 3.2. Both GAD and GABA-T have been purified from various vertebrate sources. However, in the invertebrates, since the early studies in lobster, there has been only a limited amount of work on GABA-T and marginally more on GAD, largely in insect tissues. Recently, though, both enzymes have been purified from insect nervous tissue.

The synthesis of GABA

Frontali (1961) detected the presence of GAD in insect nervous tissue and demonstrated that this enzyme required pyridoxal phosphate (PLP, $4 \times 10^{-4}-10^{-3}$ M) for optimal activity. Furthermore the insect GAD exhibited two pH optima of 7.2 and 8.0. Glutamate has also been shown to be converted to GABA in the CNS of locust (Bradford et al., 1969; Sugden and Newsholme, 1977), the moth, Manduca sexta (Maxwell et al., 1978) and cockroach thoracic ganglion (Huggins et al., 1967). Working with extracts of lobster, Molinoff and Kravitz (1968) obtained GAD activity with a single pH optimum of 8.0; potassium ions and 2-mercaptoethanol were essential for enzyme activity and it was competitively inhibited by GABA, implying a possible role of GAD in regulating GABA levels. Furthermore GAD activity was found to be 100-fold higher in inhibitory than excitatory lobster axons (Kravitz et al., 1965). Glutamate has also been shown to be converted to GABA in the neuronal somata of the lobster (Otsuka et al., 1967; Potter, 1968) and the ventral nerve photoreceptors of the horseshoe crab (Battelle et al., 1979).

Wu et al. (1973) purified GAD from mouse brain by ammonium sulphate fractionation, gel filtration, and calcium phosphate gel and DEAE-sephadex chromatography. By high-speed sedimentation equilibrium centrifugation the native enzyme exhibited a molecular weight of 85000. The denatured enzyme had a molecular weight of 44000, and hence the mouse brain GAD was thought to be a dimer. GAD had been purified from a variety of other vertebrate sources including human brain (Blinderman et al., 1978). This enzyme was purified by DEAE-cellulose, hydroxyapatite, phenyl-sepharose, QAE-sepharose and Ultrogel chroma-

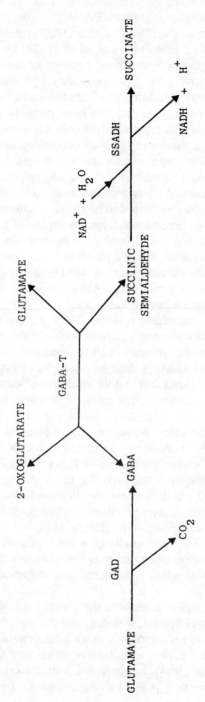

Figure 3.2: Metabolic pathway of GABA in mammalian nervous tissue

tography. The human enzyme has also been shown, by SDS–polyacryl-amide gel electrophoresis (SDS–PAGE), to be dimeric, and to have a molecular weight of 140 000. The pure GAD protein from catfish brain (Su *et al.*, 1979) had a molecular weight of 90 000 by PAGE. However, the denatured protein exhibited three peptides of molecular weights 22 000, 40 000 and 90 000, raising the possibility that the enzyme is tetrameric rather than dimeric. A number of studies have been made of GAD in insect tissues, and the enzyme has been partially purified from housefly (Chude *et al.*, 1979). Very recently, GAD has been purified 700-fold from locust brain by a fast protein liquid chromatography method (Stapleton, 1986). The native enzyme had a molecular weight of 97 000 as estimated by gradient PAGE. Under denaturing electrophoretic conditions, locust brain GAD consisted of two subunits of molecular weights 44 000 and 51 000 and therefore appeared to be dimeric as described for the vertebrate enzyme. The two-dimensional tryptic mapping of these subunits showed a considerable degree of homology between the respective separated peptides. Furthermore, this enzyme has been used to raise antibodies for mapping invertebrate neurones for GAD activity (Stapleton, 1986). The various studies on invertebrate GAD are compared with some of the data for the vertebrate enzyme in Table 3.2.

All the GAD activities shown in Table 3.2 are similar in that they are inhibited by carbonyl trapping agents such as aminooxyacetic acid and they require pyridoxal phosphate (PLP) for maximal activity (though note that mammalian brain GAD is thought to consist of two species, one of which has PLP tightly bound and one which does not such that there is a hetero-geneous PLP requirement for maximal activity: Tapia, 1983). However, the vertebrate and invertebrate GAD activities do differ in their pH optima and their affinity for glutamate. Until purified by Stapleton (1986), invertebrate GAD consistently exhibited a lower affinity for glutamate than vertebrate GAD. Such differences in K_m may be a function of the different purities of the preparations, such that crude preparations contain enzymes other than GAD which compete for glutamate, though the affinity of GAD for glutamate was decreased rather than increased on partial purification from mouse brain (Susz *et al.*, 1966). However, though both insect and crustacean GAD have, if anything, a lower affinity for glutamate than the mammalian enzyme, insect nervous tissues have a higher activity of GAD ($\approx$ 10-fold) than either crustacean or vertebrate CNS (Baxter and Torralba, 1975).

Insect GAD differs from both the vertebrate and crustacean enzymes in possessing two pH optima, whereas lobster, crayfish and mammalian GAD (either purified or measured in a tissue homogenate) have only one, though note that in early experiments lobster nerve-cord GAD exhibited two pH optima (Frontali, 1964). Furthermore, both the crude and purified enzyme from locust brain only possess a single optimum pH. It is unlikely that two

Table 3.2: Comparison of GAD activity in vertebrate and invertebrate tissues

Tissue	Purity	K_m Glutamate (mM)	PLP (μM)	pH optimum[a]	Reference
Mouse brain	Homogenate	3		6.4	Susz et al. (1966)
	158×	8		7.2	
Mouse brain	700×	0.7	0.05	7.0	Wu et al. (1973)
Human brain	8000×	1.3	0.13	6.8	Blinderman et al. (1978)
Calf brain	850×			7.9	Heinamaki et al. (1983)
Lobster ganglia	Crude extract	26		8.0	Molinoff and Kravitz (1968)
Crayfish brain	Homogenate	14		7.1	Wu et al. (1976)
Honeybee brain	Homogenate	48		6.8–7.8 and 8–8.5	Fox and Larsen (1972)
Drosophila larval brain	Crude extract			7.2 and 7.8	Chen and Widner (1968)
Cockroach metathoracic ganglion	Crude extract	28		7.1	Baxter and Torralba (1975)
Locust head thoracic ganglion	Homogenate	12		6.3 7.6	Breer and Heilgenberg (1986)
Whole housefly	Crude extract	60		4.8–5.2	Chude et al.
	600×	11		7.5	(1979)
Locust brain	Crude extract			7–7.5	Stapleton (1986)
	700×	2.7		7–7.4	

[a]Lower pH optimum is a plateau rather than a peak. Crude extract generally. The supernatant from a single centrifugation of a homogenate.

pH optima represent membrane-bound and non-bound GAD since detergents were used in the production of both honeybee and cockroach enzymes (Fox and Larsen, 1972; Baxter and Torralba, 1975). Mammals were thought to possess a second GAD activity ('GAD II') in nonneuronal tissues such as kidney, glial cells and developing chick-brain mitochondria (Haber et al., 1969, 1970a, b). However, more recently the existence of GAD II has been questioned since its measurement was based on the measurement of labelled CO_2 from [1-^{14}C]glutamate, and Wu et al. (1978) have demonstrated that this is an invalid means of measuring GAD activity in crude preparations because pathways other than GAD can cause the same CO_2 production. By analogy with this, a possible explanation of two pH optima for insect GAD activity is that one pH optimum represents the production of $^{14}CO_2$ by a pathway other than the decarboxylation of [1-^{14}C]glutamate to $^{14}CO_2$ and GABA by GAD. Fox and Larsen (1972) claimed that this is unlikely because they obtained similar GAD activities when measuring the production of labelled GABA from [U-^{14}C]glutamate as when measuring $^{14}CO_2$ production from [1-^{14}C]glutamate. However, White (1981) reported that most of the GABA synthesis in peripheral tissues does not occur via a direct decarboxylation of glutamic acid. Hence glutamate dehydrogenase could convert glutamate to 2-oxoglutarate, which in turn could be converted to SSA (for instance via pyruvate decarboxylase) with concomitant CO_2 release. SSA could be converted to GABA by GABA-T. The problem with this theory is that pyruvate decarboxylase is a fermentative rather than aerobic enzyme, and, even if there was significant activity in brain tissue, it is very dubious as to whether 2-oxoglutarate would serve as a good substrate. However, a pathway may exist for the production of GABA from glutamate other than via GAD, which may explain the second pH optimum observed for the conversion of glutamate to GABA in some invertebrate preparations. An alternative explanation of the two pH optima of insect GAD might be that insect nervous tissues could possess two isoenzymes, one of which is lost on purification of the enzyme (Chude *et al.*, 1979). This is only speculative though, since attempts to demonstrate two isoenzymes with cockroach GAD were unsuccessful (Baxter and Torralba, 1975), and Stapleton (1986) did not detect two pH optima in a crude locust GAD preparation. Alternatively, when varying the pH with a fixed substrate concentration in a crude preparation, there are several basic parameters, such as rate-limiting steps which may shift with change in pH, resulting in two or more apparent pH optima.

A further difference between insect, crustacean and vertebrate GAD is the effect of potassium and chloride ions on GAD activity. Both insect GAD (Baxter and Torralba, 1975; Breer and Heilgenberg, 1985) and mammalian GAD (Susz *et al.*, 1966; Blinderman *et al.*, 1978) are inhibited by chloride ions, whereas the lobster enzyme is not (Molinoff and Kravitz, 1968). Baxter and Torralba (1975) have suggested that chloride ion inhibition of GAD has evolved only in terrestrial species which were not faced, like the lobster, with a high chloride ion concentration in their environment. Conversely, insect GAD is similar to lobster and crayfish GAD in being stimulated by potassium ions (Molinoff and Kravitz, 1968; Baxter and Torralba, 1975; Wu *et al.*, 1976; Chude *et al.*, 1979; Breer and Heilgenberg, 1986; Stapleton, 1986). This effect has not been reported for vertebrate GAD.

GAD is thought to be involved in limiting the rate of GABA metabolism and, therefore, to be important in the control of tissue concentrations of GABA. A refinement of this control is the possibility of feedback inhibition of GAD by GABA, since competitive inhibition of GAD by GABA has been reported in honeybee (Fox and Larsen, 1972), cockroach (Baxter and Torralba, 1975), locust (Breer and Heilgenberg, 1986), lobster (Molinoff and Kravitz, 1968) and crayfish (Wu *et al.*, 1976). Baxter and Torralba (1975) have suggested that this feedback inhibition is a characteristic of the arthropod GAD which is not shared by the vertebrate enzyme. However, human brain GAD is weakly inhibited by GABA ($K_i = 10$ mM,

Blinderman *et al.*, 1978). The physiological relevance of these inhibition studies is difficult to assess since any inhibition *in vivo* will depend on the concentration of GABA in the relevant cellular compartment and the affinity of the enzyme for glutamate (which is higher in the case of mammalian GAD than invertebrate GAD). It is probable that insects are capable of GABA inhibition of GAD *in vivo*, but whether or not this effect is exclusive to the arthropod enzyme is unclear.

A further means of studying the similarities between enzymes from different species involves the use of antibodies. Su *et al.* (1983) have demonstrated that antibodies raised in rabbit against catfish GAD cross-react with goldfish, turtle, chick and frog GAD, but not with the mammalian enzyme. These antibodies also inhibit the GAD activity of *Drosophila* and crayfish (as well as frog, chick and goldfish), suggesting that the GAD enzymes of invertebrates and lower vertebrates are closely related to each other but not to the mammalian enzyme. In agreement with this, Breer and Heilgenberg (1986) have observed low cross-reactivity between antiserum against rat GAD and the locust enzyme, although they also observed strong cross-reactivity of the antiserum against rat GAD with trout GAD, which argues for a closer relationship between the GAD enzymes of higher and lower vertebrates than postulated by Su *et al.* (1983).

In summary, there is ample evidence, in insects and crustaceans at least, that invertebrates possess GAD activity in their nervous tissues which has overall similarities with the well-characterised mammalian enzyme. However, differences may exist between insect, crustacean and vertebrate GAD enzymes with respect to the inhibitory control of the enzyme, possible isoenzymes and the affinity of the enzyme for its substrate. Though GAD is important in the control of GABA concentration in all these tissues, the fine tuning of the system probably differs, not only between vertebrates and invertebrates, but also between different invertebrate classes.

The metabolism of GABA

In 1967 Hall and Kravitz (1967a) demonstrated GABA-T in extracts of lobster CNS. The enzyme is similar to other transaminases; it is stimulated by pyridoxal phosphate (PLP) and inhibited by carbonyl reagents such as hydroxylamine as well as high concentrations of the sustrates GABA and 2-oxoglutarate. Aminooxyacetic acid is the most potent inhibitor. The kinetic properties of the enzyme are consistent with a substituted enzyme mechanism in which the product of the first substrate must leave the enzyme before the second substrate can bind. Other amino acids such as β-alanine and γ-aminovaleric acid can act as substrates for the enzyme, but not as well as GABA.

The same workers also characterised the lobster SSADH enzyme. It has an alkaline pH optimum of about 9, requires NAD^+ (though $NADP^+$ will also serve as cofactor), and also contains an active thiol group (2-mercaptoethanol and dithiothreitol activate the enzyme equally). In these respects the lobster SSADH is similar to the mammalian brain enzyme, but it differs in being inhibited by sodium chloride (Hall and Kravitz, 1967b). Both GABA-T and SSADH have been demonstrated in honeybee extracts (Fox and Larsen, 1972), and GABA-T has also been measured in locust and cockroach CNS (Sugden and Newsholme, 1977).

GABA-T has been purified from mouse brain by ammonium sulphate fractionation, gel filtration, and calcium phosphate gel and DEAE-sephadex chromatography (Schousboe et al., 1973). The enzyme is thought to be dimeric, having a molecular weight of 109000, and has a pH optimum of 8.05. The Michaelis constants for GABA and 2-oxoglutarate are 1.1 mM and 0.25 mM, respectively, and the authors postulate therefore that the enzyme activity in vivo is probably not regulated by GABA concentration, but that 2-oxoglutarate may be important since its concentration is lower than its K_m. The mouse enzyme has a wide amino acid substrate specificity but narrow ketoacid substrate specificity. Of a range of ketoacids tested, only 2-oxoglutarate served as an amino group acceptor, but a range of amino acids were capable of being donors. The donor needs to be a neutral ω-aminocarboxylic acid with the two functional groups separated by between two and four carbon atoms.

More recently, GABA-T has been studied in the parasitic worm, Nippostrongylus brasiliensis (Watts and Atkins, 1984). The enzyme activity and K_m values (GABA, 0.33 mM; 2-oxoglutarate, 0.57 mM) are comparable with the mammalian enzyme. Glutamate inhibits the enzyme (K_i, 0.35 mM) and, as described above for purified mouse GABA-T, only 2-oxoglutarate acted as an amino group acceptor while a narrow range of amino acids were able to serve as donors. Of a range of substrate analogues tested, other than GABA, only canaline significantly inhibited the parasite enzyme, and the authors are not optimistic about the use of conventional substrates in inhibiting the enzyme. However, an unusual feature of the worm enzyme is that, in a parallel isoelectric and chromatographic study with GABA-T from rat brain, whereas the rodent enzyme exhibited a pI of 5.4, in agreement with other workers, the parasite enzyme had a pI of 10.5. This implies that GABA-T from the nematode has a quite different amino acid composition from that of the enzyme in mammalian brain.

GABA-T has been detected in locust head and thoracic ganglia (Breer and Heilgenberg, 1986). More recently the enzyme has been purified in parallel from both sheep brain and whole locust heads (D. Jeffery, personal communication). The insect enzyme was purified by acid and heat precipitation, ammonium sulphate precipitation, negative chromatography on DE52 cellulose, phenyl agarose chromatography, and chromatofocusing on

FPLC. Both the locust and sheep enzymes appeared as single proteins of identical size on native polyacrylamide gel electrophoresis (PAGE); the molecular weight of both native enzymes was 97000 by gel permeation chromatography. In SDS–PAGE both enzymes migrated as a single species of molecular weight 50000, and are therefore thought to be dimeric, in agreement with other studies on mammalian GABA-T (Schousboe et al., 1973; John and Fowler, 1976). However, when a monoclonal antibody against rabbit brain GABA-T is used in immunoblotting studies, purified rabbit GABA-T shows as a single band of molecular weight 52000 whereas unfractionated rabbit brain supernatant shows a single band of 58000 (Edwardson et al., 1985). These workers therefore suggest that the lengthy purification procedure may result in limited proteolysis of the enzyme, and by analogy it is possible that the authentic insect GABA-T may have a marginally larger subunit size than reported by D. Jeffery (personal communication).

In the parallel study of Jeffery (personal communication), GABA-T from locust has a pH optimum of 8.3 and the K_m values for GABA and 2-oxoglutarate are comparable for the locust (0.8 mM and 0.27 mM, respectively) and sheep (1.68 mM and 0.22 mM) enzymes. Also, for both enzymes, 2-oxoglutarate is the only effective amino group acceptor, as described above for GABA-T from other sources. Furthermore gabaculline is a potent suicide substrate ($K_i = 1$–2 mM) for GABA-T from both sheep and locust. However, the enzymes have different isoelectric points (locust, 6.65; sheep, 5.50), and antibodies raised in rabbit against sheep GABA-T show no cross-reactivity with the locust enzyme in several different immunological studies. None of these antibodies interacts with the active site of the enzyme though. Therefore, it seems possible that the GABA-T enzymes from insect and mammalian CNS have a very similar overall size and active sites, but that the amino acid composition of the enzymes differs, though this theory awaits the definitive evidence of amino acid analyses of the enzymes.

Within the invertebrates there is now very good evidence that insects, crustaceans and nematodes possess the enzymic machinery for the metabolism of GABA. Furthermore, the major metabolic enzyme, GABA-T, appears very similar in size, subunit composition, and catalytic characteristics in both invertebrate and vertebrate CNS, though major differences exist in the amino acid composition of the enzyme from different species.

STUDIES ON GABA UPTAKE

Once GABA has been released from the nerve terminal and interacted with its postsynaptic receptor, it is necessary that the resulting signal be terminated. Though nervous tissues are known to contain the enzymes necessary for the metabolism of GABA, it is thought that GABA-T is not directly responsible for the removal of GABA from the synaptic cleft. Because it requires 2-oxoglutarate as a substrate, GABA-T is unlikely to be located outside the nerve endings, and, furthermore, GABA-T inhibitors do not have any prolonging effect on neuromuscular transmission in the lobster (Hall and Kravitz, 1967a). Instead, GABA is thought to be removed from the synapse by neuronal and glial uptake.

One of the first demonstrations of such uptake was in an invertebrate nerve–muscle preparation. Iversen and Kravitz (1966, 1968) found that on incubating the ventral superficial abdominal muscles of the lobster in 20 nM [^{3}H]GABA, the preparation accumulated four to five times the concentration of radioactivity in the medium, and that 90 per cent of this was still in the form of GABA. This uptake was sodium-dependent, exhibited a K_m of 6×10^{-5} M, and was inhibited by β-guanidinopropionic acid, γ-hydroxy GABA, desipramine and chlorpromazine (all at 10^{-3} M). Furthermore, these workers demonstrated that GABA could be released by stimulation of a single inhibitory axon innervating lobster muscle (Kravitz et al., 1968). More recently, Craelius and Fricke (1981) have shown that the inhibitory neurones innervating the muscle receptor organ of crayfish, following exposure to [^{3}H]GABA, will release the radioactivity in response to low-frequency electrical stimulation in a Ca^{2+}-dependent manner. Subsequent to the early work of Iversen and Kravitz (1966), Iversen and Neil (1968) demonstrated the active accumulation of GABA by slices of rat cerebral cortex, and 70 per cent of [^{3}H]GABA uptake by slices and homogenates of various brain regions was shown by autoradiography to be localised in the nerve terminals (Iversen and Bloom, 1972). Sodium-dependent GABA uptake was also demonstrated in synaptosomes from mouse brain (Kuriyama et al., 1969) and rat brain (Martin and Smith, 1972). Indeed, the availability of mammalian synaptosomes opened the doors to several extensive studies of GABA uptake such that it became widely accepted that neurones (and glial cells, see Enna and Snyder, 1975; De Feudis, 1979) of mammalian CNS are capable of GABA uptake. Subsequently it was possible to obtain both Ca^{2+}-dependent and Ca^{2+}-independent, K^+-evoked GABA release from synaptosomes. However, there is still controversy as to whether or not Ca^{2+}-dependent release is from cytosolic (e.g. De Belleroche and Bradford, 1977) or non-cytosolic (Haycock et al., 1978) compartments.

It should be noted that in the early studies it was thought that GABA uptake was responsible for the termination of the response. However, Eccles and Jaeger (1958) had demonstrated mathematically that the time course of diffusion of transmitter from the synaptic cleft was sufficiently rapid to account for transmitter inactivation. Evidence has recently been provided in support of this for GABA inactivation in crayfish stretch receptor neurones (Deisz et al., 1984). Bath-applied GABA causes a conductance increase in voltage-clamped neurones which is enhanced in magnitude, but not in time course, by the uptake inhibitor, nipecotic acid. Furthermore, rapid conductance increases were only observed at high GABA concentrations, leading the authors to postulate that the prominent action of GABA uptake is to maintain a low cleft concentration of GABA rather than to terminate the response.

Following the demonstration of GABA uptake in lobster muscle (Iversen and Kravitz, 1966, 1968), whereas work on mammalian synaptosomal GABA uptake and release was making rapid progress, invertebrate investigators had to wait fourteen years before they had synaptosomes available to them. However, it was possible to make autoradiographic and electrophysiological studies of invertebrate [^{3}H]GABA uptake.

Frontali and Pierantoni (1973) incubated brain slices from the cockroach, *Periplaneta americana*, with [^{3}H]GABA and demonstrated that GABA accumulated over certain neurones and/or their processes or endings, while much longer times of autoradiograph exposure were required to detect low levels of glial uptake. In an electrophysiological study on the abductor muscle of crayfish, Horwitz and Orkand (1979) reported a GABA uptake mechanism which was blocked by nipecotic acid and L-2,4-diaminobutyric acid. However, this mechanism was not Na$^+$-dependent, an observation which the authors attributed to the relatively high concentrations (10^{-5} M) of GABA employed in the study. In a more recent autoradiographic study (Hue *et al.*, 1982), the two last abdominal ganglia from the cockroach were incubated in a saline solution containing [^{3}H]GABA. The transmitter accumulated at the periphery of the ganglia, and uptake of GABA into the terminal ganglion was partially inhibited by sodium-free medium. Furthermore, since the uptake was also inhibited by 10 mM β-alanine, this observed GABA uptake mechanism was thought to be at least partially glial, though more recently Griffiths *et al.* (1986) have shown that β-alanine is not specific for glial GABA uptake systems. Inhibitory motor neurones of the leech CNS have been shown to take up [^{3}H]GABA, and this uptake is stimulated by electrical activity of the neurones (Cline *et al.*, 1985). Van Marle *et al.* (1985) have made an autoradiographic investigation of GABA uptake into two morphologically distinct insect neuromuscular synapses; the common inhibitor innervated slow muscle exhibits [^{3}H]GABA uptake by both glia and terminal axons, whereas the slow extensor tibia that is not innervated exhibits only glial uptake. Shepherd and Tyrer (1985) have also demonstrated a GABA uptake mechanism electrophysiologically in insect skeletal muscle, and they conclude that this may be glial though like many invertebrate studies, this conclusion is based largely on mammalian pharmacology.

Crayfish muscle plasma membrane vesicles have been shown to accumulate GABA at 0°C in a process dependent on sodium ions and membrane integrity (Meiners *et al.*, 1979). The process was saturable with

103

a K_m of 0.55 µM, which is much lower than that reported for intact lobster muscle at 20°C (K_m=60 µM, Iversen and Kravitz, 1966) or crayfish muscle slices at 22°C (22 µM, Olsen et al., 1975). These differences may be due to the use of isolated vesicles as against slices of or intact neuromuscular tissue. Alternatively, an additional low-affinity uptake system may exist in the vesicular preparation. The vesicular uptake of [³H]GABA (Meiners et al., 1979) is inhibited by both nipecotic acid (IC_{50}=0.45 µM) and slightly more potently by the mammalian neuronal inhibitor 2,4-diaminobutyric acid (DABA) (IC_{50}=0.3 µM). However, even at 1 µM neither of these ligands inhibited the uptake by more than 80 per cent, supporting the possibility of a secondary uptake system. It therefore seemed likely that uptake was involved in the removal of GABA from the crustacean neuromuscular junction, but claims for the existence of any similar uptake system in the invertebrate CNS relied solely on the autoradiographic evidence described above.

In 1980 Breer and Jeserich published a method for the production of synaptosomes from insect CNS. Gordon et al. (1982) used this method to produce synaptosomes from locust nervous tissue which possessed an active GABA transport mechanism. To characterise this mechanism the synaptosomes were osmotically shocked to produce membrane vesicles. These vesicles had a saturable high-affinity uptake mechanism, with a K_m of 3 µM, which was dependent on the simultaneous presence of sodium and chloride ions. Also, when the sodium gradient (outside > inside) was collapsed with the ionophore nigericin, the uptake was inhibited. More recently, intact locust synaptosomes have been shown to possess high-affinity (K_m=1.7 µM) and low-affinity (K_m=19 µM) GABA uptake systems (Breer and Heilgenberg, 1985). The transport of GABA is optimal at 150 mM Na⁺, is blocked by replacement of Na⁺ with K⁺, and is stimulated by Cl⁻ ions. The uptake was inhibited by DABA, muscimol, chlorpromazine and imipramine (IC_{50} values of 50, 70, 90 and 200 µM, respectively). Synaptosomes derived from cockroach CNS have also been shown to possess an active GABA uptake system; however, after 5 min of uptake the apparent uptake decreases, which the authors attribute to endogenous GABA-T activity (Whitton et al., 1986). Robinson and Lunt (1986) have reported a method for the rapid production of a preparation possessing a high density of synaptosomes. This preparation possesses a GABA uptake system which is blocked by 50 mM K⁺ but is unaffected by either of the GABA-T inhibitors gabaculline or aminooxyacetic acid. The uptake is potently inhibited by 0.1 mM DABA, but 0.1 mM nipecotic acid is much less potent. Preliminary results suggest that GABA can be released from this preparation by 50 mM K⁺ in a partially Ca²⁺-dependent manner (Robinson and Lunt, unpublished observation). Though all these studies are on synaptosomal preparations, one should bear in mind that glial uptake of GABA cannot be ruled out as a possible component of the observed uptake. Lees and Beadle (1986) have reported a Na^+-dependent, high-affinity uptake system for GABA in cockroach neuronal cultures. Using autoradiography these workers have shown this uptake to be highly specific and limited to only 10–15 per cent of the neurones. This may explain some of the difficulties in working with insect synaptosomal GABA uptake in that only a small proportion of any synaptosome preparation may possess a GABA uptake capacity.

It is evident from the limited work described above that the use of synaptosomes in the study of invertebrate GABA uptake is very much in its infancy. However, evidence is beginning to emerge that, in addition to invertebrate neuromuscular junctions, insect (and annelid) CNS possesses GABA uptake systems which may well play a role in the removal of GABA from the synapse. We are only just beginning to gain information on the pharmacology of these uptake systems, and on the relative contributions of glial and neuronal uptake. Let us hope that our knowledge in this area continues to grow and that with improved invertebrate synaptosome technology we might be able to study GABA release in continuous perfusion systems.

A GABA RECEPTOR COMPLEX

GABA receptors

Much of the early evidence for GABA receptors in nervous tissues came from electrophysiological studies in invertebrate systems (see Gerschenfeld, 1973). Owing to such factors as the ease of obtaining large quantities of tissue, the subsequent biochemical characterisation of the GABA receptors involved in neurotransmission was carried out with mammalian brain. However, as we will see later, in the last five years there has been an upsurge of interest in invertebrate GABA receptors.

The Na^+-independent, receptor-specific binding of [^{3}H]GABA to mammalian brain fractions was first demonstrated by Zukin *et al.* (1974), and was subsequently characterised by several groups (e.g. Enna and Snyder, 1975; Greenlee *et al.*, 1978; Krogsgaard-Larsen *et al.*, 1979). It is now known that there are at least two classes of GABA receptor, $GABA_A$ and $GABA_B$, in mammalian brain. $GABA_A$ receptors activate Cl^- channels and are sensitive to bicuculline, while $GABA_B$ receptors inactivate Ca^{2+} channels, or activate K^+ channels, may be linked to G-proteins and adenylate cyclase, and are sensitive to baclofen (see Bowery *et al.*, 1984; Hill *et al.*, 1984). The $GABA_A$ receptor is in fact a well-characterised complex of a chloride channel associated with interacting binding sites for GABA, the benzodiazepines, and picrotoxin, cage convulsants and the barbiturates (Olsen, 1981, 1982; Braestrup and Nielsen, 1983; Haefely,

1984; Tallman and Gallager, 1985). This complex is depicted in a highly simplified model in Figure 3.3 (after Olsen, 1981), though it should be noted that more complex models have since been postulated (e.g. Haefely, 1984).

All the binding activities of the mammalian GABA$_A$ receptor have been co-purified as one complex from both bovine and porcine brain (Sigel *et al.*, 1983; Olsen *et al.*, 1984; Sigel and Barnard, 1984; Kirkness and Turner, 1986). The purified complex has a molecular weight of approximately 200000 and contains two different species of subunit of approximate molecular weights 50000 and 55000. Polyclonal (Stephenson *et al.*, 1986) and monoclonal (Schoch *et al.*, 1985) antibodies have been raised to the purified complex. The monoclonal antibodies have been used to demonstrate that the receptor complex has a mixed subunit composition such that each complex contains both (50000 and 55000) subunits.

Figure 3.3: Simplified model of the mammalian GABA$_A$ receptor complex, after Olsen (1981), showing interacting binding sites for GABA, the benzodiazepines (BZ) and picrotoxin (PTXN) and cage convulsants (TBPS) associated with a Cl$^-$ ion channel

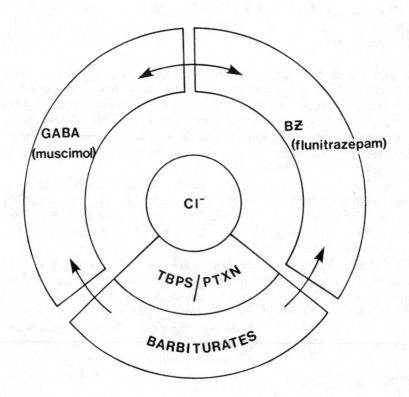

Furthermore, the benzodiazepine and high- and low-affinity GABA binding sites are all precipitated by monoclonal antibodies to different subunits and hence all the binding sites reside on the same structural complex (Haring *et al.*, 1985).

Despite early electrophysiological evidence for GABA receptors in some invertebrate nervous tissues (Gerschenfeld, 1973; Walker *et al.*, 1975), for a long time the only complementary biochemical evidence was that of De Robertis and Fiszer de Plazas (1974). These workers extracted a proteolipid from shrimp muscle that bound [^{14}C]GABA with a K_D of 8 × 10^{-6} M. At 10^{-4} M, bicuculline was the most potent displacing ligand, while picrotoxin (10^{-4} M) and the mammalian GABA agonist muscimol (4 × 10^{-4} M), were equally potent. However, such proteolipids are difficult to work with, and no further work has been published in this area.

The first invertebrate GABAergic binding studies were performed in crayfish muscle by Meiners *et al.* (1979) who demonstrated specific, saturable Na$^+$-independent binding of [^{3}H]muscimol. However, it was not until 1983 that a GABAergic binding activity was reported in insect tissue. Abalis *et al.* (1983) detected the binding of the benzodiazepine [^{3}H]flunitrazepam (FNZP) to housefly thoracic membranes (though they could not detect GABA binding in the same preparations). This binding was enhanced by GABA in a dose-dependent manner, optimal at 10^{-8} M GABA, which in turn was blocked by the GABA antagonist bicuculline at 1.7 × 10^{-5} M. Though this preparation would contain a proportion of thoracic ganglionic membranes, the authors attributed the observed binding to muscle membranes. Numerous electrophysiological studies (e.g. Nistri and Constanti, 1979) have reported that GABA-evoked responses in crustaceans are reduced in the presence of bicuculline. This observation, together with those of Abalis *et al.* (1983), suggests that invertebrate muscle possesses GABA receptors which show some sensitivity to bicuculline, though the IC_{50} for bicuculline was only 3 × 10^{-3} M in the [^{3}H]muscimol binding study of Meiners *et al.* (1979) in crayfish muscle. More recently there has been an increase in interest in invertebrate CNS GABA receptors, and several groups have probed insect CNS with both [^{3}H]muscimol and [^{3}H]GABA (Lummis and Sattelle, 1985a; Lunt *et al.*, 1985; Tanaka and Matsumura, 1985; Abalis and Eldefrawi, 1986; Breer and Heilgenberg, 1986). A common denominator in all these studies is a virtual lack of any sensitivity of insect CNS GABA receptors to bicuculline. The characteristics of these various invertebrate binding studies are compared in Table 3.3 with the data from a selection of mammalian GABAergic binding studies.

The lack of bicuculline sensitivity in invertebrate GABAergic binding studies compared with the mammalian CNS is in agreement with Beadle *et al.* (1985) who reported that 10^{-5} M bicuculline failed to block GABA-evoked conductance increases measured in current- and voltage-clamped

Table 3.3: Comparison of GABAergic binding studies in invertebrate and vertebrate nervous tissues

Reference	1	2	3	4	5	6	7	8	9	10
K_D (nM)	10	40	30	3 and 144	384	100	9	5 and 30	370	100
B_{max} (fmol/mg protein)	70	20	200	50 and 500	1420	2200	500	500 and 900	680	—
Ligand [^{3}H]	Muscimol	Muscimol	GABA	Muscimol	GABA	GABA	Muscimol	Muscimol	GABA	GABA
Tissue	Locust ganglia	Housefly head	Locust ganglia	Honeybee brain	Cockroach CNS	Locust ganglia	Crayfish muscle	Rat brain	Rat brain	Rat brain
IC$_{50}$(M)										
muscimol	3×10^{-8}	3×10^{-8}	1×10^{-7}	6×10^{-9}	7×10^{-7}	9×10^{-9}	2×10^{-8}	1×10^{-8}	—	4×10^{-8}
GABA	1×10^{-7}	4×10^{-8}	9×10^{-8}	4×10^{-8}	1×10^{-7}	—	2×10^{-7}	4×10^{-8}	4×10^{-7}	2×10^{-7}
Isoguvacine	2×10^{-7}	1×10^{-7}	5×10^{-5}	—	5×10^{-4}	—	6×10^{-8}	—	—	2×10^{-7}
3-APS	3×10^{-6}	3×10^{-7}	—	3×10^{-5}	4×10^{-4}	NS	3×10^{-6}	3×10^{-7}	3×10^{-7}	6×10^{-7}
Imidazole acetate	—	6×10^{-7}	—	1×10^{-6}	—	—	2×10^{-7}	2×10^{-7}	2×10^{-7}	9×10^{-7}
Bicuculline	$>10^{-3}$	4×10^{-4}	$>10^{-3}$	$>10^{-3}$	$>10^{-3}$	NS	3×10^{-3}	6×10^{-6}	4×10^{-6}	4×10^{-6}
Chlorpromazine	—	—	—	—	—	1×10^{-7}	—	—	2×10^{-4}	—
Imipramine	—	—	—	—	—	1×10^{-7}	—	—	—	—
2,4-DABA	—	$>10^{-3}$	—	$> 10^{-3}$	—	NS	—	$>10^{-5}$	$>10^{-3}$	$>10^{-4}$
Nipecotic acid	—	$>10^{-4}$	$>10^{-3}$	$>10^{-4}$	$>10^{-3}$	—	$>10^{-3}$	$>10^{-5}$	$>10^{-4}$	$>10^{-4}$
β-Alanine	—	—	$>10^{-3}$	5×10^{-6}	$>10^{-3}$	—	—	$>10^{-5}$	8×10^{-5}	4×10^{-5}

Sources: 1, Robinson (1986). 2, Lunt et al. (1985). 3, D. Jeffery, (personal communication). 4, Abalis and Eldefrawi (1986). 5, Lummis and Sattelle (1985a). 6, Breer and Heilgenberg (1986); 'NS' denotes 10% displacement of binding, maximum concentration of ligand not given. 7, Meiners et al. (1979). 8, Williams and Risley (1979). 9, Enna and Snyder (1975). 10, Greenlee et al. (1978b).

cultured locust neurones. However, apart from the lack of bicuculline sensitivity, the invertebrate binding studies are conflicting on several points. All the studies except that of Breer and Heilgenberg (1985) are in agreement with respect to the potencies of GABA and muscimol, irrespective of the radioactive ligand used, and this characteristic is similar to that of the mammalian $GABA_A$ receptor. Furthermore, Beadle et al. (1985) have reported that 1 mM GABA and muscimol both cause the desensitisation of cultured insect neurones, and that cross-desensitisation is also evident, implying that these ligands act on a common receptor population. The binding of [³H]muscimol in locust and housefly CNS (Lunt et al., 1985) and crayfish muscle (Meiners et al., 1979) is also similar to the mammalian GABA receptor in being sensitive to isoguvacine, 3-aminopropane sulphonic acid (3-APS), and imidazole acetate, whereas [³H]GABA binding in cockroach CNS (Lummis and Sattelle, 1985a) is relatively insensitive to either isoguvacine or 3-APS. The binding of [³H]muscimol to honeybee brain (Abalis and Eldefrawi, 1986) is also only weakly displaced by 3-APS, though imidazole acetate is reasonably potent.

A further discrepancy lies in the 10-fold higher number of binding sites observed when using [³H]GABA compared with [³H]muscimol. There is a danger (as discussed by Meiners et al., 1979) that when working with [³H]GABA the observed binding can be contaminated by a contribution from GABA uptake sites, a problem probably not encountered with nanomolar concentrations of [³H]muscimol. Though the [³H]GABA binding activity of locust CNS (Breer and Heilgenberg, 1986) is insensitive to the neuronal GABA uptake inhibitor, DABA, it is also insensitive to muscimol and strongly inhibited by both the uptake inhibitors chlorpromazine and imipramine. Although these binding studies were carried out in Na^+-free medium, the membranes did not undergo the same extensive washing as those utilised by D. Jeffery (personal communication). Therefore, it is possible that the membrane preparation contains sufficient residual Na^+ ions to permit GABA uptake into vesiculated membranes or that in any case [³H]GABA can bind to the recognition site of its transporter, resulting in a contribution from GABA uptake systems. The uptake of GABA may also contribute somewhat to the data of Lummis and Sattelle (1985a), because their observed [³H]GABA binding is decreased by freezing, washing or detergent treatment of the membrane preparation prior to assay; these treatments elevate mammalian GABAergic binding activity by removal of endogenous GABA (Greenlee et al., 1978). The data in Table 3.3 do not agree on the level of GABAergic binding activity in insect CNS, but, from the considerations outlined above, it seems likely that the concentration of GABA receptors in insect CNS is less than picomolar. As for the affinities of binding, Abalis and Eldefrawi (1986) have demonstrated the presence of two different affinity binding sites, comparable with the mammalian CNS. Robinson and Lunt (unpublished results have also observed low-affinity (K_D 70 nM, B_{max} 400 fmol/mg protein) binding of [^{3}H]muscimol to locust brain.

There is therefore growing biochemical evidence for the presence of GABA receptors in insect CNS which exhibit a pharmacology comparable to but subtly different from that of their mammalian counterparts. An obvious progression from these observations is to ask whether these central insect GABA receptors possess associated receptors for benzodiazepines and cage-convulsant/picrotoxin as described above for mammalian GABA$_A$ receptors. The question has two components: first, does insect CNS have benzodiazepine or cage-convulsant/picrotoxin binding sites, and secondly, if so, are these sites linked to GABA receptors?

Benzodiazepine receptors

In 1978 it was reported that benzodiazepine binding sites were present in a wide range of vertebrate species but absent in the nervous system of five invertebrate species studied (earthworm, squid, woodlouse, lobster and locust) (Nielsen et al., 1978). The authors suggested, therefore, that brain-specific benzodiazepine receptors had a late evolutionary appearance.

However, Abalis et al. (1983) demonstrated the presence of [^{3}H]FNZP binding sites in a housefly thoracic preparation, and though these sites were more susceptible to the peripheral benzodiazepine Ro5-4864 than the central ligand clonazepam, they were affected by GABA. Tanaka and Matsumura (1985) observed binding of [^{3}H]diazepam in cockroach ganglia, and more recently, using a filtration assay, Robinson et al. (1986) have characterised a heterogeneous population of benzodiazepine receptors in locust CNS which also show a greater susceptibility to Ro5-4864 than clonazepam. Lummis and Sattelle (1986) have demonstrated [^{3}H]FNZP binding to cockroach brain membranes which is more potently displaced by Ro5-4864 than by clonazepam, and it may be that, contrary to the suggestion of Bolger et al. (1986) that peripheral-type benzodiazepine receptors appeared very late in evolution, they do in fact have an evolutionary old history. Robinson et al. (1986) have reported that it is necessary to replace the physiological concentrations of calcium that are removed by washing during the membrane preparation to obtain binding of [^{3}H]FNZP to ganglionic membranes, implying a difference in cation requirements from the mammalian central benzodiazepine receptor. Though the invertebrate benzodiazepine binding sites appear to have a peripheral-type pharmacology, a proportion of the receptors in locust CNS can be photoaffinity labelled with [^{3}H]FNZP (Robinson et al., 1986) as described for mammalian central GABAergic benzodiazepine receptors (Mohler et al., 1980). Mammalian peripheral benzodiazepine receptors can not be photoaffinity labelled (Thomas and Tallman, 1981). In locust ganglionic membranes two major and two minor bands are labelled, and this pattern is different from any published for mammalian receptors (Robinson *et al.*, 1986) though it is most similar to that of fish (Hebebrand *et al.*, 1986). What is clear is that insect CNS possesses benzodiazepine receptors which are analogous in some respects to mammalian central receptors, and that GABAergic benzodiazepine receptors may have an earlier evolutionary origin than first suggested by Nielsen *et al.* (1978).

Cage-convulsant/picrotoxin receptors

Tanaka *et al.* (1984) have reported the specific binding of picrotoxin to a preparation from cockroach CNS which by Scatchard analysis exhibited a B_{max} of 1.4 pmol/mg protein and a relatively low affinity constant, K_D, of 800 nM. More recently Cohen and Casida (1986) have reported binding of the cage convulsant, [^{35}S]t-butylbicyclophosphorothionate (TBPS) to a preparation from whole housefly thorax and abdomen. This binding does not require Cl^- ions for optimal activity, has a B_{max} of 2.5 pmol/mg protein, and a K_D of 210 nM, and is only weakly inhibited by picrotoxin (IC_{50}=3.28 × 10^{-5} M). These data are in contrast with the low concentrations of high-affinity [^{35}S]TBPS binding measured in membranes from locust CNS (K_D=10 nM, B_{max} approximately 200 fmol/mg protein; M. Brown, personal communication), housefly head (K_D=51 nM, B_{max}=220 fmol/mg protein) and crayfish muscle (K_D=23 nM, B_{max}90 fmol/mg protein, Szamraj *et al.*, 1986) and cockroach CNS (K_D=18 nM, B_{max}=180 fmol/mg protein, Lummis and Sattelle, 1986). These binding sites were inhibited weakly by picrotoxin (10–30 μm). Furthermore, the binding in housefly head is not dependent upon, but stimulated by, Cl^- ions. The binding of [^{35}S]TBPS has been observed in *Torpedo* electroplax, which are known not to possess any GABA receptors (Abalis *et al.*, 1985a). Therefore the possibility arises that [^{35}S]TBPS binding can be contaminated by some non-GABAergic component. Such a situation might be more likely in the study of Cohen and Casida (1986) which uses a very heterogeneous tissue source. Evidently both invertebrate muscle and CNS can possess binding sites for cage convulsants, though their relationship to GABA receptors requires more study. One observation common to all the binding studies described above is that picrotoxin is not very potent at invertebrate TBPS binding sites.

The interaction of sites in an invertebrate GABA receptor complex

The evidence points to both invertebrate muscle and CNS possessing binding sites for GABA and muscimol, benzodiazepines and cage convulsants.

The binding of [^{3}H]FNZP is enhanced by GABA in both insect muscle (Abalis *et al.*, 1983) and CNS (Robinson *et al.*, 1986) though the characteristics of enhancement are different. Furthermore, the locust central benzodiazepine binding is enhanced 20 per cent by 0.1 mM pentobarbital (M. Brown and T.N. Robinson, unpublished observation), and similarly the binding of [^{3}H]GABA in the same tissue is enhanced by 20 per cent by 0.1 mM pentobarbital (D. Jeffery, personal communication). The binding of [^{3}H]muscimol in housefly heads is enhanced 25 per cent by 1 mM pentobarbital (R.W. Olsen, unpublished observation). The [^{35}S]TBPS binding observed by Cohen and Casida (1986) is enhanced by 40 per cent by 10–100 μM GABA, in contrast with that observed by Lummis and Sattelle (1986) which is inhibited by 1 μM GABA, and [^{35}S]TBPS binding in housefly heads is insensitive to GABA or muscimol (0.1–1000 μM: Szamraj *et al.*, 1986). In other words we are just beginning to gain biochemical evidence that a portion of these different binding activities may be linked to each other in a complex somewhat analogous to the mammalian GABA$_A$ receptor. This has been borne out by the work of Beadle and colleagues on current- and voltage-clamped cultured locust neurones. GABA-evoked responses in these cultures are enhanced by 10^{-6} M FNZP (Lees *et al.*, 1985) and also by 10^{-4}–10^{-5} M pentobarbital (Beadle *et al.*, 1985). However, the GABA responses are also completely blocked by 10^{-6} picrotoxin (Lees *et al.*, 1985) in contrast with the weak efficacy of this ligand described above in biochemical assays.

Therefore, contrary to the conclusions of Simmonds (1983) that invertebrates possess 'simple' GABA receptors which lack any associated allosteric sites, it appears that the CNS of at least insects, and probably other invertebrates, possesses GABA receptor complexes which are similar in overall organisation to their mammalian counterparts but with subtle differences, as postulated by Robinson *et al.* (1986). We can only speculate whether such receptor complexes appeared in insect brain at the same time as the vertebrates evolved, or whether the evolutionary ancestors of today's insects also possessed GABA receptor complexes. Hebebrand *et al.* (1986) reported that phylogenetically older species, as far back as the fishes, have fewer central benzodiazepine receptors, compared with mammalian brain. It is evident that any insect central GABA receptor complexes are present in greatly reduced numbers compared with their counterparts in vertebrate brain. One must therefore postulate some evolutionary pressure leading to an increased density of central complex GABA receptors which occurred with the arrival of the vertebrates. Whether or not this reflects the evolution of higher brain functions is unknown.

With continued work in this field the details of the pharmacology and stoichiometry of such invertebrate GABA receptor complexes will hopefully be elucidated. The logical progression, by analogy with mammalian CNS, would be to attempt to solubilise the receptor out of the membrane with a view to purifying it. Unfortunately, however, it seems that these receptors are present in relatively low concentrations and so attempts to purify invertebrate GABA receptors by conventional means may prove extremely frustrating. In the long term it may prove more fruitful to apply the strategies and probes of molecular biology, presently being developed by mammalian workers, to the study of invertebrate GABA receptors.

THE ACTION OF PESTICIDES

The chlorinated hydrocarbons are a major class of insecticide whose actions may involve perturbation of GABAergic transmission. Cyclodienes (such as endrin) and γ-hexachlorocyclohexane (lindane) have been shown to affect mammalian GABA receptors by potently displacing [^{3}H]dihydro-picrotoxinin ([^{3}H]DPTXN) binding (Matsumura and Ghiasuddin, 1983) and [^{35}S]TBPS binding in rat brain membranes (Abalis et al., 1985b). With the availability of comparable binding assays in invertebrate nervous tissues it is now possible to compare the effects of these compounds in binding and physiological studies on the same tissues. Preliminary work in this direction has shown some chlorinated hydrocarbons to be active in displacing [^{3}H]DPTXN and [^{35}S]TBPS binding from insect membranes, though the relative potencies are different from those reported in analogous vertebrate studies (Tanaka et al., 1984; Cohen and Casida, 1986; Szamraj et al., 1986). Hopefully further work utilising the invertebrate techniques now available will lead to the elucidation of the mode of action of these insecticides.

A second class of pesticides which has received a lot of attention recently is the avermectins (AVMs). These macrocyclic lactones are extremely potent against a variety of parasitic nematodes and arthropods. Electrophysiological evidence for their acting at GABA receptor chloride channels came from work on lobster (Fritz et al., 1979) and nematodes (Kass et al., 1980). However, biochemical evidence for a GABAergic mode of action was obtained in mammalian preparations. AVMs were shown to increase the release of GABA from rat brain synaptosomes (Pong et al., 1980), enhance the in vivo muscle-relaxant effects of benzo-diazepines in mice (Williams and Yarborough, 1979), modulate the binding of TBPS and GABA (Pong and Wang, 1982; Drexler and Sieg-hart, 1984a, b; Calcott and Fatig, 1984; Olsen and Snowman, 1985) and enhance benzodiazepine binding in dog and rat brain preparations (Williams and Yarborough, 1979; Paul et al., 1980; Pong et al., 1981, 1982; Supavilai and Karobath, 1981; Williams and Risley, 1982, 1984; Drexler and Sieghart, 1984c). Comparable data in invertebrate preparations are limited though. AVMs have been shown to potently inhibit honeybee brain [^{3}H]muscimol binding (IC$_{50}$ 3 nM, Abalis and Eldefrawi,

113

1986), but at 0.1–5 μM they show enhancement of [³H]GABA binding in cockroach CNS (Lummis and Sattelle, 1985b). No effect on housefly or crayfish [³⁵S]TBPS binding was observed by Szamraj *et al.* (1986), nor was there any modulation by AVM of [³H]muscimol binding in these tissues (Olsen, unpublished observations). No effect of AVM was found in cockroach ganglia on the binding of [³H]DPTXN, [³H]muscimol or [³H]diazepam (Tanaka and Matsumura, 1985). [³H]Ivermectin (5 nM) exhibits specific binding to locust ganglionic membranes while the unlabelled ligand enhances [³H]FNZP binding in the same tissue (A. Scott, T.N. Robinson and G.G. Lunt, unpublished observation). However, in a recent review of the mode of action of the AVMs, Wright (1986) concluded from the electrophysiological evidence that AVMs had two distinct actions in vertebrates and invertebrates. These two postulated actions include a reversible effect at concentrations in the 10 nM range which appears specific to GABA receptors, whereas an irreversible effect is suggested to occur at higher concentrations. This latter might involve glycine (inhibitory) responses in vertebrates and glutamatergic (excitatory) responses in invertebrates. This conclusion is consistent with the more recent data of Beadle and Lees (1986) obtained from locust neuronal cultures in which AVM acts as a GABA-mimetic, but it may act on two separate populations of channels. Also its action is partially an irreversible one on a population of channels that do not respond to GABA. Furthermore, although Mellin *et al.* (1983) demonstrated that AVM opened a picrotoxin-sensitive chloride ion channel in lobster muscle, Chalmers *et al.* (1986) showed that AVM potentiated GABA responses in crayfish muscle but not in stretch receptor neurones, despite being able to increase chloride conductance on its own in both tissues. Tanaka and Matsumura (1985) demonstrated in cockroach leg muscle that AVM increased chloride permeability, apparently independently of GABA receptors, and Duce and Scott (1985) showed that AVM activated chloride channels in a variety of insect muscles including some that lacked inhibitory GABAergic innervation. Abalis *et al.* (1986) have reported that, in rat brain, AVMs do open a chloride channel by binding to a GABA receptor and acting as a partial agonist, but that they also open voltage-dependent chloride channels which are totally insensitive to GABA. In conclusion, it would appear that the mode of action of the avermectins is only partially GABAergic.

As our biochemical knowledge of invertebrate GABAergic transmission increases, so our practical ability to measure its various parameters increases. It is to be hoped that this will not only help us in our understanding of the mode of action of existing pesticides, as outlined above, but perhaps more importantly might open new doors for the rational design of novel, specific pesticides.

CONCLUSION

In 1973 Gerschenfeld concluded from the available, largely electrophysio-logical data that GABA was probably the universal inhibitory transmitter of the invertebrate neuromuscular junction, excluding molluscs, and that the evidence was particularly convincing in the arthropods. The situation in invertebrate central neurotransmission was less clear; GABA was probably involved in crustacean and insect CNS, but its role in molluscan CNS was dubious to say the least. Certainly there was scant knowledge of any of the biochemical components of any invertebrate GABA system (Gerschenfeld, 1973).

We hope that this chapter has shown that our knowledge of GABAergic transmission in insect CNS and crustacean muscle has come a long way since 1973 due to a quite recent growth of interest in these fields, and that we are beginning to see a picture of a GABAergic system in the arthropods which in overall organisation is similar to that outlined in Figure 3.1 for the mammalian nervous system. However, the biochemical evidence for GABA being a ganglionic transmitter in any other invertebrate phylum is still limited, though what evidence there is implicates GABA as a neuro-transmitter in nematodes, annelids and molluscs. Let us hope that this is confirmed in the near future.

We are starting to see the mapping of invertebrate GABAergic neurones both with antibodies to GAD (Stapleton, 1986) and with mono-clonal antibodies to GABA (Meyer *et al.*, 1986). In the future these approaches should provide a more precise picture of GABA's role in invertebrate nervous tissues.

Throughout any comparisons between the emerging insect GABA system and its established mammalian counterpart, one sees an overall similarity but also subtle differences. Such differences in GABAergic enzymes are being characterised, but the insect uptake system and recep-tors are much harder to study. In the future, insect synaptosomal prepar-ations of greater stability than are presently available might permit not only the full characterisation of GABA uptake but also investigation of the mechanisms of insect neuronal GABA release. Greater exploitation of invertebrate neuronal (and glial) cell cultures would not only allow the pharmacological characterisation of GABA receptors but also provide an alternative vehicle for uptake studies.

In the future, then, we hope to see the full characterisation of inverte-brate GABAergic neurotransmission. This is not only of applied import-ance in the development of pesticides, but is also essential from a compar-ative standpoint in our understanding of the evolution of such a trans-mission system.

REFERENCES

Abalis, I.M. and Eldefrawi, A.T. (1986) [³H]Muscimol binding to a putative GABA receptor in honeybee brain and its interaction with avemectin B_{1a}. *Pest. Biochem. Physiol. 25*, 279-87

Abalis, I.M., Eldefrawi, M.E. and Eldefrawi, A.T. (1983) Biochemical identification of putative GABA/benzodiazepine receptors in housefly thorax muscles. *Pest. Biochem. Physiol. 20*, 39-48

Abalis, I.M., Eldefrawi, M.E. and Eldefrawi, A.T. (1985a) Binding of GABA receptor channel drugs to a putative voltage-dependent Cl⁻ channel in torpedo electric organ. *Biochem. Pharmacol. 34*, 2579-82

Abalis, I.M., Eldefrawi, M.E. and Eldefrawi, A.T. (1985b) High affinity stereospecific binding of cyclodiene insecticides and γ-hexachlorocyclohexane to γ-aminobutyric acid receptors of rat brain. *Pest. Biochem. Pharmacol. 24*, 95-102

Abalis, I.M., Eldefrawi, A.T. and Eldefrawi, M.E. (1987) Actions of AVM_{B1a} on $GABA_A$ receptor and chloride channels in rat brain. *Biochem. Toxicol.* in press

Ackerman, D. and Kutscher, F. (1910) Uber aporrhegmen. *Z. Physiol. Chem. 69*, 1265-72

Aprison, M.H., McBride, W.J. and Freeman, A.R. (1973) The distribution of several amino acids in specific ganglia and nerve bundles of the lobster. *J. Neurochem. 21*, 87-95

Awapara, J., Landua, A.J., Fuerst, R. and Seale, B. (1950) Free gamma-aminobutyric acid in brain. *J. Biol. Chem. 187*, 35-9

Battelle, B.A., Kravitz, E.A, and Stieve, H. (1979) Neurotransmitter synthesis in *Limulus* ventral nerve photoreceptors. *Experientia 35* 77-8

Baxter, C.F. and Torralba, G.F. (1975) γ-Aminobutyric acid glutamate decarboxylase (*L*-glutamate l-carboxy-lyase, E.C.4.1.1.15) in the nervous system of the cockroach, *Periplaneta americana*, 1. Regional distribution and properties of the enzyme. *Brain Res. 84*, 383-97

Bazemore, A.K., Eliott, A.C. and Florey, E. (1957) Isolation of factor I. *J. Neurochem. 1*, 334-9

Beadle, D.J., Benson, J.A., Lees, G. and Neuman, R. (1985) Flunitrazepam and pentobarbital modulate GABA responses of insect neuronal somata. *J. Physiol. 2731*, 273p

Beadle, D.J. and Lees, G. (1986) Insect neuronal cultures — a new tool in insect neuropharmacology. In: Ford, M.G., Lunt, G.G., Reay, R.C. and Usherwood, P.N.R. (eds) *Neuropharmacology and pesticide action*, pp. 425-44. Ellis Horwood, Chichester

Blinderman, J-M., Maitre, M., Ossola, L. and Mandel, P. (1978) Purification and some properties of *L*-glutamate decarboxylase from human brain. *Eur. J. Biochem. 86*, 143-52

Bolger, G.T., Weissman, B.A., Lueddens, H., Basile, A.S., Mantione, C.R., Barrett, I.E., Witkin, J.M., Paul, S.M. and Skolnick, P. (1986) Late evolutionary appearance of peripheral-type binding sites for benzodiazepines. *Brain Res. 338*, 366-70

Bowery, N., Price, G.W., Hudson, A.L., Hill, D.R., Wilkin, G.P. and Turnbull, M.J. (1984) GABA receptor multiplicity — visualization of different receptor types in the mammalian CNS. *Neuropharmacol. 23* 219-31

Bradford, H.F. (1986) *Chemical neurobiology*. W.H. Freeman, New York, 229-42

Bradford, H.F., Chain, E.B., Cory, H.T. and Rose, S.P.R. (1969) Glucose and amino acid metabolism in some invertebrate nervous systems. *J. Neurochem. 16*, 969-78

Braestrup, C. and Nielsen, M. (1983) Benzodiazepine receptors. In: Iversen, L.L., Iversen, S.D. and Snyder, S.H. (eds) *Handbook of pharmacology*, vol. 17, pp. 285-384, Plenum, New York

Breer, H. and Heilgenberg, H. (1986) Neurochemistry of GABAergic activities in the central nervous system of *Locusta migratoria. J. Comp. Physiol. A 157*, 343-54

Breer, H. and Jeserich, G. (1980) A microscale flotation technique for the isolation of synaptosomes from nervous tissue of *Locusta migratoria. Insect Biochem. 10*, 457-63

Calcott, P.H. and Fatig, R.O. (1984) Avermectin modulation of GABA binding to membranes of rat brain, brine shrimp and a fungus, *Mucor miehei. J. Antibiot. 37*, 253-9

Callec, J.J. (1974) Synaptic transmission in the central nervous system of insects. In: Treherne, J.E. (ed.) *Insect neurobiology*, pp. 119-78. Elsevier North-Holland, Amsterdam

Chalmers, A.E., Miller, T.A. and Olsen, R.W. (1987) The action of avermectin on crayfish nerve and muscle. *Eur. J. Pharmacol.* in press

Chen, R.S. and Widner, B. (1968) Content and synthesis of GABA in the larval brain of *Drosophila melanogaster. Experientia, 24*, 516-17

Chude, O., Roberts, E. and Wu J-Y. (1979) Partial purification of *Drosophila* glutamate decarboxylase. *J. Neurochem. 32*, 1409-15

Cline, H.T., Nusbaum, M.P. and Kristan, W.B. (1985) Identified GABAergic inhibitory motor neurones in the leech central nervous system take up GABA. *Brain Res. 384*, 359-62

Cohen, E. and Casida, J.E. (1986) Effects of insecticides and GABAergic agents on a housefly [^{35}S]t-butylbicyclophosphorothionate binding site. *Pest. Biochem. Physiol. 25*, 63-72

Cory, H.T. and Rose, S.P.R. (1969) Glucose and amino acid metabolism in octopus optic and vertical lobes *in vitro. J. Neurochem. 16*, 979-88

Craelius, W. and Fricke, R.A. (1981) Release of [^{3}H]gamma-aminobutyric acid (GABA) by inhibitory neurones of the crayfish. *J. Neurobiol. 12*, 249-58

De Belleroche, J.S. and Bradford, H.F. (1977) On the site of origin of transmitter amino acids release by depolarization of nerve terminals *in vitro. J. Neurochem. 29*, 335-43

De Feudis, F.V. (1979) Binding and ionophoretic studies on centrally active amino acids — a search for physiological receptors. *Int. Rev. Neurobiol. 21*, 155-77

Deisz, R.A., Dose, M. and Lux, H.D. (1984) The time course of GABA action on the crayfish stretch receptor: evidence for a saturable GABA uptake. *Neurosci. Lett. 47*, 245-50

De Robertis, E. and Fiszer de Plazas, M. (1974) Isolation of hydrophobic proteins binding neurotransmitter amino acids: γ-aminobutyric acid receptor of the shrimp muscle. *J. Neurochem. 23*, 1121-5

Drexler, G. and Seighart, W. (1984a) Properties of a high affinity binding site for [^{3}H]avermectin$_{B1a}$. *Eur. J. Pharmacol. 99*, 269-77

Drexler, G. and Seighart, W. (1984b) [^{35}S]t-butylbicyclophosphorothionate and avermectin bind to different sites associated with the gamma-aminobutyric acid–benzodiazepine receptor complex. *Neurosci. Lett. 50*, 273-7

Drexler, G. and Seighart, W. (1984c) Evidence for association of a high affinity avermectin binding site with the benzodiazepine receptor. *Eur. J. Pharmacol. 101*, 201-7

Duce, I.R. and Scott, R.H. (1985) Actions of avermectin$_{B1a}$ on insect muscle. *Br. J. Pharmacol. 85*, 395-401

Eccles, J.C. and Jaeger, J.C. (1958) The relationship between the mode of operation and the dimensions of the junctional regions at synapses and motor end organs. *Proc. Roy. Soc. (Lond.) 148*, 38-56

Edwardson, J.M., Phillips, N.I., Kirby, N. and Fowler, L.J. (1985) A monoclonal antibody to rabbit brain GABA transaminase. *J. Neurochem 44*, 1679-84

Enna, S.J. and Snyder, S.H. (1975) Properties of GABA receptor binding in rat brain synaptic membrane fractions. *Brain Res. 100*, 81-97

Florey, E. (1954) An inhibitory and excitatory factor from mammalian brain and their action on a single sensory neurone. *Arch. Int. Physiol. 62*, 33-53

Fox, P.M. and Larsen, J.R. (1972) Glutamic acid decarboxylase and the GABA shunt in the supraoesophageal ganglion of the honeybee, *Apis mellifera. J. Insect Physiol. 18*, 439-57

Fritz, L.C., Wang, C.C. and Gorio, A. (1979) Avermectin$_{Bla}$ irreversibly blocks postsynaptic potentials at the lobster neuromuscular junction by reducing muscle membrane resistance. *Proc. Natl Acad. Sci. USA 76*, 2062-6

Frontali, N. (1961) Activity of glutamic acid decarboxylase in insect nervous tissue. *Nature (Lond.) 191* 178-9

Frontali, N. (1964) Brain glutamic acid decarboxylase and synthesis of γ-amino-butyric acid in vertebrate and invertebrate species. In: Richter, D. (ed.) *Comparative neurochemistry*, pp. 185-92. Pergamon Press, Oxford

Frontali, N. and Pierantoni, R. (1973) Autoradiographic localisation of [^{3}H]GABA in the cockroach brain. *Comp. Biochem. Physiol. 44A*, 1369-72

Gerschenfeld, H.M. (1973) Chemical transmission in invertebrate central nervous systems and neuromuscular junctions. *Phys. Rev 53* 1-119

Gordon, D., Zlotkin, E. and Kanner, B. (1982) Functional membrane vesicles from the nervous system of insects, 1. Na$^+$ and Cl$^-$ dependent GABA transport. *Biochim. Biophys. Acta 688*, 229-36

Greenlee, D.V., Van Ness, P.C, and Olsen, R.W. (1978) Endogenous inhibitor of GABA binding in mammalian brain. *Life Sci. 22*, 1653-62

Griffiths, R., Larsson, O.M., Allen, I.C. and Schousboe, A. (1987) Reassessment of β-alanine as a selective inhibitor of GABA uptake by glia. *Biochem. Soc. Trans.* in press.

Haber, B., Kuriyama, K. and Roberts, E. (1969) Decarboxylation of glutamate by tissues other than brain, non-identity with CNS GAD. *Fed. Proc. 28*, 577

Haber, B., Kuriyama, K. and Roberts, E. (1970a) *L*-Glutamic acid decarboxylase: a new type from glial cells and human brain gliomas. *Science 168*, 598-9

Haber, B., Kuriyama, K. and Roberts, E. (1970b) An anion stimulated *L*-glutamic acid decarboxylase in non-neuronal tissues. *Biochem. Pharmacol. 19*, 1119-36

Haefely, W. (1984) Benzodiazepine interactions with GABA receptors. *Neurosci. Lett. 47*, 201-6

Hall, Z.W. and Kravitz, E.A. (1967a) The metabolism of γ-aminobutyric acid (GABA) in the lobster nervous system. I. GABA-glutamate transaminase. *J. Neurochem. 14*, 45-54

Hall, Z.W. and Kravitz, E.A. (1967b) The metabolism of γ-aminobutyric acid (GABA) in the lobster nervous system. II. Succinic semialdehyde dehydro-genase. *J. Neurochem. 14*, 55-61

Haring, P., Stahli, C., Schoch, P., Takacs, B., Staehelin, T. and Mohler, H. (1985) Monoclonal antibodies reveal structural homogeneity of γ-aminobutyric acid/benzodiazepine receptors in different brain regions. *Proc. Natl Acad. Sci. USA 82*, 4837-41

Haycock, J.W., Levy, W.B., Denner, L.A. and Cotman, C.W. (1978) Effects of elevated [K$^+$] on the release of neurotransmitters from cortical synaptosomes:

efflux or secretion? *J. Neurochem. 30*, 1113-25

Hebebrand, J., Friedl, W., Lentes, K-L. and Propping, P. (1986) Qualitative variations of photolabelled benzodiazepine receptors in different species. *Neurochem. Int. 7*, 267-71

Heinemaki, A.A., Malila, S.I., Tolonen, K.M., Valkonen, K.H. and Piha, R.S. (1983) Resolution and purification of taurine- and GABA-synthesising decarboxylases from calf brain. *Neurochem. Res. 8*, 207-18

Hill, D.R., Bowery, N.G. and Hudson, A.L. (1984) Inhibition of GABA$_B$ binding by guanyl nucleotides. *J. Neurochem. 42*, 632-57

Horwitz, I.S. and Orkand, R.K. (1979) GABA inactivation at the crayfish neuromuscular junction. *J. Neurobiol. 11*, 447-58

Hue, B., Gabriel, A. and Le Patezour, A. (1982) Localization of [^{3}H]GABA accumulation in the sixth abdominal ganglion of the cockroach, *Periplaneta americana. J. Insect Physiol. 28*, 753-9

Huggins, A.K., Rick, J.T. and Kerkut, G.A. (1967) A comparative study of the intermediate metabolism of *L*-glutamate in muscle and nerve tissue. *Comp. Biochem. Physiol. 21*, 23-30

Iversen, L.L. and Bloom, F.E. (1972) Studies on the uptake of [^{3}H]GABA and [^{3}H]glycine in slices and homogenates of rat brain and spinal cord by electron microscopic autoradiography. *Brain. Res. 41*, 131-43

Iversen, L.L. and Kravitz, E.A. (1966) Uptake of γ-aminobutyric acid (GABA) in a lobster nerve–muscle preparation. *Fed. Proc. 25*, 714

Iversen, L.L. and Kravitz, E.A. (1968) The metabolism of γ-aminobutyric acid (GABA) in the lobster nervous system: uptake of GABA in nerve–muscle preparations. *J. Neurochem. 15*, 609-20

Iversen, L.L. and Neil, M.J. (1968) The uptake of [^{3}H]GABA by slices of rat cerebral cortex. *J. Neurochem. 15*, 1141-9

John, R.A. and Fowler, L.J. (1976) Kinetic and spatial properties of rabbit brain 4-aminobutyrate aminotransferase. *Biochem. J. 155*, 645-51

Kass, I.S., Wang, C.C., Walrond, J.P. and Stretton, A.O.W. (1980) Avermectin$_{Bla}$, a paralysing antihelminthic that affects interneurones and motorneurones in *Ascaris. Proc. Natl. Acad. Sci. USA 77*, 6211-15

Kirkness, E.F. and Turner, A.J. (1986) The γ-aminobutyrate/benzodiazepine receptor from pig brain. *Biochem. J. 233*, 265-70

Kravitz, E.A. (1967) Acetylcholine, gamma-aminobutyric acid and glutamic acid: physiological and chemical studies related to their role as neurotransmitter agents. In: Quarton, G.C., Melnechuk, T. and Schmitt, F.O. (eds) *The neurosciences: a study program*, pp. 433-44. Rockefeller University Press, New York

Kravitz, E.A., Iversen, L.L., Otsuka, M. and Hall, Z.W. (1968) Gamma-aminobutyric acid in the lobster nervous system: release from inhibitory nerves and uptake into nerve–muscle preparations. In: von Euler, C., Skoglund, S. and Soderberg, U. (eds) *Structure and function of inhibitory neuronal mechanisms*, pp. 371-6. Pergamon Press, Oxford

Kravitz, E.S., Molinoff, P.B. and Hall, Z.W. (1965) A comparison of the enzymes and substrates of gamma-aminobutyric acid metabolism in lobster excitatory and inhibitory axons. *Proc. Natl Acad. Sci. USA 54*, 778-82

Kravitz, E.A., Potter, D.D. and Van Gelder, N.M. (1962) Gamma-aminobutyric acid and other blocking substances extracted from crab muscle. *Nature (Lond.) 194*, 382-3

Krogsgaard-Larsen, P., Hjeds, H., Curtis, D.R., Lodge, D. and Johnstone, G.A.R. (1979) Dihydromuscimol, thiomuscimol and related heterocyclic compounds as GABA analogues. *J. Neurochem. 32*, 1717-24

Kuriyama, K., Weinstein, H. and Roberts, E. (1969) Uptake of GABA by mitochondrial and synaptosomal membrane fractions from mouse brain. *Brain Res.* *16*, 479-92

Lees, G. and Beadle, D.J. (1986) Dihydroavermectin B_1: actions on cultured neurones from the insect central nervous system. *Brain Res. 366*, 369-72

Lees, G., Neuman, R., Beadle, D.J. and Benson, J.A. (1985) Flunitrazepam enhances GABA- and muscimol-induced responses in freshly dissociated locust neuronal somata. *Pest. Sci. 16*, 534

Lummis, S.C.R. and Sattelle, D.B. (1985a) Insect central nervous system γ-aminobutyric acid. *Neurosci. Lett. 60*, 13-18

Lummis, S.C.R. and Sattelle, D.B. (1985b) GABA and benzodiazepine binding sites in insect CNS. *Pest. Sci. 16*, 695-7

Lummis, S.C.R. and Sattelle, D.B. (1987) Binding sites for [^{3}H]GABA, [^{3}H]flunitrazepam and [^{3}H]TBPS in insect CNS. *Neurochem. Int. 9*, 287-93

Lunt, G.G., Robinson, T.N., Miller, T., Knowles, W.P. and Olsen, R.W. (1985) The identification of GABA receptor binding sites in insect ganglia. *Neurochem. Int. 7*, 751-4

McAdoo, D.J. and Coggeshall, R.E. (1976) Gas chromatographic–mass spectrophotometric analysis of biogenic amines in identified neurones and tissues of *Hirudo medicinalis. J. Neurochem. 26*, 163-7

Martin, D.L. and Smith III, A.A. (1972) Ions and the transport of GABA by synaptosomes. *J. Neurochem. 19*, 841-55

Matsumura, F. and Ghiasuddin, S.M. (1983) Evidence for similarities between cyclodiene type insecticides and picrotoxin in their action mechanisms. *J. Environ. Sci. Health B18*, 1-14

Maxwell, G.D., Tait, J.F. and Hildebrand, J.G. (1978) Regional synthesis of neurotransmitter candidates in the CNS of the moth, *Manduca sexta. Comp. Biochem. Physiol. 61C*, 109-19

Meiners, B.M., Kehoe, P., Shaner, D.M. and Olsen, R.W. (1979) GABA receptor binding and uptake in membrane fractions of crayfish muscle. *J. Neurochem. 32*, 979-90

Mellin, T.N., Busch, R.D. and Wang, C.C. (1983) Postsynaptic inhibition of invertebrate neuromuscular transmission by avermectin$_{B1a}$. *Neuropharmacology 22*, 89-96

Meyer, E.P., Matute, C., Streit, P. and Nassel, D.R. (1986) Insect optic lobe neurones identifiable with monoclonal antibodies to GABA. *Histochemistry 84*, 207-16

Mohler, H., Battersby, M.K. and Richards, J.G. (1980) Benzodiazepine receptor protein identified and visualized in brain tissue by photoaffinity labelling. *Proc. Natl Acad. Sci. USA 77*, 1666-70

Molinoff, P.B. and Kravitz, E.A. (1968) The metabolism of gamma-aminobutyric acid (GABA) in the lobster nervous system — glutamic decarboxylase. *J. Neurochem. 15*, 391-409

Nielsen, M., Braestrup, C. and Squires, R.F. (1978) Evidence for a late evolutionary appearance of brain-specific benzodiazepine receptors: an investigation of 18 vertebrate and 5 invertebrate species. *Brain Res. 141*, 342-6

Nistri, A. and Constanti, A. (1979) Pharmacological characterization of different types of GABA and glutamate receptors in vertebrates and invertebrates. *Proc. Neurobiol. 13*, 117-235

Olsen, R.W. (1981) GABA–benzodiazepine–barbiturate receptor interactions. *J. Neurochem. 37*, 1-13

Olsen, R.W. (1982) Drug interactions at the GABA–receptor–ionophore complex.

Ann. Rev. Pharmacol. Toxicol. 22, 245-77

Olsen, R.W., Fischer, J.B., King, R.G., Ransom, J.Y. and Stauber, G.B. (1984) Purification of the GABA/benzodiazepine/barbiturate receptor complex. *Neuropharmacol. 23*, 853-4

Olsen, R.W., Lee, J.M. and Ban, M. (1975) Binding of γ-aminobutyric acid to crayfish muscle and its relationship to receptor sites. *Mol. Pharmacol 11*, 566-77

Olsen, R.W. and Snowman, A.M. (1985) Avermectin$_{Bla}$ modulation of γ-aminobutyric acid benzodiazepine receptor binding in mammalian brain. *J. Neurochem. 44*, 1074-82

Osborne, N.N., Briel, G. and Neuhoff, V. (1972) Distribution of GABA and other amino acids in different tissues of the gastropod mollusc, *Helix pomatia*, including *in vitro* experiments with ^{14}C-glucose and ^{14}C-glutamic acid. *Int. J. Neurosci. 1*, 265-72

Otsuka, M.E., Kravitz, E.A. and Potter, D.D. (1967) Physiological and chemical architecture of a lobster ganglion with particular reference to gamma-aminobutyrate and glutamate. *J. Neurophysiol. 30*, 725-52

Paul, S.M., Skolnick, P. and Zats, M. (1980) Avermectin Bla: an irreversible activator of the GABA−benzodiazepine−chloride ionophore receptor complex. *Biochem. Biophys. Res. Comm. 96*, 632-8

Pichon, Y. (1974) The pharmacology of the insect nervous system. In: Rockstein, M. (ed.) *The physiology of insects*, pp. 102-74. Academic Press, New York

Pong, S.S., De Haven, R. and Wang, C.C. (1981) Stimulation of benzodiazepine binding to rat brain membranes and solubilized receptor complex by avermectin Bla and 4-aminobutyric acid. *Biochim. Biophys. Acta 646*, 143-50

Pong, S.S., De Haven, R. and Wang, C.C. (1982) A comparative study of avermectin Bla and other modulators of the 4-aminobutyric acid receptor chloride ion channel complex. *J. Neurosci. 2*, 966-71

Pong, S.S. and Wang, C.C. (1982) Avermectin Bla modulation of 4-aminobutyric acid receptors in rat brain. *J. Neurochem. 38*, 375-9

Pong, S.S., Wang, C.C. And Fritz, L.C. (1980) Studies on the mechanism of action of avermectin Bla: stimulation of release of 4-aminobutyric acid from brain synaptosomes. *J. Neurochem. 34*, 351-8

Potter, D.D. (1968) The chemistry of inhibition in crustaceans with special reference to gamma-aminobutyric acid. In: von Euler, C., Skoglund, S. and Soderberg, U. (eds) *Structure and function of inhibitory neuronal mechanisms*, pp. 359-70. Pergamon Press, Oxford

Ray, J.W. (1964) The free amino acid pool of the cockroach (*Periplaneta americana*) central nervous system and the effects of insecticides. *J. Insect Physiol. 10*, 587-97

Roberts, E. (1964) Comparative aspects of the distribution of ninhydrin-reactive constituents in nervous tissue. In: Richter, D. (ed.) *Comparative neurochemistry*, pp. 401-8. Pergamon Press, Oxford

Roberts, E. and Frankel, S. (1950) γ-Aminobutyric acid in the brain; its formation from glutamic acid. *J. Biol. Chem. 187*, 55-63

Robinson, T.N. (1986) PhD Thesis, University of Bath

Robinson, T.N. and Lunt, G.G. (1986) A rapid technique for the production of a locust synaptosomal preparation. *Proc. 17th FEBS Meeting biol. Chem Hoppe Seiler. 367*, suppl. 298

Robinson, T.N., MacAllan, D., Lunt, G.G. and Battersby, M. (1986) The GABA receptor complex of insect CNS: characterization of a benzodiazepine binding site. *J. Neurochem. 47*, 1955-62

Sargent, P.B. (1977) Synthesis of acetylcholine by excitatory motorneurones of the leech. *J. Neurochem. 40*, 453-60

Schoch, P., Richards, J.G., Haring, P., Takacs, B., Stahli, C., Staehelin, T., Haefely, W. and Mohler, H. (1985) Co-localization of $GABA_A$ receptors and benzodiazepine receptors in the brain shown by monoclonal antibodies. *Nature (Lond.) 314*, 168-71

Schousboe, A., Wu, J-Y. and Roberts, E. (1973) Purification and characterization of 4-aminobutyrate-2-ketoglutarate transaminase from the mouse brain. *Biochemistry 12*, 2868-73

Shepherd, D. and Tyrer, N.M. (1985) Inhibition of GABA uptake potentiates the effects of exogenous GABA on locust skeletal muscle. *Comp. Biochem. Physiol. 82C*, 315-21

Sigel, E. and Barnard, E.A. (1984) A GABA/benzodiazepine receptor complex from bovine cerebral cortex. *J. Biol. Chem. 259*, 7219-23

Sigel, E., Mamalaki, C. and Barnard, E.A. (1985) Reconstitution of the aminobutyric acid–benzodiazepine receptor complex from bovine cerebral cortex into phospholipid vesicles. *Neurosci. Lett. 61*, 165

Sigel, E., Stephenson, F.A., Mamalaki, C. and Barnard, E.A. (1983) GABA/benzodiazepine receptor complex of bovine cerebral cortex–purification and partial characterization. *J. Biol. Chem. 258* 6965-71

Simmonds, M.A. (1983) Multiple GABA receptors and associated regulatory sites. *TINS 6* 279-81

Stapleton, A.G. (1986) Studies on *L*-glutamic acid decarboxylase from locust brain. Ph.D. Thesis, UMIST, England

Stephenson, F.A., Casalotti, S.O., Mamalaki, C. and Barnard, E.A. (1986) Antibodies recognizing the $GABA_A$/benzodiazepine receptor including its regulatory sites. *J. Neurochem. 46*, 854-61

Su, Y.Y.T., Wu, J-Y. and Lam, D.M.K. (1979) Purification of *L*-glutamic acid decarboxylase from catfish brain. *J. Neurochem. 33*, 169-79

Su, Y.Y.T., Wu, J-Y. and Lam, D.M.K. (1983) Species specificities of *L*-glutamic decarboxylase: immunochemical comparison. *Neurochem. Int. 5*, 587-92

Sugden, P.H. and Newsholme, E.A. (1977) Activities of ChAT, AChE, GAD, GABA-T and carnitine acetyl transferase in nervous tissue from some vertebrates and invertebrates. *Comp. Biochem. Physiol. 56C*, 89-94

Supavilai, P. and Karobath, M. (1981) *In vitro* modulation by avermectin Bla of the GABA/benzodiazepine receptor complex of rat cerebellum. *J. Neurochem. 36*, 798-803

Susz, J.P., Haber, B. and Roberts, E. (1966) Purification and some properties of mouse brain *L*-glutamic acid decarboxylase. *Biochemistry 5*, 2870-7

Szamraj, O.I., Miller, T. and Olsen, R.W. (1986) Cage convulsant [^{35}S]TBPS binding to GABA receptor–chloride channel complex in invertebrates. *Soc. Neurosci. 12*, 656

Tallan, H.H. (1962) A survey of amino acids and related compounds. In: J.T. Holden (ed.) *Nervous tissue, amino acid pool*, pp. 156-84. Elsevier, Amsterdam and London

Tallman, J.F. and Gallager, D.W. (1985) The GABA-ergic system: a locus of benzodiazepine action. *Ann. Rev. Neurosci. 8*, 21-44

Tanaka, K. and Matsumura, F. (1985) Action of avermectin Bla on the leg muscles and the nervous system of the American cockroach. *Pest. Biochem. Physiol. 29*, 124-35

Tanaka, K., Scott, J.G. and Matsumura, F. (1984) Picrotoxin receptor in the central nervous system of the American cockroach: its role in the action of cyclodiene insecticides. *Pest. Biochem. Physiol. 22*, 117-27

Tapia, R. (1983) Regulation of GAD activity. In: Hertz, L., Kramme, E., McGeer, E.G. and Schousboe, A. (eds) *Glutamine, glutamate and GABA in the central nervous system*, p. 113. Alan Liss, New York

Thomas, J.W. and Tallman, J.F. (1981) Characterization of photoaffinity labelling of benzodiazepine binding sites. *J. Biol. Chem. 256*, 9838-42

Tsukada, Y., Uemura, K., Hirano, S. and Nagata, Y. (1964) Distribution of, amino acids in the brain of different species. In: Richter, D. (ed) *Comparative neurochemistry*, pp. 179-83. Pergamon Press, Oxford

Udenfriend, S. (1950) Identification of γ-aminobutyric acid in brain by the isotope derivative method. *J. Biol. Chem. 187*, 33-59

Van Marle, J., Piek, T., Lammerste, Th., Lind, A. and Van Weeren-Kramer, J. (1985) Selectivity of the uptake of glutamate and GABA into two morphologically distinct insect neuromuscular synapses. *Brain Res. 348*, 107-11

Walker, R.J., Azanza, G.A. and Woodruff, G.N. (1975) The action of γ-aminobutyric acid (GABA) and related compounds on two identifiable neurones in the brain of the snail, *Helix aspersa. Comp. Biochem. Physiol. 50C*, 147-54

Watts, S.D.M. and Atkins, A.M. (1984) Kinetics of 4-amino-2-oxoglutarate aminotransferase from *Nippostrongylus brasiliensis. Mol. Biochem. Parasitol. 12*, 207-16

White, H.L. (1981) Glutamate as a precursor of GABA in rat brain and peripheral tissue. *Mol. Cell. Pharmacol. 39*, 253

Whitton, P.S., Nicholson, R.A. and Strang, R.H.C. (1986) Metabolism of taurine by insect synaptosomes. *Biochem. Soc. Trans. 14*, 609

Williams, M. and Risley, E.A. (1979) Characterization of the binding of [^{3}H]muscimol, a potent γ-aminobutyric acid agonist, to rat brain synaptosomal membranes using a filtration assay. *J. Neurochem. 32*, 713-18

Williams, M. and Risley, E.A. (1982) Interaction of avermectins with [^{3}H]-carboline-3-carboxylate ethyl ester and [^{3}H]diazepam binding sites in rat brain cortical membranes. *Eur. J. Pharmacol. 77*, 307-12

Williams, M. and Risley, E.A. (1984) Avermectin interactions with benzodiazepine receptors in rat cortex and cerebellum *in vitro. J. Neurochem. 42*, 745-53

Williams, M. and Yarborough, G.G. (1979) Enhancement of *in vitro* binding and some of the pharmacological properties of diazepam by a novel antihelminthic agent, avermectin Bla. *Eur. J. Pharmacol. 56*, 273-6

Witte, P.U. and Matthaei, H. (1980) *Mikrochemische Methoden für neurobiologische Unterschungen*, pp. 99-101. Springer, Berlin, Heidelberg, New York

Wright, D.J. (1986) Biological activity and mode of action of avermectins. In: Ford, M.G., Lunt, G.G., Reay, R.C. and Usherwood, P.N.R. (eds) *Neuropharmacology and pesticide action*, pp. 174-202. Ellis & Horwood, Chichester

Wu, J-Y., Chude, O., Weber, B., Driskell, J. and Roberts, E. (1976) Properties of L-glutamate decarboxylase from crayfish. *Trans. Am. Soc. Neurosci. 7*, 190

Wu, J-Y., Chude, O., Wein, J., Roberts, E., Saito, K. and Wong, K. (1978) Distribution and tissue specificity of glutamate decarboxylase (EC 4.1.1.15) *J. Neurochem. 30*, 849-59

Wu, J-Y., Matsuda, T. and Roberts, E. (1973) Purification and characterization of glutamate decarboxylase from mouse brain. *J. Biol. Chem. 248*, 3029-34

Zukin, S.R., Young, A.B. and Snyder, S.H. (1974) Gamma-aminobutyric acid binding to receptor sites in rat central nervous system. *Proc. Natl. Acad. Sci. USA 71*, 4802-7

4

Amine Transmitters and their Associated Second Messenger Systems

Peter F.T. Vaughan

The biogenic amines form an important group of compounds, based on the phenylethylamine or indolamine structures (Figure 4.1). Catecholamines are widely distributed throughout the animal kingdom, as they have been detected in all the major groups of invertebrates and vertebrates (see Welsh, 1972). Serotonin has been detected in all classes of invertebrates and vertebrates, with the exception of the Echinoderms (Robertson and Juorio, 1976).

An interesting pattern occurs in the distribution of the phenylethylamines. Thus the invertebrate phyla, comprising the protostomia (with the exception of the cephalopods), have high levels of octopamine relative to noradrenaline. In contrast the phyla belonging to the deuterostomia have low octopamine and high noradrenaline levels (Robertson and Juorio, 1976). In relation to this it is of interest that in the rat brain the ratio of octopamine to noradrenaline decreases from 1.9 at day 16 of gestation to 0.04 at birth, compared with a ratio of 0.01 in adult brain. This is associated with a decrease in octopamine levels from 30 ng/g wet weight at day 16 of gestation to 6.24 ng/g wet weight at birth, and an increase in noradrenaline content from 16.4 ng/g wet weight at day 16 to 156 ng/g wet weight at birth (Coyle and Henry, 1973).

A sensitive histochemical technique (the Falck–Hillarp) has been widely used to study the localisation of catecholamines and indolamines in specific regions of invertebrate (Klemm, 1976) and vertebrate (Ungerstedt, 1971) CNS. This is based on the reaction of dopamine, noradrenaline and 5HT with formaldehyde vapour to form a highly fluorescent condensation product (Falck and Owman, 1965). The regional localisation of catecholamines and indolamines provided strong evidence that biogenic amines might function as neurotransmitters in the CNS of a wide range of animals (Cooper et al., 1982).

The main aim of this chapter is to discuss those aspects of monoamine biochemistry which are consistent with their role as either neurotransmitters or neuromodulators. In this context (Evans, 1980) a neuro-

Figure 4.1: Structure of the major biogenic amines

i) Phenylethylamines

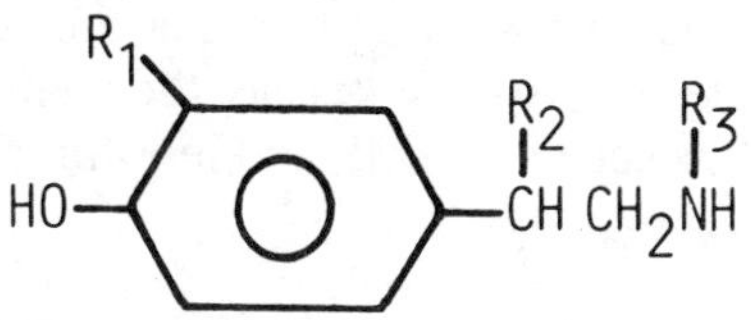

Compound	Substituent		
	R_1	R_2	R_3
1. Tyramine	H	H	H
2. Octopamine	H	OH	H
3. Dopamine	OH	H	H
4. Noradrenaline	OH	OH	H
5. Adrenaline	OH	OH	CH_3

ii) Indolamines

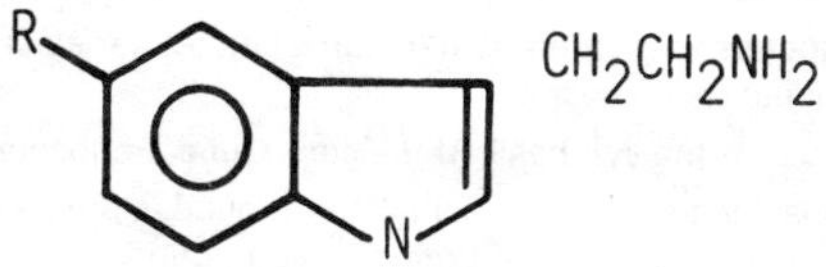

	R
1. Tryptamine.	H
2. 5-hydroxytryptamine (serotonin)	OH

transmitter refers to a compound that is released at a synapse and diffuses across the synaptic cleft to act on a receptor located on the membrane of a postsynaptic cell, which may be another neurone, a muscle cell or a special-ised gland cell. The specificity of neurotransmitter action is therefore partly dependent on the anatomical distribution of synapses throughout the CNS. The term 'neuromodulator' refers to the release of a compound, within a localised region of the CNS, the receptor for which is not necessarily sited on an anatomically apposed postsynaptic cell. Thus a neuromodulator may affect several postsynaptic cells. In this case specificity is conferred mainly by the distribution of receptors. No specific synaptic junctions occur, and neuromodulators are postulated to modify the basic rapid switching on and off of neuronal activity. Current views on the mammalian CNS suggest that the basic pattern of neuronal activity is mediated by glutamate (excitatory) and GABA (inhibitory), and is due to changes in ion flux. It has been proposed (see, for example, Iversen, 1985) that the other neurotransmitter systems, e.g. acetylcholine, catecholamines and neuropeptides, can be regarded as neuromodulators which modify this basic pattern. The action of neuromodulators is relatively slow and long lasting, compared with the effect of neurotransmitters on ion fluxes, as their effect is mediated by second messenger systems such as cAMP and inositol trisphosphate.

CRITERIA FOR A SUBSTANCE ACTING AS A NEUROTRANSMITTER OR NEUROMODULATOR

It will be helpful, in the subsequent discussion, to list a number of criteria that are required to be fulfilled before a substance can be regarded as acting as a neurotransmitter or a neuromodulator. These are summarised in Table 4.1. In a discussion on neurotransmission it is convenient to group these criteria under two main headings.

Table 4.1: Criteria for a substance acting as a neurotransmitter (neuromodulator)

---

(1) Occurs in specific neurones. Concentrated in the nerve terminal. Usually stored in vesicles.

(2) Liberated in response to membrane depolarisation (either electrical or chemical), by means of a Ca^{2+}-dependent mechanism.

(3) Enzymes required for the synthesis of the compound are located in specific neurones.

(4) Mechanism of inactivation (or removal) of compound is present. This can be enzymic, but more common method is active re-uptake of compound into nerve terminal.

(5) Specific receptors for compound are located in nervous system. In the case of neurotransmitters, receptors are located in close proximity to site of release. This is not necessarily so for neuromodulators.

(6) Addition of compound to nervous system (iontophoresis(should mimic a physiological response.

I Presynaptic

This includes the storage, synthesis and release of the neurotransmitter or neuromodulator.

II Postsynaptic

This includes binding to a specific receptor and receptor-mediated responses, e.g. second messenger systems. Receptor stimulation can be mediated in two ways: either via ion channels, in which case there is usually

126

a rapid response, e.g. GABA receptor linked to Cl⁻ channel, or glutamate and acetylcholine (nicotinic) linked to Na^+ channel; activation of specific enzymes with a subsequent alteration in the level of second messengers, e.g. adenylate cyclase/cAMP or phospholipase C/inositol trisphosphate–diacylglycerol. The activation of these second messenger systems is usually associated with the mediation of slower effects. However, note that:

(1) Inactivation mechanisms do not fall neatly into this classification. Thus high-affinity uptake of neurotransmitters has been reported for both neurones and glial cells (particularly glutamate and GABA). Furthermore, enzymes that inactivate neuotransmitters are distributed both presynaptically and postsynaptically. However, since the major mechanism for termination of action of most neurotransmitters is re-uptake into neurones, inactivation will be considered with presynaptic mechanisms.

(2) Receptors for most neurotransmitters are also located on presynaptic nerve terminals where they act as 'auto' receptors regulating the release of transmitter. For example, stimulation of adrenergic α_2 receptors in vertebrate CNS inhibits the release of NA from nerve endings in the occipital cortex and hypothalamus (Chesselet, 1984).

In spite of these limitations it is intended to discuss amine transmitters and their associated second messenger systems in invertebrates under these two main headings.

PRESYNAPTIC MECHANISMS

Distribution

Histochemical studies

Much useful information on the distribution of monoamines has been provided by the application of the histochemical methods of Falk and Hillarp to nervous tissue. Klemm (1976), in a most valuable and detailed review, has summarised the results of applying this technique to the cerebral ganglia of several insects. He concluded that three categories of insect brain structure can be recognised in relation to the distribution of monoamines (Figure 4.2).

(1) Structures in which no biogenic amines occur, e.g. the primary photoreceptors, the afferent and efferent fibres of the antennal nerve, and the fibre bundles of the tractus olfactorio-globularis (Figure 4.2).

(2) Structures which contain biogenic amines in every species studied, e.g.

127

Figure 4.2: Diagram of an 'insect' brain

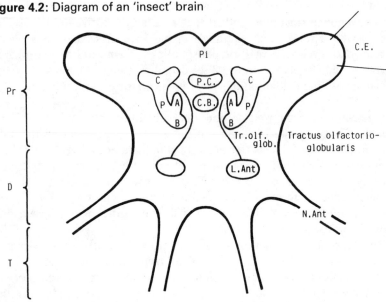

Pr = protocerebrum, D = dentocerebrum, T = tritocerebrum, A — α lobe, B — β lobe, C — calyx, P — peduncle (of mushroom body), C.B. — central body, C.E. — compound eye, N.Ant. — antennal nerve, Pi — pars intercerebralis, P.C. — pons cerebralis, L.Ant. — antennal lobe.

cell bodies in the posterior region of the pars intercerebralis, the α- and β-lobes of the mushroom body, the central body complex and the tritocerebrum.

(3) Structures in which the presence of monoamines varies between insects, e.g. cell bodies in the anterior region of the pars intercerebralis, in the optic ganglia, the calyx of the mushroom body, the peduncle, the pons cerebralis and the antennal lobe.

Klemm (1976) concluded that there was no relationship between the distribution and occurrence of monoamines (a) in the visual centres of insects and their visual abilities, (b) in the mushroom body and the 'intelligence' of the insect, and (c) in the antennal lobes and the functional development of the antennae. Further support for the uneven distribution of monoamines in insect nervous tissue was provided by Evans and O'Shea (1977). These workers reported that octopamine, based on the staining by neutral red of non-fluorescent neurones, was localised in a group of neurones, the dorsal unpaired medial neurones (DUM), in the thoracic ganglia of the locust, *Schistocerca gregaria.*

An uneven distribution of monoamines has also been reported in other invertebrates. Thus Stefano and Aiello (1975) found that although sero-

128

tonin predominated in the cerebral cortex and dopamine in the visceral ganglia of the bivalve mollusc *Mytilus edulis*, most of the cells in these ganglia were non-fluorescent. Furthermore, octopamine has been located in individual neurones in *Aplysia* (Saavedra *et al.*, 1974), and dopamine occurs in single neurones in *Planorbis* (Powell and Cottrell, 1974) and serotonin is located in the Retzius cell of the leech (Lent, 1982). Thus monoamines are unevenly distributed throughout both invertebrate and vertebrate nervous systems.

Quantitative studies

The early work on the levels of monoamines in invertebrate nervous system has been well reviewed by Robertson and Juorio (1976) and Evans (1980). These studies (e.g. Hiripi and S-Rosa, 1973) used fluorimetric methods which were of limited sensitivity and required heroic amounts of insect nervous tissue. Hiripi and S-Rosa (1973) used 300 to 400 locusts to determine the catecholamine content in the nervous system.

Progress in the measurement of monoamines was accelerated by the development of radioenzymic methods in which the [^{3}H-CH$_2$-] group from S-adenosyl methionine was incorporated into either the 3-hydroxyl group of catecholamines or the amino group of octopamine (Evans *et al.*, 1985). These methods are capable of measuring picogram amounts of noradrenaline, dopamine and octopamine, thus allowing accurate determination of these monoamines in small amounts of tissues from both vertebrates and invertebrates. Evans and co-workers (see Evans, 1980) have used the radioenzymic method to measure the distribution of biogenic amines in locust and cockroach CNS. These studies confirm earlier observations that (a) adrenaline could not be detected, and (b) the ratio of octopamine to noradrenaline in *P. americana* is > 1, ranging from 2.7 in the cerebral ganglion to 33.3 in the metathoracic ganglion.

A further advance in the analysis of biogenic amines in invertebrate nervous tissue has been the use of high-performance liquid chromatography linked to electrochemical detection (HPLC–ED; Evans *et al.*, 1985). The application of this technique to crustacean (Elofsson *et al.*, 1982; Laxmyr, 1984) and insect (Nässel and Laxmyr, 1983) CNS has confirmed earlier reports (Robertson and Juorio, 1976; Evans, 1980) that insect nervous tissue contains low concentrations of noradrenaline (Table 4.2). Crustacea also have higher levels of octopamine than noradrenaline in their nervous systems (Table 4.2), in agreement with previous reports that the protostomia have higher levels of octopamine than noradrenaline.

These studies using HPLC–ED have also provided evidence for the occurrence of DOPA in the CNS of arthropods (Table 4.2), which is consistent with its role as an intermediate in the biosynthesis of dopamine

Table 4.2: Biogenic amines and dopa in the CNS of arthropods: HPLC–ED measurements.
(1) = pmol/brain; (2) = μg g wet wt

Insect tissue	OA (1)	OA (2)	DA (1)	DA (2)	NA (1)	NA (2)	DOPA (1)	DOPA (2)	5HT (1)	5HT (2)
Calliphora[a] (Blowfly)										
mid brain	2.1	1.3	2.1	1.3	ND	—	0.76	—	1.8	1.2
Optic lobes (per pair)	1.6	0.55	0.92	0.33	ND	—	0.31	—	0.78	0.31
Whole brain	4.0	0.65	2.4	0.39	ND	—	0.29	—	1.8	0.33
Ventral ganglia	1.3	0.6	1.4	0.67	ND	—	0.54	—	1.1	0.58
Periplaneta americana										
Cerebral ganglion			21.5[d] 42.0[e]	—	5.1[e]		—		28.0[e]	—
Crustacean species (brain)										
Carcinus maenas[b]	9.8	0.33	6.3	0.17	1.2	0.05	4.5	0.16	2.80	0.09
Eupagurus berhnardos[b]	5.0	0.2	10.0	0.27	3.1	0.11	4.5	0.13	5.2	0.24
Homarus vulgaris[b]	38.0	0.29	14.0	0.11	13.0	0.14	5.3	0.06	18.0	0.15
Leandes squilla[b]	—	—	1.8	0.19	0.45	0.05	2.0	0.29	—	—
Pandalus borealis[b]	16	0.71	7.4	0.27	2.3	0.1	0.17	0.01	4.7	0.24
Pacifastacus leniusculus[c]	35.9	0.55	13.7	0.2	7.4	0.125	10.2	0.2	11.3	0.2

[a]Nässel and Laxmyr (1983).
[b]Laxmyr (1984).
[c]Elofsson et al. (1982). Values for pmol per brain calculated on the assumption that average weight of brain = 10 mg
[d]Murdock and Omar (1981).
[e]Sloley and Owen (1982).
N.D. = None detectable.

and noradrenaline. This point will be returned to in a later section when the biosynthesis of biogenic amines will be discussed.

It would appear, therefore, that the concentration of monoamines in invertebrate nervous tissue is relatively high as it is of the order of micrograms per gram wet weight of tissue. This value remains relatively constant throughout the Animal Kingdom. For example, the noradrenaline content of the echinoderm *Pycnopodia helianthoides* has been reported as 2.1 μg/g wet weight (Robertson and Juorio, 1976), compared with the noradrenaline content of adult human hypothalamus of 1–2 μg/g wet weight (Hornykiewicz, 1981). The corresponding values for dopamine are 6.0 μg/g wet weight in *P. helianthoides* and 2–5 μg/g wet weight in adult human caudate nucleus (Hornykiewicz, 1981).

Storage and release

A considerable amount of work has been carried out on the storage and release of monoamines from vertebrate preparations (see Chesselet, 1984; Cooper and Meyer, 1984). In vertebrates there is good evidence (Nelson and Molinoff, 1976) that biogenic amines are stored in vesicles located in nerve endings. Electron microscopic studies suggest that noradrenaline is contained in at least two types of membrane-bound granules, small dense-cored vesicles (30–60 nm in diameter) and large dense-cored vesicles (60–150 nm in diameter; see Cooper *et al.*, 1982). The noradrenaline in these granules is associated with ATP (in a molar ratio of noradrenaline: ATP of about 4:1 in adrenal granules together with a soluble protein, chromogranin A, which is thought to be involved in the storage of monoamines). The release of catecholamines from the adrenal medulla and adrenergically innervated peripheral tissues (e.g. the hypogastric nerve vas deferens preparation) has been well studied, and is used as a model for the release of catecholamines from nerve endings in the CNS. In the adrenal medulla, acetylcholine released from the preganglionic fibres combines with receptors on the plasma membrane of the chromaffin cells with a consequent influx of Ca^{2+} and other ions. The influx of Ca^{2+} is regarded as the main stimulus responsible for the mobilisation of catecholamines and their release. There is good evidence, based on release of ATP, chromogranin A and dopamine β-hydroxylase together with catecholamines, that release from chromaffin granules occurs by exocytosis.

It is much more difficult, however, to study the release of neurotransmitter from nerve endings in the central nervous system, due to the complexity of the brain. It is thus well established that exocytosis is the mechanism by which catecholamines are released from sympathetic or central nerve endings. Since exocytosis implies that the entire content of the vesicles be released (i.e. catecholamine, ATP and soluble protein), and as the nerve terminal does not synthesise protein, a high rate of axonal flow would be required to replenish the protein lost by exocytosis. Furthermore there is evidence that newly synthesised transmitter is released preferentially, which implies that neurotransmitter is stored in more than one pool. Both these points provide evidence against the view that release from central neurones occurs via exocytosis.

Considerably less work has been carried out on the storage and release of monoamines by invertebrates. Electron microscopic studies have shown that putative aminergic neurones in invertebrates contain dense-cored vesicles (50–200 nm in diameter; see Evans, 1980). In several cases aldehyde–osmium fixation also shows the presence of smaller, clear vesicles, 20–60 nm in diameter (Hoyle et al., 1974; Robertson, 1974; Oertel et al., 1975; Maxwell, 1978). The function of these clear vesicles is not known. However, the presence of clear vesicles (250 nm in diameter) in the corpora cardiaca of S. gregaria, which are neurohumoral organs, suggests that they may store neurohormones. Alternatively the small clear vesicles may represent a stage in regeneration of the vesicles following release of neurotransmitter.

Evidence that dense-cored vesicles in mammalian neurones contain catecholamines has been provided by studies in which treatment with reserpine resulted in loss of such vesicles (Hökfelt, 1966; Nelson and Molinoff, 1976). Studies with reserpine have also provided evidence for the localization of amines in vesicles in insects (Frontali, 1968; Rutschke and Thomas, 1975; Robertson, 1976). However, a relatively large amount of reserpine (50–60 µg/g body weight) had to be used before loss of dense-cored vesicles could be observed. Furthermore, in one study (Mancini and Frontali, 1970) a concentration of reserpine which completely abolished catecholamine histofluorescence had little effect on the size, appearance or number of vesicles in cockroach brain (see also Sloley and Owen, 1982).

Subcellular fractions containing nerve terminals (synaptosomes) have provided a useful preparation for the study of neurotransmitter release from vertebrate CNS (Chesselet, 1984; Cooper and Meyer, 1984). Relatively few studies using synaptosomes have been reported for invertebrates. However, recent work (Donnellan et al., 1976; Breer and Knipper, 1985) has shown that it is possible to prepare synaptosomes from insect nervous system. Most work to date has been carried out on acetylcholine, and it has been shown that this is located, synthesised and released from insect synaptosomes (Breer and Knipper, 1985). There is no reason, however, why insect synaptosomes should not prove useful for the study of the uptake, synthesis and release of monoamines. Evidence for the Ca^{2+}-dependent release of octopamine from lobster second thoracic nerve roots has been provided by Evans et al. (1975, 1976) following prelabelling with either [^{3}H]tyrosine or [^{3}H]tyramine.

132

Biosynthesis

The biosynthetic pathways for catecholamines and indoleamines have been extensively studied in vertebrates, and the enzymes catalysing each step have been isolated and partially characterised (Cooper *et al.*, 1982). Both pathways start from aromatic amino acids; tyrosine in the case of the catecholamines and tryptophan for the synthesis of serotonin. There are other similarities between the pathways as they both involve a hydroxylation followed by a decarboxylation (Figure 4.3). The key enzymes regulating the activity of these pathways are tyrosine hydroxylase and tryptophan hydroxylase (Kaufman, 1974). The affinity of these enzymes for their substrates (particularly reduced pteridine) and hence their overall activity are altered by their level of phosphorylation (Lazar *et al.*, 1982). Although tyrosine hydroxylase can be phosphorylated by both cAMP and Ca^{2+}/ calmodulin-dependent protein kinases, the current view is that the depolarisation-induced activation of both tyrosine hydroxylase (El Mestikawy *et al.*, 1983) and tryptophan hydroxylase (Hamon *et al.*, 1979) is mediated by Ca^{2+}/calmodulin-dependent protein kinases. In contrast to tyrosine and tryptophan hydroxylases, which are very specific, the enzyme catalysing the decarboxylation of L-dopa and 5-hydroxytryptophan is not very specific and is referred to as aromatic amino acid decarboxylase. For example, tyrosine, tryptophan and histidine are substrates in addition to L-dopa and 5-hydroxytryptophan. Aromatic amino acid decarboxylase is present in excess in nervous tissue, so little dopa or 5-hydroxytryptophan accumulates in the CNS.

Dopamine β-hydroxylase is a mixed-function oxidase which catalyses the conversion of dopamine to noradrenaline. It is located in noradrenergic neurones, probably in the vesicles associated with the storage of noradrenaline. Sufficient dopamine β-hydroxylase is present in noradrenergic neurones so that little dopamine accumulates. Dopamine β-hydroxylase is not absolutely specific; thus tyramine can be β-hydroxylated to form octopamine.

The limited specificity of aromatic amino acid decarboxylase and dopamine β-hydroxylase could account for the occurrence of low levels of octopamine in vertebrate CNS (Molinoff and Axelrod, 1972).

Relatively little work has been carried out on the biosynthesis of monoamines in invertebrates. Most studies have incubated nervous tissue (for individual neurones, e.g. serotonin synthesis in *Helix*, Osborne, 1973) with radioactive precursors, usually [³H]tyrosine or [³H]tryptophan, followed by separation of monoamines by TLC (Osborne *et al.*, 1975b; Vaughan and Neuhoff, 1976), electrophoresis (Hildebrand *et al.*, 1971) or HPLC (Mir and Vaughan, 1981a). Hoyle and Barker (1975) demonstrated the conversion of [³H]tyrosine to tyramine and octopamine in DUM cell bodies and nerve cord of the lubber grasshopper, *Romalea microptera*, and

Figure 4.3: Steps in the biosynthesis of monoamines

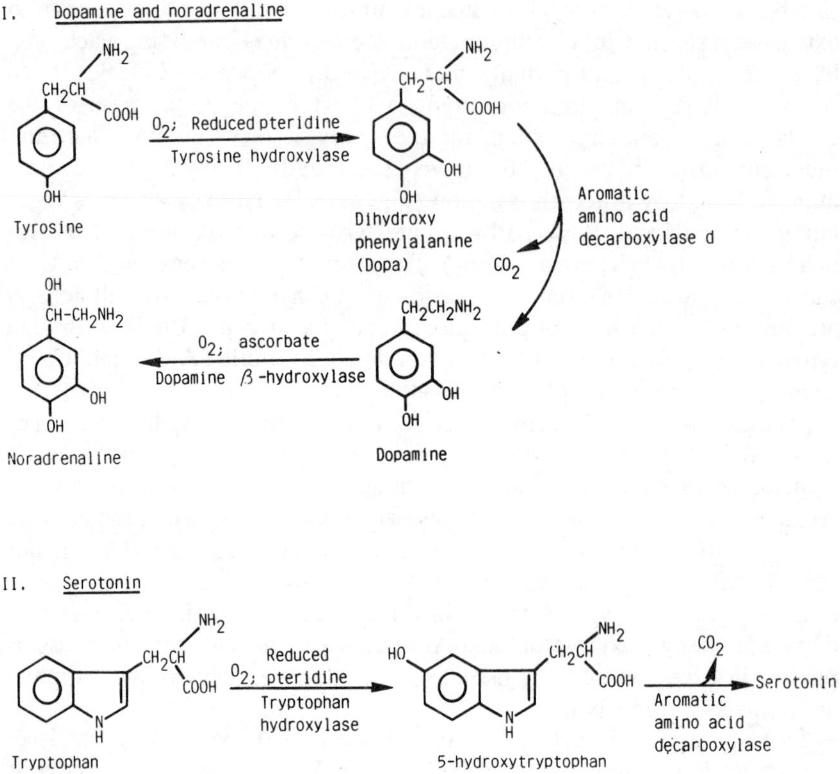
I. Dopamine and noradrenaline
NH2
CH2CH
COOH
OH
Tyrosine
O2; Reduced pteridine
Tyrosine hydroxylase
NH2
CH2-CH
COOH
OH
OH
Dihydroxy
phenylalanine
(Dopa)
Aromatic
amino acid
decarboxylase d
CO2
CH2CH2NH2
OH
OH
Dopamine
O2; ascorbate
Dopamine β-hydroxylase
OH
CH-CH2NH2
OH
OH
Noradrenaline
II. Serotonin
NH2
CH2CH
COOH
N
H
Tryptophan
Reduced
O2; pteridine
Tryptophan
hydroxylase
HO
NH2
CH2CH
COOH
N
H
5-hydroxytryptophan
CO2
Serotonin
Aromatic
amino acid
decarboxylase

in the metathoracic ganglion of the locust *Schistocerca gregaria.* Low rates of synthesis were attained, as only 230 fmol of octopamine were synthesised in 3.5 h per metathoracic ganglion in *S. gregaria.* No radioactivity was incorporated into dopamine or noradrenaline. The conversion of [^{3}H]tyrosine to octopamine and dopamine has been recorded for the thoracic and abdominal ganglia of *Manduca sexta* (Maxwell *et al.*, 1978) and for *Drosophila melanogaster* larval CNS (Wu *et al.*, 1983) and adult brain (Livingston and Tempel, 1983). Similarly, Mir and Vaughan (1981a) found that incubation of *S. gregaria* cerebral ganglia with either [^{3}H]tyrosine or [^{3}H]tyramine resulted in only 2–3 per cent of radioactivity being incorporated into either octopamine or dopamine. In contrast approximately 20 and 90 per cent of radioactivity from [^{3}H]tyrosine and [^{3}H]tyramine, respectively, was recovered in *N*-acetylated monoamines. This provides evidence that *N*-acetylation of monoamines is a major reaction in *S. gregaria* nervous tissue.

A number of points need to be made about these studies on the bio-synthesis of monoamines by invertebrate CNS.

(1) In general the amounts of monoamine that accumulate are quite low. Thus Maxwell *et al.* (1978) reported that 5–75 fmol dopamine; 5–90 fmol octopamine and 10–350 fmol tyramine were formed per structure after 19 h incubation of different regions of *Manduca sexta* CNS with [^{3}H]tyrosine. This low variable conversion corresponds to 5–16 nmol per 19 h per gram of protein. In contrast dopamine synthesis by rat striatal synaptosomes can achieve 20 nmol per minute per gram of protein (Saller and Salaman, 1984). A similar picture was obtained when thoracic and cerebral ganglia of the desert locust *S. gregaria* were incubated with [^{3}H]tyrosine, as only about 2 per cent of the label was recovered in tyramine and octopamine (Mir and Vaughan, 1981a). When ganglia were incubated with [^{3}H]tyramine, approximately 3 per cent of the label could be recovered in octopamine and dopamine.

(2) The procedure developed by Hildebrand *et al.* (1971), which has been widely used to assess monoamine synthesis in different vertebrate and invertebrate preparations (Kingan and Hildebrand, 1985), does not separate uncharged metabolites. This is particularly relevant in the case of insects, since insect nervous tissue contains a very active *N*-acetylase which catalyses the transfer of the acetyl group from acetyl-CoA to the amino group of amines (Dewhurst *et al.*, 1972; Evans and Fox, 1975; Hayashi *et al.*, 1977). If one assumes that *N*-acetylated monoamines are the normal metabolites of biogenic amines in insects, then the accumulation of label in *N*-acetylated monoamines implies that synthesis of monoamine has occurred. Mir and Vaughan (1981a) found that when [^{3}H]tyrosine was incubated with thoracic ganglia of *S. gregaria*, approximately 4 and 9 per cent of the label was incorporated into *N*-acetyloctopamine and *N*-acetyldopamine, respectively. This provided evidence that conversion of tyrosine to dopamine had occurred. However, since thoracic ganglia of *S. gregaria* convert *N*-acetyltyramine to *N*-acetyldopamine (Mir and Vaughan, 1981b), it is not known how much *N*-acetyldopamine is formed from dopamine or whether the *o*-hydroxylation of *N*-acetyltyramine represents an important pathway *in vivo*.

(3) The nature of the *o*-hydroxylation step required to form dopamine from tyrosine (or tyramine) remains uncertain. In spite of reports that label from tyrosine is incorporated into dopamine (Maxwell *et al.*, 1978), no convincing evidence for the conversion of tyrosine to *L*-dopa has been obtained for insect nervous tissue. Thus Mir and Vaughan (1981a) were unable to detect incorporation of label from [^{3}H]tyrosine into *L*-dopa in *S. gregaria* thoracic ganglia, even in the presence of the aromatic amino acid decarboxylase inhibitor NSD

135

1015, which has been shown to lead to an accumulation of L-dopa in mammalian nervous tissue (Carlsson *et al.*, 1972). In contrast, when [^{3}H]tyrosine was incubated with *S. gregaria* haemolymph, approximately 7 per cent of the label was recovered in L-dopa and this was increased to 17.5 per cent in the presence of NSD 1015 (Mir and Vaughan, 1981a).

The failure to detect dopa formation in insect CNS is particularly surprising in view of the ready conversion of tyrosine to L-dopa in insect haemolymph. In the haemolymph, dopa acts as an intermediate in the formation of *N*-acetyldopamine, which is an important protein cross-linking agent in the formation of the insect cuticle (exoskeleton) (Sekeris and Karlson, 1966; Murdock, 1971). It should be noted, however, that the hydroxylation of tyrosine to dopa in insect haemolymph is catalysed by a tyrosinase (Sekeris and Karlson, 1966; Pau and Kelly, 1975). In view of the failure to detect L-dopa as an intermediate in the synthesis of dopamine, a possible pathway for the formation of dopamine and octopamine by *S. gregaria* nervous system in which tyramine is an intermediate is outlined in Figure 4.4. However, two recent studies suggest that L-dopa may be an intermediate in the synthesis of dopamine. Thus Nässel and Laxmyr (1983), using HPLC–ED, detected L-dopa in *Calliphora* CNS. Livingston and Tempel (1983), using mutants of *Drosophila melanogaster* in which monoamine synthesis was abnormal, found that there are two different aromatic amino acid decarboxylases, one for L-dopa and 5-hydroxytryptophan and the other for tyrosine. On the other hand, support for the suggestion that insect CNS does not contain tyrosine hydroxylase is provided by Omar *et al.* (1982) who found that α-methyltyrosine does not decrease the amount of dopamine in *Periplaneta americana* cerebral ganglia. In view of the ready conversion of monoamines to *N*-acetyl compounds, a method of inhibiting the active *N*-acetylase will be a necessary prerequisite before a detailed study can be made of the synthesis of monoamines by insect nervous tissue.

Relatively few studies appear to have been carried out on monoamine biosynthesis in other invertebrates. Stuart *et al.* (1974) reported the incorporation of radioactivity into dopamine and octopamine when leech ganglia were incubated with [^{14}C]tyrosine. However, the amounts incorporated were very low (2–300 cpm) and the incubation times excessively long (3 days), and no evidence was obtained for L-dopa as an intermediate.

In contrast to the failure to obtain evidence for tyrosine hydroxylase in insect CNS, the conversion of radioactive tyrosine to L-dopa has been demonstrated in snail, *Planorbis corneus* (Osborne *et al.*, 1975b) and lobster *Homarus americanus* (Barker *et al.*, 1972) nervous tissue.

Several studies have shown that invertebrate tissue contains appreciable

Figure 4.4: Suggested pathway for the conversion of tyrosine to dopamine and octopamine in insects

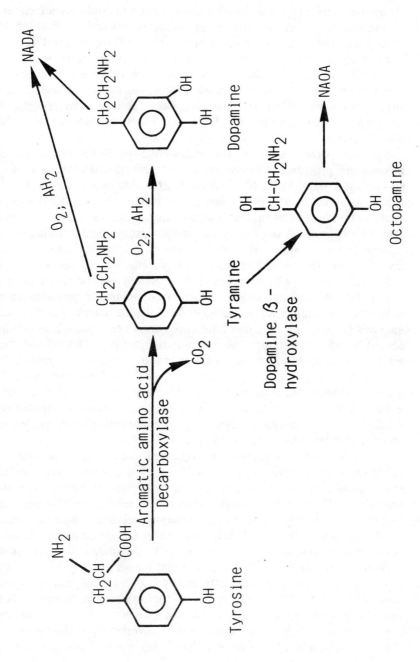

amounts of aromatic amino acid decarboxylase. For example, Murdock *et al.* (1973) reported relatively high rates of *L*-dopa decarboxylation by homogenates of brain of *Locusta migratoria*, *S. gregaria* and the crayfish *Astacus astacus*. The dopa decarboxylase activity in crude homogenates of *L. migratoria* cerebral ganglia is approximately 30 and 10 times that in homogenates of pig and rat brain, respectively. *L*-Tyrosine was decarboxylated at approximately 1.6 per cent of the rate for *L*-dopa, which is of interest in view of the relative ease with which tyrosine is converted to octopamine compared with dopamine by insect nervous tissue. A nonspecific aromatic amino acid decarboxylase has also been reported for *Drosophila* (Dewhurst *et al.*, 1972) and locust (Emson *et al.*, 1974) nervous system.

The incorporation of label from radioactive tyrosine and tyramine into octopamine provided evidence for a tyramine β-hydroxylase in insect (Hoyle and Barker, 1975; Maxwell *et al.*, 1978; Mir and Vaughan, 1981a), lobster (Barker *et al.*, 1972) and leech (Stuart *et al.*, 1974) nervous systems. It is of interest that, although mammals and invertebrates contain aromatic amino acid decarboxylase and dopamine β-hydroxylase, octopamine and not noradrenaline is the major β-hydroxylated monoamine in invertebrates. This would appear to be due to the absence of tyrosine hydroxylase from octopaminergic neurones and the presence of a β-hydroxylase in octopaminergic but not in dopaminergic neurones, rather than to a major difference in specificity between the mammalian and insect enzymes (Evans, 1980). Some differences do occur, however, as Wallace (1976) found that although the tyramine β-hydroxylase from lobster had similar cofactor requirements for Cu^{2+} and ascorbate as mammalian dopamine β-hydroxylase, high levels of fumarate inhibited the lobster but stimulated the mammalian enzyme. Furthermore the ratio of tyramine to dopamine β-hydroxylation was 4.8 in crude homogenates, compared with 4.5 in the partially purified enzyme. Thus both activities reside in the same protein, with tyramine acting as the preferred substrate.

Little work appears to have been carried out on the biosynthesis of serotonin in invertebrates. Such evidence as is available indicates that serotonin is synthesised in invertebrates by a pathway similar to that in vertebrates (Figure 4.3). Thus the serotonergic cerebral cells of gastropod ganglia convert tryptophan to 5-hydroxytryptophan (Eisenstadt *et al.*, 1973; Osborne, 1973) and convert 5-hydroxytryptophan to serotonin (Cottrell and Powell, 1971; Osborne, 1972). Additional support for this pathway is provided by the observation that the serotonergic cerebral cells of *Aplysia* and *Tritonia* contain relatively high levels of aromatic amino acid decarboxylase (Weinreich *et al.*, 1973). Osborne and Neuhoff (1974) provided evidence for this pathway in insects when they found that cerebral ganglia from *Periplaneta americana* converted [³H]tryptophan to 5-hydroxytryptophan and serotonin. They also reported that the incorporation of label into 5-hydroxytryptophan and serotonin was inhibited by *p*-chlorophenylalanine, an inhibitor of tryptophan hydroxylase.

Inactivation

The major mechanism for the termination of monoamine action in vertebrates is the active re-uptake of the transmitter by nerve terminals. Two uptake systems have been described: uptake I, which has a high affinity for monoamines, is sodium dependent and is found in neurones, and uptake II, which is non-neuronal and has a lower affinity for monoamines (see Cooper *et al.*, 1982).

High-affinity uptake systems have been observed in a number of invertebrates. Thus Osborne *et al.* (1975a,b) reported the high-affinity uptake of biogenic amines of *Helix pomatia* nervous tissue, and for dopamine in *Planorbis corneus*. In another study Klemm and Schneider (1975) provided histochemical evidence, using both fluorimetric and electron microscopic techniques, for the active uptake of 6-hydroxytryptamine into neurones of *S. gregaria*, which was blocked by chlorimipramine. Similarly Evans (1978) reported that *P. americana* nervous system contains a high-affinity sodium-dependent uptake system for octopamine. Thus the available data suggest that the action of monoamines in invertebrates may be terminated by a mechanism similar to that proposed for vertebrates. Metabolism of monoamines by monoamine oxidase (MAO) and catechol-*o*-methyl transferase (COMT) is a second, more widely distributed mechanism for the inactivation of these compounds in vertebrates. The combined action of these enzymes gives rise to the metabolites homovanillic acid, 3-methoxy-4-hydroxyphenylethylene glycol and 5-hydroxyindole acetic acid from dopamine, noradrenaline and serotonin, respectively (Figure 4.5).

Insect nervous tissue appears to contain little, if any, monoamine oxidase and catechol-*o*-methyltransferase. Thus Dewhurst *et al.* (1972) reported that MAO and COMT in *Drosophila* brain were 50- and 500-fold less active, respectively, than *N*-acetyltransferase activity. In support of this low level of MAO, Evans and Fox (1975) were unable to detect 5-hydroxy indoleacetic acid when serotonin was incubated with homogenates of honeybee brain. Furthermore no radioactivity was incorporated into MAO or COMT products when cerebral and thoracic ganglia of *S. gregaria* were incubated with [^{14}C]tyrosine, [^{14}C]tyramine or [^{14}C]*L*-dopa (Vaughan and Neuhoff, 1976).

Houk and Beck (1978), however, detected MAO activity in the brain of European corn borer larvae, *Ostrinia nubilalis*, using 2-(2'-benzothiozolyl)-5-stryl-3-(4'-phthalhydrazidyl) tetrazolium salt, which forms an electron-dense product with tryptamine. This histochemical method

139

Figure 4.5: Metabolism of biogenic amines in vertebrate CNS

1.   Dopamine

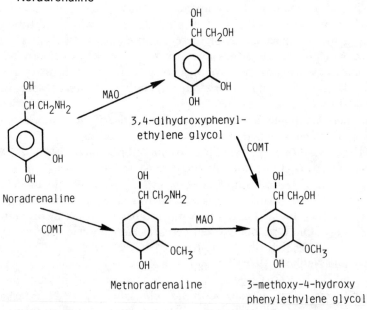

2.   Noradrenaline

Figure 4.5: (continued)

3.   Serotonin

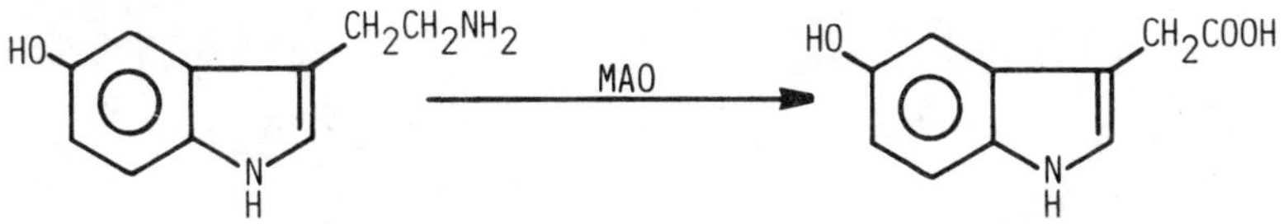

Serotonin (5-hydroxytryptamine)   5-hydroxyindole acetic acid

detected MAO reaction product in the outer membrane of mitochondria within the perineural type I cells, and the reaction was sensitive to the MAO inhibitors tranylcypromine and nialamide. A disadvantage of this histochemical approach is that it is not quantitative, thus the amounts of MAO present are unknown. Furthermore, in a later study, Evans *et al.* (1980) were unable to detect MAO products when tryptamine was incubated with either neuronal or non-neuronal tissue of *Ostrinia nubilalis.*

The very low levels of MAO reported for insect nervous tissue are interesting, in view of the importance of MAO in vertebrates. MAO would appear to have an irregular distribution throughout the Animal Kingdom as it occurs in some invertebrates particularly molluscs and echinoderms (Blaschko and Hope, 1957; Osborne *et al.*, 1975b).

In contrast to the low levels of MAO and COMT, insects possess a very active *N*-acetylase which transfers acetyl groups from acetyl coenzyme A to the amino group on biogenic amines (Figure 4.6). This enzyme has been detected in many insects including *Drosophila* (Dewhurst *et al.*, 1972; Maranda and Hodgetts, 1977), honeybee (Evans and Fox, 1975), locust (Vaughan and Neuhoff, 1976; Hayashi *et al.*, 1977) and the European corn borer, *Ostrinia nubilalis* (Evans *et al.*, 1980). It is of considerable interest that the insect CNS contains an *N*-acetylase which leads to the accumulation of *N*-acetylmonoamines. *N*-Acetylated monoamines are also formed in the haemolymph, where they act as precursors for the exoskeleton. It is tempting to speculate, therefore, that an enzyme required at one stage of the insect life cycle, i.e. formation of the exoskeleton, has been adapted for the inactivation of monoamines, thus providing a 'chemical barrier' to protect the nervous system from biogenic amines formed in the haemolymph.

In view of the low levels of MAO and COMT in insect nervous tissue it is of interest that Kennedy (1978) failed to detect these enzymes in lobster nervous tissue. This study also showed that sulphate conjugates were the major metabolites of octopamine, dopamine and serotonin in the lobster nervous system. Kennedy (1978) reported another metabolite which could also form a conjugate with sulphate (see also Evans *et al.*, 1976). Although

141

Figure 4.6: *N*-acetyl and *O*-sulphate compounds of the common biogenic amines

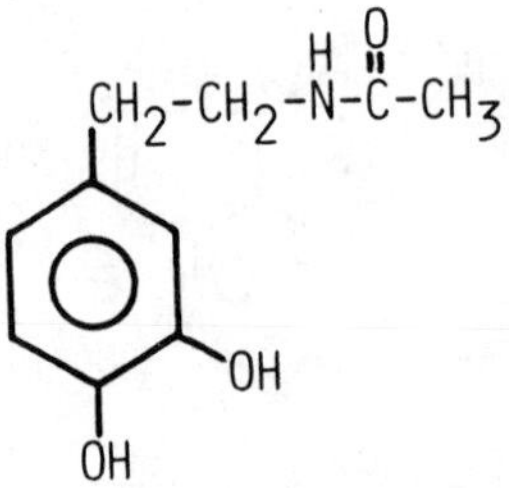

N-acetyl dopamine

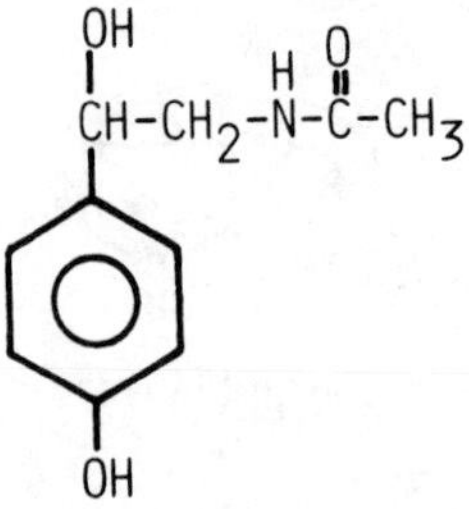

N-acetyl octopamine

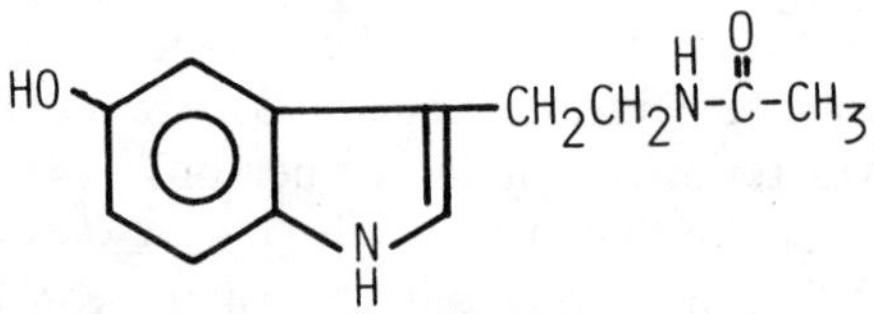

N-acetyl serotonin

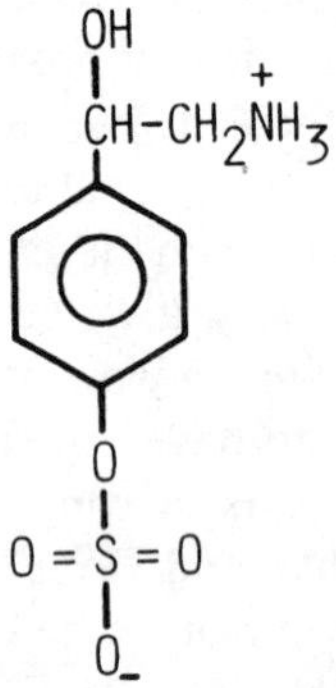

Octopamine O-sulphate

this metabolite was not identified, its electrophoretic behaviour is consistent with its being an *N*-acetylated monoamine.

Summary

The main features of monoamine presynaptic mechanisms appear to be the same in invertebrates and vertebrates. However three major differences are:-

(1) In invertebrates octopamine occurs in larger amounts than noradrenaline.
(2) The lack of monoamine oxidase and catechol-*o*-methyl transferase in arthropods. This observation, together with the occurrence of an active *N*-acetylase and sulphation reaction, suggests that *N*-acetylation or *O*-sulphation, may represent major pathways for enzyme inactivation in some invertebrates.

142

(3) The failure to find L-dopa as an intermediate in the biosynthesis of dopamine in insects. This suggests that the pathway for dopamine formation in insects may involve decarboxylation followed by hydroxylation. This pathway has also been suggested by Whitehead (1969) for N-acetyl dopamine (NADA) formation in the haemolymph.

POSTSYNAPTIC MECHANISMS

Three stages in the postsynaptic action of neurotransmitters and neuro-modulators will be discussed: (1) binding to specific receptors; (2) effect on second messenger formation; and (3) second messenger dependent protein phosphorylation.

1. Receptor binding studies

The binding of a neurotransmitter, hormone or modulator to its specific receptor is the first stage in a sequence of molecular interactions which culminates in an appropriate physiological response, e.g. change in membrane permeability, modulation of neurotransmitter or hormone release, muscle contraction or alteration in gene expression (see, for example, Berridge and Heslop, 1982; Evans, 1982).

Progress in the study of the binding of neurotransmitters to receptors depended on the development of rapid and efficient methods for the separation of bound from free ligand and the availability of ligands with high biological and radioactive specificity. Homogenates of nervous or other tissue are equilibrated with a suitable radioactive ligand, followed by rapid filtration to separate bound from free ligand. Separation by filtration is only suitable if (a) the receptor is membrane bound, and (b) the ligand has a high affinity for its receptor, e.g. a K_D in the nM range or less. Other methods that have been used to separate bound from free ligand include centrifugation (especially if the ligand has lower affinity), or in the case of soluble receptors adsorption of free ligand by charcoal or gel exclusion chromatography.

One of the major problems associated with receptor binding studies is to distinguish specific binding (i.e. to the receptor) from non-specific binding (which includes binding to membrane lipids, glassware and filters). Thus binding of radioactive ligand to its receptor is carried out in the presence and absence of sufficient unlabelled ligand (preferably not the same compound as the radioactive ligand) to displace specific but not non-specific binding. Specific binding is defined, therefore, as the difference between the total radioactivity bound and the radioactivity bound in the presence of unlabelled ligand (i.e. non-specific binding).

143

Binding studies provide information about the dissociation constant, K_d, the number of binding sites, B_{max}, and whether or not cooperativity occurs for a particular ligand–receptor interaction. A valuable chapter providing details of techniques used to study the binding of neuroactive compounds to receptors in insects has appeared recently (Dudai, 1985), and some of the problems associated with the analysis of binding data in insect nervous tissue have been discussed by Lunt (1985).

Receptors for several neurotransmitters have been detected in insects by binding studies (Dudai, 1985). Two of these, which have been well studied in *Drosophila*, are for serotonin and octopamine. Two classes of binding site were detected for serotonin, a high-affinity site, $K_d=1.4$ nM, and a lower affinity site with K_d approximately 130 nM. The low-affinity site was present at approximately eight times the concentration of the high-affinity site (Dudai and Zvi, 1984b). The binding of [³H]serotonin to its high-affinity sites was displaced by low concentrations ($IC_{50} = 10^{-8}-10^{-7}$ M) of ergot alkaloids, tryptamine derivatives (such as 5,6-dihydroxytryptamine) and the neuroleptic drugs haloperidol and chlorpromazine. [³H]Serotonin binding was not displaced by phentolamine, propranolol and dopamine. Thus the serotonergic binding site in *Drosophila* has similar pharmacological properties to the serotonergic receptor in molluscs (Drummond *et al.*, 1980a,b) and mammalian brain (Peroutka and Snyder, 1979).

Dudai and Zvi (1984a) also reported the presence of high-affinity binding sites for octopamine in homogenates from *Drosophila melanogaster* heads. These sites had an apparent dissociation constant of $6.0 \pm 0.9 \times 10^{-9}$ M for octopamine and occurred at a concentration of 0.5 ± 0.1 pmol/mg protein. Octopamine binding was displaced by low concentrations of dihydroergotamine, phentolamine, clonidine, chlorimipramine and chlorpromazine, but not by serotonin and propranolol. Thus the octopamine binding site in *Drosophila* appears to be distinct from the serotonergic site also reported by Dudai and Zvi (1984b). Furthermore, in agreement with the other invertebrate octopaminergic receptors (Evans, 1980), the *Drosophila* octopaminergic binding site has pharmacological properties which resemble α-adrenergic rather than β-adrenergic receptors in vertebrates.

2. Second messenger systems

The detection of a high-affinity binding site for a putative neurotransmitter is only the first step in establishing the role of that compound in the CNS. An important next step is to link the binding site to a biochemical change. Receptors for some neurotransmitters are linked to ion channels, such as the nicotinic cholinergic receptor and GABAergic receptors (Enna, 1981).

However, considerable evidence, obtained during the past 20 years, has shown that many hormones and neurotransmitters mediate their effects by means of second messenger systems. Two important second messenger systems are linked to biogenic amines. The first one, which has been studied for over 20 years, has cyclic nucleotides, particularly cyclic-3′,5′-adenosine monophosphate (cAMP) as the second messenger (Ross and Gilman, 1980; Stadel et al., 1982). The second example, which has only been clearly established in the past few years, links receptor stimulation to changes in inositol trisphosphate and diacylglycerol levels (Berridge and Irvine, 1984). These second messenger systems have been established in a large number of tissues and appear to be widely distributed throughout the Animal Kingdom.

The general model for receptor-linked second messenger systems requires that the hormone, or neurotransmitter, bind to its receptor, which is located on the outer surface of the cell's membrane. The ligand–receptor complex, in association with a guanine nucleotide regulatory protein, alters the activity of adenylate cyclase in the case of cAMP-mediated effects (Figure 4.7), or phospholipase C for inositol-1,4,5-trisphosphate diacylglycerol second messenger systems (Figure 4.8).

Figure 4.7: Second messenger mediated effects of receptor stimulation by biogenic amines

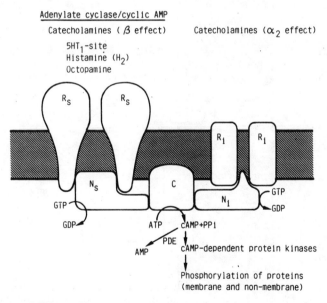

R_s = Receptor dimer for stimulatory agonists of adenylate cyclase; R_i = Receptor dimer for inhibitory agonists of adenylate cyclase; N_s = Guanine nucleotide binding protein associated with stimulation of adenylate cyclase; N_i = Guanine nucleotide binding protein associated with inhibition of adenylate cyclase; C = Calatytic unit of adenylate cyclase complex; PDE = Phosphodiesterase catalysing hydrolysis of cAMP.

Figure 4.8: Second messenger mediated effects of receptor stimulation by biogenic amines phosphatidyl inositol/inositol triphosphate (RPI = receptor for agonists stimulating phosphatidyl inositol hydrolysis; PLC = phospholipase C (phosphodiesterase); $N_{PI?}$ = putative guanine nucleotide binding protein associated with stimulation of phospholipase C)

Phosphatidyl inositol / Inositol trisphosphate

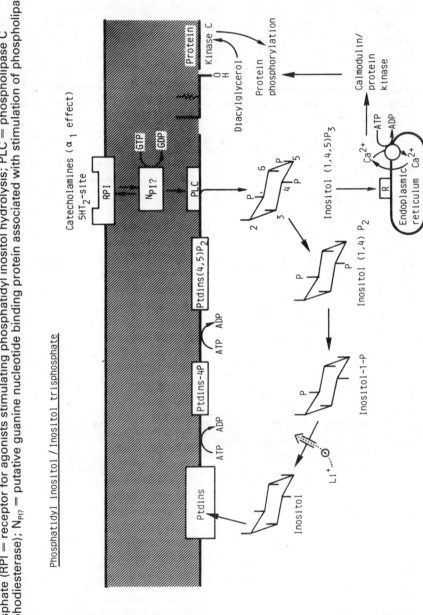

The change in cAMP levels alters the activity of cAMP-dependent protein kinase(s) with the consequent alteration to the phosphorylation state, and hence activity, of a variety of proteins. One of the best studied examples of cAMP-mediated effects is the hormonal control of glycogen synthesis and breakdown, in which the activity of the key enzymes glycogen synthetase and glycogen phosphorylase is regulated by their level of phosphorylation (see Krebs and Beavo, 1979). The role of cAMP as a second messenger, linking changes in neuronal activity with protein phosphorylation, has been clearly established in the vertebrate nervous system (see Drummond, 1985).

Hormonal stimulation of phosphatidyl inositol hydrolysis, although well documented for non-neuronal tissue (Berridge and Irvine, 1984), has been less well established for the CNS. Both products of phosphatidylinositol-4,5-bisphosphate hydrolysis, inositol-1,4,5-trisphosphate (IP_3) and diacylglycerol (Figure 4.8) have a second messenger function. Inositol trisphosphate mobilises intracellular calcium from its storage site in the endoplasmic reticulum whereas diacylglycerol is postulated to activate a membrane-associated protein kinase C (Nishizuka, 1984) with the subsequent phosphorylation of membrane proteins. Furthermore, since calcium/calmodulin dependent protein kinases are found in brain (Cheung, 1981) increases in inositol trisphosphate levels can also lead to stimulation of protein phosphorylation. Thus protein phosphorylation would appear to represent a common stage in the mediation of the effects of receptor stimulation, whether via cAMP, inositol trisphosphate or diacylglycerol.

The binding of catecholamines (to β-adrenergic receptors), 5HT (to $5HT_1$ receptors) and octopamine to its receptor all lead to the stimulation of adenylate cyclase. Catecholamines, acting at $α_2$-adrenergic receptors, lead to an inhibition of adenylate cyclase, whereas catecholamines acting at $α_1$-adrenergic receptors and 5HT acting at $5HT_2$ receptors have been found to stimulate phosphatidylinositol-4,5-bisphosphate hydrolysis. These actions of biogenic amines are in no sense unique and represent one example of the general way in which binding of hormones and neurotransmitters to their receptors can lead to changes in cellular activity.

It is now intended to discuss some of the evidence which associates these second messenger systems with biogenic amines in invertebrates and in vertebrate nervous tissue.

Adenylate cyclase/cAMP

Several studies have shown that biogenic amines can stimulate adenylate cyclase in particulate preparations from invertebrates. Thus octopamine stimulates adenylate cyclase in homogenates from brains of *Periplaneta americana* (Nathanson and Greengard, 1973; Harmar and Horn, 1977). *Schistocerca gregaria* (Kilpatrick *et al.*, 1980; Morton, 1984), the moth *Mamestra configurata* (Bodnaryk, 1979a), *Drosophila melanogaster*

(Uzzan and Dudai, 1982), *Aplysia* (Levitan *et al.*, 1974) and the annelid *Lumbricus terrestris* (Robertson and Osborne, 1979).

Stimulation of adenylate cyclase by dopamine has also been reported in preparations of cerebral ganglia from *S. gregaria* (Kilpatrick *et al.*, 1980), *Locusta migratoria* (Lafon-Cazal and Bockaert, 1984), *M. configurata* (Bodnaryk, 1979b), *Aplysia* (Drummond *et al.*, 1980b) and the mollusc *Helix pomatia* (Osborne, 1977).

The stimulation of adenylate cyclase by serotonin in insect CNS is less consistent. Thus, although Nathanson and Greengard (1974) reported that low concentrations of serotonin stimulated adenylate cyclase 160 per cent, in homogenates of *P. americana* cerebral ganglia Harmar and Horn (1977) were unable to repeat this observation. Furthermore Kilpatrick *et al.* (1980) obtained only a 20 per cent stimulation of adenylate cyclase in *S. gregaria* cerebral ganglia by 10^{-5} M serotonin, and serotonin did not have any effect on adenylate cyclase in *M. configurata* (Bodnaryk, 1979b), *Ceratitis capitata* (Garcia *et al.*, 1981), or locust and tick (Morton, 1984) CNS.

Serotonin stimulation of adenylate cyclase has, however, been reported for other invertebrates. Thus serotonin has been observed to stimulate adenylate cyclase in annelid CNS (Robertson and Osborne, 1979), liver fluke (McNall and Mansour, 1985) and *Aplysia* nervous tissue (Drummond *et al.*, 1980b).

The stimulation of adenylate cyclase by dopamine and octopamine was found to be additive in *P. americana* (Harmar and Horn, 1977) and *S. gregaria* (Kilpatrick, 1981) cerebral ganglia. This suggested the presence of separate receptors for these monoamines.

Pharmacological evidence has also provided support for the hypothesis that dopamine and octopamine stimulate adenylate cyclase in invertebrates via separate receptors. For example, stimulation of *S. gregaria* adenylate cyclase by octopamine is inhibited by phentolamine and cyproheptadine, but not by propranolol nor chlorpromazine (Kilpatrick *et al.*, 1980). Stimulation by dopamine, however, was hardly affected by phentolamine but was inhibited by chlorpromazine. Similarly, Harmar and Horn (1977) reported that phentolamine and cyproheptadine were more potent inhibitors of the octopamine stimulation of adenylate cyclase of *P. americana* than either the neuroleptic α-flupenthixol or the β-adrenergic antagonist propranolol. In contrast α-flupenthixol was a more potent inhibitor of the dopamine-sensitive adenylate cyclase than cyproheptadine. Other insects for which it has been found that octopamine stimulation of adenylate cyclase is blocked by phentolamine but not by propranolol include *M. configurata* (Bodnaryk, 1979b) and *D. melanogaster* (Uzzan and Dudai, 1982). These studies suggest that the insect octopaminergic receptor has pharmacological properties more similar to those of an α-adrenergic receptor than a β-adrenergic receptor in vertebrates. More recent studies (Evans, 1981; see below) suggest that octopamine receptor subtypes exist, with differing pharmacological and biochemical properties. Thus at least one of the octopamine receptor subtypes resembles a β-adrenergic receptor as its occupancy is accompanied by activation of adenylate cyclase.

Two important questions that arise in the study of receptor function relate to (a) the number of receptor types which can be defined pharmacologically, and (b) the relationship of binding sites for a given neurotransmitter to its second messenger system. The following discussion on the invertebrate octopamine receptor may help to illustrate these points. Evans (1981) used pharmacological criteria to distinguish two classes of octopamine receptor in the locust extensor tibiae muscle. Class 1 receptors have a high affinity for octopamine ($EC_{50}=1.5 \times 10^{-8}$ M). They are antagonised by both chlorpromazine ($EC_{50}=2.6 \times 10^{-8}$ M) and cyproheptadine ($EC_{50}=3.7 \times 10^{-8}$ M), and do not appear to be linked to adenylate cyclase. Class 2 receptors on the other hand have a lower affinity for octopamine ($EC_{50}=1.5–3.5 \times 10^{-6}$ M), and are linked to adenylate cyclase. They are not very sensitive to chlorpromazine.

Morton (1984) made a detailed comparison between the octopamine stimulated adenylate cyclase in homogenates of the cerebral ganglion of the locust *S. gregaria americana* and the tick *Boophilus microplus*. He reported maximal stimulation of the locust adenylate cyclase with 10^{-5} M octopamine and noted that this stimulation was inhibited by phentolamine, chlorpromazine and cyproheptadine, but not by propranolol. Synephrine and the formamidine, dimethylchlordimeform (DCDM) were both as potent agonists as octopamine for the locust adenylate cyclase, but DCDM was a less potent agonist than octopamine for the tick adenylate cyclase. Stimulation by formamidines is of particular interest as these compounds have acaricidal and insecticidal properties and have been shown to mimic the actions of octopamine in a number of different insects (Hollingworth, 1976; Murdock and Hollingworth, 1980; Singh *et al.*, 1981). Thus the general properties of the locust and tick octopamine-stimulated adenylate cyclase agree well with those reported by other workers.

On the basis of agonist potency Morton (1984) concluded that the octopamine receptor linked to adenylate cyclase in locust was similar to class 2A of Evans (1981). However, a discrepancy between Morton's study and the classification proposed by Evans relates to the potency of chlorpromazine as an antagonist. Thus Evans (1981) did not find that class 2 octopamine receptors (linked to adenylate cyclase) were sensitive to chlorpromazine whereas Morton (1984) reported that chlorpromazine inhibited the octopamine stimulation of adenylate cyclase in both locust and tick. In agreement with Evans, Kilpatrick *et al.* (1980) did not find that chloropromazine was a potent inhibitor of octopamine-stimulated adenylate cyclase in the locust. It would thus appear that the octopamine receptor linked to adenylate cyclase in insect CNS is similar to the class 2 receptor proposed by Evans. Sub-classification into class 2A or 2B is not really feasible in view of the difficulties of comparing pharmacological studies on homogenates of CNS with an intact muscle preparation.

The octopamine binding sites in *Drosophila* heads reported by Dudai and Zvi (1984b) appear to be similar to class 1 octopaminergic sites of Evans on the basis of both their agonist potency and antagonist sensitivity (see above). It is not known whether these binding sites are linked to a second messenger system, but if they do belong to class 1 subspecificity they are unlikely to be linked to adenylate cyclase.

A further complication in relating high-affinity binding sites to adenylate cyclase stimulation is that adenylate cyclase assays are carried out in the presence of guanine nucleotides which convert high-affinity binding sites to low-affinity binding sites (see e.g. Burgisser *et al.*, 1982). Thus high-affinity binding sites might not be present under the conditions used to assay adenylate cyclase in CNS homogenates, which could account for the greater concentration of octopamine required to stimulate adenylate cyclase than is suggested from its binding affinity.

The requirement for GTP for full activation of monoamine-stimulated adenylate cyclase(s) in invertebrates (Harmar and Horn, 1977; Bodnaryk, 1979a; Kilpatrick *et al.*, 1982) suggests that the model (summarised in Figure 4.7, which was derived from studies on vertebrates) also applies to invertebrates.

There is some evidence that adrenergic receptors in vertebrates are located in glial cells as well as on neurones. Thus cultures of vertebrate astrocytes contain an adenylate cyclase which is linked to a β-adrenergic receptor (Gilman and Schrier, 1971; Harden and McCarthy, 1982; Chneiweiss *et al.*, 1984). Furthermore brain lesion studies suggest that a large proportion of the β-adrenergic receptors found in mammalian brain is located on glial cells (Chang *et al.*, 1980; Karnusina *et al.*, 1983).

Few studies appear to have been carried out on the location of adenylate cyclase in invertebrates. In one study (Taylor *et al.*, 1976) the effect of biogenic amines on the level of cAMP in neuronal perikarya and glia separated from the nerve cord of *Manduca sexta* was examined. Exposure to 5HT increased cAMP levels 33000-fold in the neuronal fraction, compared with only a 430-fold increase in glia. This implies that the major proportion of serotonergic receptors is located on neurones and not glia in *M. sexta.*

Guanylate cyclase/cGMP

Considerably less is known about the function of cGMP in nervous tissue than about cAMP. However, current views (see e.g. Drummond, 1985) suggest that cGMP has a role in modulating the actions of excitatory neurotransmitters, such as acetylcholines and glutamate, as well as monoamines. No clear mechanism appears to have been established linking receptor stimulation, by monoamines or other neuroactive compounds, to guanylate cyclase. There is, however, some evidence that the cGMP content of cells is regulated by cholinergic muscarinic receptors (El-Fakahany and Richelson, 1980). Characteristics of receptor-mediated cGMP synthesis include rapid onset and short duration of the response, together with a dependency on Ca^{2+} (Nathanson, 1977). Although there is little evidence for neurotransmitter stimulation of guanylate cyclase, in cell-free preparations of nervous tissue, glutamate (an excitatory neurotransmitter at the insect neuromuscular junction) has been reported to stimulate guanylate cyclase in a membrane fraction from the body wall of the dipteran *Sarcophaga* (Robinson *et al.*, 1982). This effect of glutamate did not appear to be Ca^{2+} dependent, although Ca^{2+} stimulated guanylate cyclase in the absence of glutamate.

Evidence for a link between monoamines and cGMP in vertebrates has arisen from a number of different sources. Thus 5-methoxy-dimethyl-tryptamine, a 5HT analogue, administered intraperitoneally caused an increase (2.5-fold) in rat cerebellar cGMP (Lykouras *et al.*, 1980). This effect is abolished by prior injection of the 5HT antagonists cyprohep-tadine, methysergide or haloperidol. Apomorphine, a dopaminergic agonist, has been reported to increase cGMP levels in pituitary, cerebellum, corpus striatum and nucleus accumbens of the rat (Kant *et al.*, 1980). Furthermore, intrastriatal or intraperitoneal injection of (−)sulpiride, a dopaminergic D_2 antagonist, resulted in a decrease in cGMP in cerebellar cortex. Thus dopaminergic D_2 receptors may be involved in regulating cerebellar cGMP (Corda *et al.*, 1979). However, the situation is even more complex since α-receptors also appear to be involved in the regulation of cGMP levels in rat cerebellum (Haidamous *et al.*, 1980). Thus noradrenaline and the α-adrenergic receptor agonists methoxamine and phenylephrine increased cGMP levels in rat cerebellum, whereas the α-adrenergic receptor antagonists phentolamine and phenoxybenzamine decreased cerebellar cGMP levels. It would therefore appear that both α-adrenergic and dopaminergic, D_2, receptors may play a role in regulating cGMP levels in the cerebellum. It is not clear, however, whether this elevation of cGMP is caused by an influx of Ca^{2+} stimulated by monoamines or by a change in intracellular Ca^{2+} as a consequence of changes in inositol trisphosphate levels (see below). Some evidence for the former suggestion is provided by the observation that in muscle fibres of the giant barnacle the increase in cyclic GMP levels after potassium depolarisation or nerve stimulation is dependent on the entry of Ca^{2+}. Pre-incubation of muscle fibres in the absence of Ca^{2+} abolishes the rise in cGMP (Beam *et al.*, 1977).

Phosphatidylinositol/inositol trisphosphate

Progress has been made during the past five years in establishing the hydrolysis of phosphatidylinositides (Pt Ins $(4,5)P_2$) as an important response to the binding of some hormones and neurotransmitters to their receptors. The two hydrolysis products, inositol 1,4,5-triphosphate ($InsP_3$) and diacylglycerol (DAG), can both act as second messengers for a variety of cellular processes, including secretion, phototransduction and cell proliferation. The phosphatidylinositol/inositol trisphosphate system is widely distributed (see Berridge and Irvine, 1984) in many tissues, including brain (Berridge et al., 1983), and has been shown to be linked to a wide range of receptors, including α_1-adrenergic, cholinergic, muscarinic and serotonergic $5HT_2$ receptors. The current view is that $InsP_3$ plays a key role in regulating the intracellular level of free Ca^{2+} by altering the efflux of Ca^{2+} from its store in the endoplasmic reticulum (see e.g. Berridge and Irvine, 1984).

One example that has provided much information on the mechanisms by which agonists stimulate $InsP_3$ formation is the effect of serotonin on secretion of saliva in *Calliphora* (Berridge, 1983; Berridge et al., 1984). Important steps in establishing a role for $InsP_3$ in saliva secretion were: (1) the demonstration that the formation of $InsP_3$ precedes that of inositol monophosphate (which was the expected product of phosphatidylinositol hydrolysis) in *Calliphora* salivary gland; and (2) $InsP_3$ formation in *Calliphora* salivary glands is stimulated by serotonin with no apparent lag, whereas the onset of the calcium-dependent physiological response is delayed by at least 1 s. Increases in $InsP_3$ following cell stimulation also precede physiological responses in liver (Thomas et al., 1984) and GH_3 cells (Drummond, 1985). These considerations suggest that $InsP_3$ formation is sufficiently rapid for it to function as a second messenger, and provide evidence for the scheme outlined in Figure 4.8.

One attractive hypothesis to account for the coupling between receptor occupancy and Pt $Ins(4,5)P_2$ phosphodiesterase (phospholipase C) stimulation is that the linkage may be controlled by a guanine nucleotide binding protein. Thus GTP reduces the affinity of noradrenaline for α-adrenergic receptors (Goodhardt et al., 1982) and of carbachol for muscarinic receptors (Hulme et al., 1981). Furthermore, whereas pertussis toxin prevents the modulatory effect of GTP on α_2-adrenoreceptors, which act by inhibiting adenylate cyclase, it has no effect on Ca^{2+} mobilisation regulated by α_1-adrenoreceptors. This implies α_1- and α_2-adrenoreceptors are associated with different nucleotide binding proteins (Boyer et al., 1984). The enhancement of the effect of Ca^{2+} on secretion when GTP or its analogues GTPγs or GppNHp are added to permeabilised blood platelets provides additional support for the role of nucleotide binding proteins in Ca^{2+}-mediated processes. Furthermore a guanine nucleotide binding protein has been identified in association with squid and octopus photoreceptors where light stimulates the hydrolysis of Pt $Ins(4,5)P_2$ (Brown et al., 1984; Vandenberg and Montal, 1984). The need for a second messenger in phototransduction is emphasised by the observation that a single photon can open approximately 1000 sodium channels in *Limulus* photoreceptors (Fein *et al.*, 1984). Injection of InsP$_3$ into the ventral photoreceptors of *Limulus* stimulates an inward current, carried mainly by sodium ions, which has a reversal potential identical with that induced by light. This provides further support for the suggestion that InsP$_3$ acts as a second messenger in phototransduction in invertebrates.

3. Second messenger dependent protein phosphorylation

At least four different second messenger dependent protein kinases have been detected in the nervous system: (a) cAMP-dependent; (b) cGMP-dependent, (c) Ca^{2+}/calmodulin-dependent, and (d) protein kinase C. Some of the evidence which suggests that these protein kinases have a function in the nervous system will now be discussed.

cAMP-dependent protein kinases

cAMP-dependent protein kinases were discovered by Krebs and his co-workers (see Krebs and Beavo, 1979). However, the suggestion that the regulatory functions of cAMP may be explained by cAMP-dependent protein-kinase-catalysed phosphorylation of specific cellular proteins was expounded by Greengard (see, for example, Greengard, 1978a,b).

Two forms of membrane-bound cAMP-dependent protein kinases have been described, which are tetramers consisting of two regulatory and two catalytic subunits. The catalytic subunit is identical in both kinases, which differ in the molecular weights of their regulatory subunits, 47000 (type I) and 52000–55000 (type II, Walter and Greengard, 1981). Activation of both types of protein kinase is due to the dissociation of the regulatory subunit from the enzyme following the binding of cAMP.

cAMP-dependent protein kinases are widely distributed as they are present in every phylum of the Animal Kingdom (Table 4.3). The highest specific activities of both types of cAMP-dependent protein kinase are found in synaptic membranes (Walter *et al.*, 1978).

The cAMP-dependent phosphorylation of protein in the mammalian CNS has been well studied (Nestler and Greengard, 1983). For example, in synaptic membrane preparations, cAMP stimulated the incorporation of ^{32}P mainly into two proteins: protein I, which occurs as a doublet (Ia and Ib), and protein II, with molecular weights of 86000, 80000 and 59000, respectively (Ueda *et al.*, 1973). Maximal phosphorylation of these proteins was achieved with 5μM cAMP in 5 s. Protein I has been purified to homogeneity from bovine brain (Ueda and Greengard, 1977), and appears to be associated with almost all synaptic vesicles in all nerve terminals. It represents approximately 6 per cent of the total protein present in highly

153

Table 4.3: Cyclic AMP-dependent protein kinase activity in different phyla of the Animal Kingdom

| Phylum | Name (Genus) | Protein kinase μmol P_i/min mg protein | |
		$-$cAMP	$+$cAMP
Protozoa	Paramecium (Paramecium)[a]	0.28	0.58
Porifera	Sponge (Haliclona)[a]	0.15	0.30
Coelenterata	Jellyfish (Cyanea)[a]	3.72	5.15
Nematoda	Roundworm (Golfingia)[a]	2.10	3.18
Annelida	Sandworm (Nereis)	0.76	1.74
Mollusca			
Class Pelecypoda	Clam (Mya)[a]	0.87	2.13
Class Cephalopoda	Squid (Loligo)[b]	3.45	6.01
Arthropoda	Lobster (Homarus)[b]	0.54	6.72
Echinodermata	Starfish (Asterias)[a]	0.20	0.74
Chordata	Fish (Cyprinus)[b]	0.55	1.15

[a]Whole animal or [b]muscle tissue used to measure protein kinase activity. See Greengard (1978b).

purified preparations of synaptic vesicles, where it occurs as an extrinsic membrane protein located on their outer or cytoplasmic surfaces (Huttner *et al.*, 1983), and it has been renamed as synapsin I.

Phosphorylation of synapsin I has been found to be regulated by serotonin (Dolphin and Greengard, 1981), dopamine (Nestler and Greengard, 1982a) and noradrenaline, in several regions of the central and peripheral nervous systems. In each of these regions the respective neurotransmitter stimulates the phosphorylation of synapsin I, apparently via an increase in cAMP levels. These observations have led to the suggestion by Greengard that phosphorylation of synapsin I plays a role in modulating the release of neurotransmitters (Nestler and Greengard, 1983).

Further evidence for the importance of synapsin I in neuronal function is provided by the observation that nerve impulse conduction, at physiological frequencies, increases the phosphorylation of about 80 per cent of synapsin I in nerve terminals of the rabbit superior cervical ganglion (Nestler and Greengard, 1982b) via an influx of calcium into the nerve terminals.

In related studies Kelly *et al.* (1979) observed that cAMP stimulated the phosphorylation of four proteins, M_r 85 000, 82 000, 78 000 and 55 000, in brain synaptic junctions. The first two proteins correspond to synapsin Ia and Ib and the 55 000-M_r component is probably a phosphorylated regulatory subunit of protein kinase and is similar to protein II of Ueda and Greengard (1977). Phosphorylation of synapsin I has also been observed in post-mortem samples of human brain from individuals 3 to 82 years of age (Routtenberg *et al.*, 1981).

The activity of tyrosine hydroxylase, the rate-limiting enzyme in the biosynthesis of catecholamines, is regulated by its phosphorylation state. Phosphorylation appears to be catalysed by both cAMP-dependent and calcium/calmodulin-dependent protein kinases (El Mestikawy et al., 1983). Bustos and Roth (1979) provide evidence which suggests that K^+ depolarisation-induced activation of tyrosine hydroxylase in crude striatal extracts is mediated by the calcium/calmodulin-dependent kinase, and not the cAMP-dependent protein kinase. The significance of the cAMP-dependent activation of tyrosine hydroxylase in vivo is less certain. One possibility, however, is that presynaptic regulation of tyrosine hydroxylase may be mediated by cAMP (Pollock et al., 1981). Phosphorylation of one of the microtubule-associated proteins (MAP-2), thought to have a role in microtubule assembly and subsequent interactions with other cellular structures, is inactivated by cAMP-dependent protein kinase. Thus recent work has provided some evidence that cAMP-dependent protein phosphorylation is important in the regulation of a number of presynaptic functions, namely neurotransmitter synthesis and release and microtubule assembly in mammals. Several studies have been carried out on cAMP-dependent phosphorylation of proteins in invertebrates, and some of these will now be summarised.

Rotondo et al. (1987b) found that maximal incorporation of ^{32}P from [^{32}P]ATP into protein in homogenates of S. gregaria cerebral ganglia was achieved by 50 μM cAMP, with an EC_{50} of 2.5 μM, within 60 s. Subcellular fractionation studies showed that cAMP stimulated phosphorylation of total protein in crude nuclear, crude mitochondrial and soluble fractions of cerebral ganglia, homogenised in either 0.25 M sucrose or 6 mM Trismaleate buffer. However, very little cAMP-dependent phosphorylation could be detected in washed pellet fractions. This suggested that the cAMP-dependent protein kinase in S. gregaria cerebral ganglia is easily solubilised. Incubation of crude homogenates with cAMP and [γ ^{32}P]ATP before preparing the washed membrane fraction resulted in incorporation of ^{32}P into membrane proteins. Thus cerebral ganglia membranes did contain substrate protein for cAMP-dependent protein kinase. Similar results were obtained with tissue from the snail Helix pomatia (Bandle and Levitan, 1977), as cAMP stimulated protein phosphorylation in the 20 000 g and 120 000 g supernatants but not in their respective pellets. If, however, phosphorylation was carried out in the 700 g supernatant prior to preparation of the 20 000 g pellet, cAMP stimulated incorporation of phosphate into proteins, which are recovered in the pellet. Similar results have been obtained with Aplysia as it was reported that cAMP stimulates the phosphorylation of 10 proteins (M_r 21 000–130 000) in the 12 000 g supernatants but not in the pellet fraction (Novak-Hover and Levitan, 1983). If, however, intact Aplysia nervous tissue was incubated with $^{32}P_i$ prior to preparation of a 15 000 g pellet, dibutyryl cAMP was observed to stimulate incorporation of phosphate into a membrane protein of M_r 118000–120000 (Levitan *et al.*, 1974).

It would appear, therefore, that the cAMP-dependent protein kinase in *S. gregaria*, *Helix* and *Aplysia* is readily solubilised, although protein substrates are found in membrane fractions. This represents a difference between invertebrates and mammals since most of the cAMP-dependent protein kinase activity in mammalian nervous tissue can be recovered in plasma membranes prepared by lysing synaptosomes in hypo-osmotic buffer prior to density gradient centrifugation (Greengard, 1978b).

Another difference between cAMP-dependent protein phosphorylation in invertebrates and mammals is that invertebrate nervous tissue does not appear to contain synaptosomal proteins with molecular weights corresponding to synapsin Ia or Ib of mammalian neurones. Thus cAMP stimulates the phosphorylation of a single protein with M_r 118000–120000 in the 15000 g pellet of *Aplysia* (Levitan *et al.*, 1974) and 20000 g pellet of *Helix* (Bandle and Levitan, 1977) nervous tissue, and of three proteins with M_r 160000, 74000 and 31000 in *Drosophila* nervous tissue (Kelly, 1981). Rotondo *et al.* (1987a) found that cAMP stimulated the phosphorylation of four proteins with M_r 51000, 45000, 39000 and 21000 in a synaptosomal fraction prepared from *S. gregaria* cerebral ganglia.

Further work is clearly required to establish whether any of these proteins is associated with synaptic vesicles and thus represents the invertebrate equivalent of synapsin I.

cGMP-dependent protein kinases

Greengard suggested that the actions of cGMP could also be mediated by a protein kinase, and a specific cGMP-dependent protein kinase has been identified (Lincoln and Carlin, 1983). The cGMP-dependent protein kinase has similar size, shape and amino acid composition to type II cAMP-dependent protein kinase, but differs from it immunologically and in its relative affinity for cAMP and cGMP (Walter *et al.*, 1980). cGMP-dependent protein kinases are unevenly distributed in the mammalian nervous system. They are enriched in the cerebellum, in agreement with the localisation of cGMP in the Purkinje cells (Cumming *et al.*, 1977).

A cGMP-dependent protein kinase has been isolated from lobster with a K_m of 0.075 μM for cGMP, compared with 3.6 μM for cAMP (Kuo *et al.*, 1971). In contrast a cAMP-dependent protein kinase has been detected in the same tissue with a K_m of 1.2 μM for cGMP and 0.018 μM for cAMP (Kuo and Greengard, 1970). This suggests that separate cGMP and cAMP-dependent protein kinases exist in invertebrates as well as vertebrates. Rotondo (1984) found that both cGMP and cAMP stimulated the phosphorylation of one protein with M_r 45000 in homogenates of *S. gregaria* cerebral ganglia. This is analogous to the situation in mammals, where it has been found that cAMP and cGMP can stimulate the phosphorylation of the same protein (Greengard, 1978b). cGMP has been reported to stimulate the phosphorylation of three proteins (M_r 92000, 68000 and 62000) in membrane fractions prepared from nerve roots of *Aplysia* (Ram and Ehrlich, 1978). cAMP can also stimulate the phosphorylation of these proteins; however, a much higher concentration is required and the maximal stimulation by cAMP is not as great as that achieved by cGMP. The effect of cGMP on protein phosphorylation in *S. gregaria* was studied in the presence of EGTA (Rotondo, 1984). This may have reduced the extent of phosphorylation since the cGMP-dependent phosphorylation of some proteins (e.g. in mouse brain) requires Ca^{2+} (Malkinson, 1975).

Calcium/calmodulin-dependent protein kinases

Calmodulin is a Ca^{2+}-binding protein, very widely distributed, which has multiple regulatory functions in nervous tissue. For example, calmodulin has been found to stimulate both cyclic nucleotide phosphodiesterase and adenylate cyclase activities in the brain (Cheung, 1980). The role of calmodulin in regulating Ca^{2+}-dependent processes as well as the synthesis and degradation of cAMP has led to suggestions that calmodulin may play a part in synaptic transmission. Evidence to support this view includes the location of a pool of calmodulin and an associated kinase in the postsynaptic density (Wood *et al.*, 1980; Grab *et al.*, 1981).

A number of studies have suggested that GTP is not required for calmodulin stimulation of adenylate cyclase either in brain membranes (Seaman and Daly, 1982) or with the partially purified enzyme (Heideman *et al.*, 1982). More direct evidence has been provided by Salter *et al.* (1981) who found that calmodulin stimulated the catalytic unit of adenylate cyclase which had been separated from the regulatory unit by gel filtration. Thus calmodulin would appear to activate brain adenylate cyclase by a direct interaction with the catalytic unit, rather than by a receptor/nucleotide binding protein complex.

Calmodulin may also be involved in neurotransmitter stimulation of adenylate cyclase. Thus both Ca^{2+} and dopamine stimulate adenylate cyclase activity in the caudate nucleus, and calmodulin reverses the Ca^{2+} inhibition of dopamine-stimulated activity and increases the sensitivity of the enzyme to dopamine (Gnegy and Treisman, 1981).

Less is known about Ca^{2+}–calmodulin-dependent protein kinases in nervous tissue than about cAMP-dependent protein kinases. However, incubation of synaptosomes with $^{32}P_i$ in the presence of veratridine, K^+ (60 mM) or the Ca^{2+} ionophore A23187 stimulated the phosphorylation of specific endogenous proteins as well as augmenting Ca^{2+} uptake (Krueger *et al.*, 1977). In particular the phosphorylation of synapsin Ia and Ib was stimulated (Sieghart *et al.*, 1979). Although synapsin I is phosphorylated

157

by both Ca^{2+}–calmodulin and cAMP-dependent protein kinases, the sites of phosphorylation are different. Thus Huttner and Greengard (1979) used proteolysis and peptide mapping studies to show that cAMP induced phosphate into only one peptide fragment of synapsin I, whereas depolarisation-induced Ca^{2+} influx into intact synaptosomes stimulated the phosphorylation not only of this peptide but also of two others. Further studies (Kennedy and Greengard, 1981) revealed that there are two distinct Ca^{2+}/calmodulin stimulated protein kinases in brain. One of these, located in the cytosol and membrane bound, phosphorylates the region of synapsin I stimulated by Ca^{2+} but not by cAMP, whereas the other, found only in the cytosol, phosphorylates the region of synapsin I that is stimulated by both Ca^{2+} and cAMP.

The studies of De Lorenzo and associates are of particular interest since these authors found that Ca^{2+}–calmodulin-dependent phosphorylation of two proteins, DPH-M (M_r 60000) and DPH-L (M_r 53000) (proteins associated with synaptic vesicles whose phosphorylation is inhibited by diphenylhydantoin), coincides with the stimulation of neurotransmitter release (DeLorenzo, 1980). It is very probable that α- and β-tubulin are the major components of DPH-M and DPH-L (Burke and DeLorenzo, 1982), which implies that α- and β-tubulin are involved in the release of neurotransmitters from synaptic vesicles.

Little work appears to have been carried out on Ca^{2+}–calmodulin-sensitive protein kinases in invertebrate nervous systems. Kelly (1981), however, found that Ca^{2+}–calmodulin stimulated the phosphorylation of several proteins, including two with M_r 55000 and 51000, possibly corresponding to DPH-M and DPH-L, in synaptosomal preparations from *Drosophila* heads. In another study Schoeffeniels and Dandrifosse (1980) observed that phosphorylation of two proteins, M_r 380000 and 530000, was increased by K^+ and decreased by Na^+ in neuronal membrane preparations of the crab *Carcinus maenas*. It was proposed that these high molecular weight proteins may be part of the Na^+ channel and that the conductance state of these channels may be controlled by a Ca^{2+}-dependent phosphorylation process.

Work in the author's laboratory (Rotondo, 1984) has shown that Ca^{2+}–calmodulin stimulates the incorporation of $^{32}P_i$ from $[\gamma\text{-}^{32}P]ATP$ into three proteins with M_r 62000, 54000 and 31000 in membrane fractions obtained from *S. gregaria* cerebral ganglia. Phosphorylation of these proteins was stimulated by Ca^{2+} in the absence of added calmodulin, thus implying that endogenous calmodulin was present. Calcium stimulated the phosphorylation of a protein with M_r 21000 from *Aplysia* nervous tissue. This phosphorylation is inhibited by *trans*-flupenthixol, which suggests that Ca^{2+}/calmodulin-dependent protein kinase is involved (Novak-Hofer and Levitan, 1983).

Protein kinase C

As mentioned above, one of the second messengers, generated by the hydrolysis of Pt Ins(4,5)P$_2$ diacylglycerol, stimulates membrane-bound protein kinase C. Protein kinase C, discovered in 1977, is a Ca^{2+}-activated, phospholipid-dependent enzyme which has its affinity for Ca^{2+} greatly increased by diacylglycerol (Nishizuka, 1983). Diacylglycerol, normally present in only very small amounts in membranes, usually contains at least one unsaturated fatty acid, frequently arachidonic acid, at position 2. Phosphatidyl serine is a requirement for activation of protein kinase C, and other phospholipids show positive or negative cooperativity with phosphatidyl serine. Triacylglycerol, monoacylglycerol and free fatty acids are totally inactive with protein kinase C. Protein kinase C is widely distributed throughout the Animal Kingdom, its activity exceeding that of cAMP-dependent protein kinases. It has a broad specificity as it catalyses the phosphorylation of seryl and threonyl but not tyrosyl residues. The most compelling evidence linking Pt Ins(4,5)P$_2$ hydrolysis with protein kinase C has come from studies with blood platelets. Thus stimulation of platelets by thrombin, collagen or platelet-activating factor leads to the rapid, extensive phosphorylation of two proteins with M_r 40000 and 20000 in association with the release of serotonin from dense bodies and acid hydrolases from lysosomes. The phosphorylation of the 20000 M_r protein, which corresponds to the light chain of myosin, is stimulated by Ca^{2+}/calmodulin, whereas the 40000 M_r protein is a substrate for protein kinase C (Sano *et al.*, 1983). Protein kinase C is directly activated by phorbal esters, which makes these tumour promoters useful tools for studying protein kinase C phosphorylations. This aspect of protein kinase C activation falls outside the scope of this chapter, but the interested reader is referred to the recent review by Nishizuka (1984).

It would appear that a number of physiological processes, such as the release of serotonin and lysosomal enzymes from platelets and the release of histamine and lysosomal enzymes from rat mast cells (Nishizuka, 1983) and neutrophils (Kajikawa *et al.*, 1983), depends on both Ca^{2+} mobilisation stimulated by InsP$_3$ and activation of protein kinase C by diacylglycerol. In addition the release of catecholamine from bovine adrenal medullary cells (Knight and Baker, 1983), aldosterone secretion from porcine adrenal glomerulosa (Kojima *et al.*, 1983) and insulin release from rat pancreatic islets (Zawalich *et al.*, 1983) have all been reported to require a synergistic interaction between protein kinase C and Ca^{2+} mobilisation for maximal effect.

Neurotransmitter-stimulated phosphorylation of proteins

An important step in establishing the role of second messenger associated protein phosphorylation in neurotransmission is to demonstrate that stimulation of whole cells, or membrane fractions, with a neurotransmitter leads to the phosphorylation of a specific protein. Three examples of phosphorylation of specific proteins stimulated by neurotransmitters will now be discussed.

Dopamine and adenosine 3′,5′ -monophosphate-regulated phosphorylation (DARPP)

A large number of neurone-specific phosphoproteins have been demonstrated in the mammalian CNS (Walaas *et al.*, 1983a,b). These authors found that, unlike synapsin I, which has a general distribution throughout the mammalian CNS, some of these phosphoproteins were restricted to specific brain regions, which implied a link with a particular neurotransmitter. One of these phosphoproteins, which is enriched in basal ganglia, has M_r 32000 and is confined to the CNS as it is not found in lung, heart, liver, skeletal muscle, kidney and spleen. Dopamine stimulates the phosphorylation of this protein (DARPP), without affecting the phosphorylation of several other proteins. Dopamine stimulation of DARPP was blocked by fluphenazine, a dopaminergic receptor antagonist, and phosphorylation was not stimulated by either noradrenaline or serotonin. Furthermore, a 30S depolarisation-induced Ca^{2+} influx into neostriatal neurones, which is expected to activate calcium-dependent protein kinases, had no effect on the phosphorylation state of DARPP-32 (Walaas *et al.*, 1983c).

Hemitransection of rat brain at the level of the globus pallidus, which destroys most of the nerve fibres passing from the striatum to the substantia nigra, resulted in an almost complete loss of DARPP-32, synapsin I and tyrosine hydroxylase in the substantia nigra. No change in caudato-putamen DARPP-32 was observed, however. Thus DARPP-32 appears to be located in neuronal fibres descending into the substantia nigra but not in neuronal fibres ascending into the caudato-putamen (Walaas and Greengard, 1984). Destruction of catecholaminergic neurones by infusion of 6-hydroxydopamine into either the substantia nigra or caudato-putamen did not alter the amounts of DARPP-32 in these regions. Thus DARPP-32 appears to be located outside dopaminergic neurones. Kainic acid, which destroys intrinsic neurones without lesioning glial cells, axon terminals of extrinsic neurones or fibres passing through the lesioned region, considerably reduced DARPP-32 when it was injected into the caudato-putamen. However, injection of kainic acid into the substantia nigra did not affect DARPP-32 in either the substantia nigra or the caudato-putamen (Walaas and Greengard, 1984). The results of lesion studies suggest that, in the caudato-putamen, DARPP-32 is present in local neurones (intrinsic neurones) while in the substantia nigra, the phosphoprotein is present in afferent neuronal axons originating in the neostriatum.

DARPP-32 is phosphorylated very rapidly. Thus incubation of cytosol from caudato-putamen with $[\gamma\text{-}^{32}P]ATP$ in the presence of cAMP resulted in half maximal phosphorylation of DARPP-32 by 5–10 s and maximal phosphorylation after 60–120 s. Neither cGMP nor Ca^{2+}/calmodulin stimulated the phosphorylation of DARPP-32 in cytosol from caudato-putamen. Furthermore, 100 μM dopamine stimulated phosphorylation of DARPP-32 in rat caudato-putamen slices maximally by 5 min.

It would appear, therefore, that DARPP-32 is not located in dopaminergic neurones, but in cells which contain dopaminergic D_1 receptors, linked to adenylate cyclase. It is therefore possible that phosphorylation of DARPP-32 may be involved in mediating or modulating dopamine-induced trans-synaptic effects linked to D_1 receptors.

Octopamine-stimulated phosphorylation

The desert locust, S. gregaria. Rotondo *et al.* (1987a) have shown that incubation of intact cerebral ganglia of *S. gregaria*, in the presence of 100 μM octopamine for 10 min, stimulated the phosphorylation of a protein with M_r 39000. After 22 h, the phosphorylation of two other proteins with M_r 21000 and 25000 was also observed. Octopamine (100 μM) also stimulated the phosphorylation of three proteins with M_r 51000, 39000 and 21000 in homogenates of *S. gregaria* cerebral ganglia. It is of interest that whereas in intact ganglia octopamine stimulated the phosphorylation of only one protein, M_r 39000, in homogenates two additional proteins became phosphorylated. This implies that there is considerable organisation in intact ganglia in terms of accessibility of protein kinases to cAMP.

Aplysia californicus. Octopamine, 100 μM, stimulates both an increase in cAMP levels and the phosphorylation of a protein with M_r 120000 in intact abdominal ganglia of *Aplysia* (Levitan and Barondes, 1974). Phentolamine inhibits both cAMP production and phosphorylation of M_r 120000, whereas the phosphorylation of this protein is stimulated by dibutyryl cAMP. It is likely, therefore, that the octopamine stimulation of phosphorylation of M_r 120000 is mediated by cAMP.

cAMP production in *Aplysia* and protein phosphorylation in *Aplysia* and *S. gregaria* CNS are biphasic. Thus a two-fold increase in cAMP levels is observed 10 min after the addition of octopamine to abdominal ganglia of *Aplysia*. cAMP levels return to control levels by 50 min but a second increase in cAMP levels is observed by 22 h, accompanied by phosphorylation of protein with M_r 120000 (Levitan *et al.*, 1974). Rotondo (1984) observed that octopamine stimulated two phases of protein phosphorylation in intact cerebral ganglia of *S. gregaria*. Thus phosphorylation of the protein fraction increased after 10 min, then decreased to control levels by 6 h, followed by a second phase of phosphorylation after 22 h. In view of the active N-acetylase present in *S. gregaria* nervous tissue (Mir and Vaughan, 1981a) it is unlikely that much unacetylated octopamine will remain by 22 h. Thus the changes in AMP and associated protein phosphorylation which occur after 22 h are possibly due to secondary effects of exposing intact ganglia to octopamine. These studies with invertebrate nervous tissue could, therefore, provide useful models in which to assess the long-term effects of neurotransmission on neuronal function.

Electrophysiological studies

One advantage of working with invertebrates is that their nervous systems possess large, relatively easily identifiable cells. For example, the marine snail *Aplysia californica* possesses relatively few, readily identifiable neurones, which are up to 1 mm in diameter. Studies on individual *Aplysia* neurones have yielded direct evidence that cAMP-induced changes in membrane properties are mediated by phosphorylation of specific neuronal proteins (Kupferman, 1980; Kandel and Schwartz, 1982). The abdominal ganglion of *Aplysia* contains an identifiable neurone, R_{15}, in which an oscillating membrane potential results in alternating periods of action potentials interspersed with hyperpolarisation (bursting neurones). Treistman and Levitan (1976) found that the phosphodiesterase inhibitor, isobutylmethylxanthine (IBMX) increased the duration and amplitude of interburst hyperpolarisation as well as the number of action potentials per burst. These changes in electrical activity were accompanied by a significant increase in cAMP content. Furthermore intracellular injection of cAMP, or the GTP analogue GppNHp resulted in a dramatic and extended hyperpolarisation of R_{15} (Treistman and Levitan, 1976). R_{15} membranes contain a 5HT-sensitive adenylate cyclase (Levitan, 1978). Drummond *et al.* (1980c) provided support for the role of 5HT and cAMP in the hyperpolarisation of neurone R_{15}. Thus (a) the 5HT-stimulated K^+ conductance was mimicked by application of phosphodiesterase-resistant cAMP analogues and intraneuronal injection of GppNHp, and (b) the pharmacological characteristics of 5HT-dependent changes in electrical activity of R_{15} and 5HT stimulation of adenylate cyclase were very similar. This provides good evidence that the 5HT-dependent changes in electrical activity of R_{15} are mediated by cAMP. Evidence for the involvement of protein phosphorylation associated with the 5HT-dependent changes in electrical activity was obtained by Adams and Levitan (1982) who found that injection of protein kinase inhibitor specifically inhibited the 5HT-induced increase in K^+ conductance in R_{15} neurone. Additional support is provided by the observation (Lemos *et al.*, 1982) that 5HT stimulates the incorporation of ^{32}P into three proteins, M_r 230000, 205000 and 135000, following the injection of $[\gamma\text{-}^{32}P]$ATP into neurone R_{15} *in vivo*.

Other studies (see Pellmar and Carpenter, 1980) found that ionophoretic application of 5HT on to LB and LC neurones (identifiable clusters of neurones, located in the left hemisphere of *Aplysia* abdominal ganglia) generates a voltage-sensitive slow inward Ca^{2+} current. It would appear, however, that this effect of 5HT is not mediated via cAMP (Pellmar, 1981). It is therefore tempting to speculate that the 5HT-induced calcium current in LB and LC neurones is mediated by changes in the level of inositol trisphosphate/diacylglycerol.

Kandel and associates have obtained evidence that implicates cyclic nucleotide-mediated protein phosphorylation in the sensitisation of the gill withdrawal reflex in *Aplysia*, which is regarded as an elementary form of learning (see Kandel and Schwartz, 1982). Electrical changes caused by nerve stimulation in isolated *Aplysia* CNS were monitored simultaneously from sensory and motor neurones. It was found that repeated stimulation of the sensory neurone resulted in a reduction in the monosynaptic e.p.s.p. evoked in the motor neurone. If, however, a sensitising stimulus (e.g. a single strong electric shock to the tail) was applied, an enhancement of the e.p.s.p. obtained by stimulation of the sensory neurone occurred. This enhancement lasted for 50 min, and quantal analysis suggested that release of transmitter was increased. Brunelli *et al.* (1976) reported that incubation of abdominal ganglia with 5HT, but neither dopamine nor octopamine, following synaptic depression caused by repeated sensory neurone stimulation, led to an enhanced e.p.s.p. in the follower neurone in the next trial. In other words, application of 5HT to abdominal ganglia mimicked the effect of stimulating the connective neurones. Furthermore, cinanserin, a 5HT receptor antagonist, reversibly blocked facilitation produced by connective nerve stimulation, and incubation of abdominal ganglia with dibutyryl cAMP or intraneuronal injection of cAMP mimicked the facilitation of e.p.s.p. evoked by 5HT or connective nerve stimulation.

Thus there is good evidence to support the hypothesis that 5HT receptor stimulation, mediated by cAMP and protein phosphorylation, plays an important role in the presynaptic facilitation underlying sensitisation.

ACKNOWLEDGEMENTS

The work referred to in this chapter, relating to monoamine-sensitive adenylate cyclase and cAMP-dependent protein phosphorylation in *S. gregaria*, was supported by the SERC CASE scheme in collaboration with Dr J.F. Donnellan, Shell Research, Sittingbourne, Kent. The author is grateful to Professor R.M.S. Smellie for providing facilities to study cyclic nucleotides in *S. gregaria* CNS.

REFERENCES

Adams, W.B. and Levitan, I.B. (1982) Intracellular injection of protein kinase inhibitor blocks the serotonin-induced increase in K$^+$ conductance in *Aplysia* neuron R15 *Proc. Natl Acad. Sci. USA 79*, 3877-80

Bandle, E.F. and Levitan, I.B. (1977) Cyclic-AMP-stimulated phosphorylation of a high molecular weight endogenous protein substrate in sub-cellular fractions of molluscan nervous systems. *Brain Res. 125*, 325-31

Barker, D.L., Molinoff, P.H. and Kravitz, E.A. (1972) Octopamine in the lobster nervous system. *Nature New Biol. 236*, 61-2

Beam, K.G., Nestler, E.J. and Greengard, P. (1977) Increased cyclic GMP levels associated with contraction in muscle fibres of the giant barnacle. *Nature (Lond.) 267*, 534-6

Berridge, M.J. (1983) Rapid accumulation of inositol trisphosphate reveals that agonists hydrolyse polyphosphoinositides instead of phosphatidylinositol. *Biochem. J. 212*, 849-58

Berridge, M.J., Buchan, P.B. and Heslop, J.P. (1984) Relationship of polyphosphoinositide metabolism to the hormonal activation of the insect salivary gland by 5-hydroxytryptamine. *Molec. Cell. Endocrinol. 36*, 37-42

Berridge, M.J., Dawson, R.M.C., Downes, C.P., Heslop, J.P. and Irvine, R.F. (1983) Changes in the levels of inositol phosphates after agonist-dependent hydrolysis of membrane phosphoinositides. *Biochem. J. 212*, 473-82

Berridge, M.J. and Heslop, J.P. (1982) Receptor mechanisms mediating the action of 5-hydroxytryptamine. In: *Neuropharmacology of insects*. Ciba Foundation Symposium, Vol. 88, pp. 260-70. Pitman, London

Berridge, M.J. and Irvine, R.F. (1984) Inositol trisphosphate a novel second messenger in cellular signal transduction. *Nature (Lond.) 312*, 315-21

Blaschko, H. and Hope, D.P. (1957) Observations on the distribution of amine oxidase in invertebrates. *Arch. Biochem. Biophys. 69*, 10-15

Bodnaryk, R.P. (1979a) Characterisation of an octopamine-sensitive adenylate cyclase from insect brain (*Mamestra configurato* Wlk). *Can. J. Biochem. 57*, 226-32

Bodnaryk, R.P. (1979b) Identification of specific dopamine- and octopamine-sensitive adenylate cyclases in the brain of *Mamestra configurata* Wlk. *Insect Biochem. 9*, 155-62

Bodnaryk, R.P. (1979c) Basal, dopamine and octopamine-stimulated adenylate cyclase activity in the brain of the moth, *Mamestra configurata*, during its metamorphosis. *J. Neurochem. 33*, 275-82

Boyer, J.L., Garcia, A., Posadas, C. and Garcia-Sainz, J.A. (1984) Differential effect of pertussis toxin on the affinity state for agonists of renal α_1- and α_2-adrenoreceptors. *J. Biol. Chem. 259*, 8076-9

Breer, H. and Knipper, M. (1985) Synaptosomes and neuronal membranes from insects. In: Breer, H. and Miller, T.A. (eds) *Neurochemical techniques in insect research*, pp. 125-55. Springer Verlag, Berlin

Brown, J.E., Rubin, L.J., Ghalayini, A.J., Tarver, A.P., Irvine, R.F., Berridge, M.J. and Anderson, R.E. (1984) Myo-inositol polyphosphate may be a messenger for visual excitation in *Limulus* photoreceptors. *Nature (Lond.) 311*, 160-2

Brunelli, M., Castellucci, V. and Kardel, E.R. (1976) Synaptic facilitation and behavioural sensitization in *Aplysia*: possible role of serotonin and cyclic AMP. *Science 194*, 1178-81

Burgisser, E., DeLean, A. and Lefkowitz, R.J. (1982) Reciprocal modulation of agonist and antagonist binding to muscarinic cholinergic receptors by guanine nucleotides. *Proc. Natl. Acad. Sci. USA* 79, 1734-6

Burke, D.E. and DeLorenzo, R.J. (1982) Ca^{2+} and calmodulin-dependent phosphorylation of endogenous synaptic vesicle tubulin by a vesicle-bound calmodulin kinase system. *J. Neurochem.* 38, 1205-17

Bustos, G. and Roth, R.H. (1979) Does cyclic AMP-dependent phosphorylation account for the activation of tyrosine hydroxylase produced by depolarization of dopaminergic neurons? *Biochem. Parmacol.* 28, 3026-8

Carlsson, A., Kerhr, W., Lindquist, M., Magnusson, T. and Atack, C.V. (1972) Regulation of monoamine metabolism in the central nervous system. *Pharmacol. Rev.* 24, 371-84

Chang, R., Tran, V.T. and Snyder, S.H. (1980) Neurotransmitter receptor localizations: brain lesion induced alterations in benzodiazepine, GABA, β-adrenergic and histamine H$_2$ receptor binding. *Brain Res.* 190, 95-110

Chesselet, M.F. (1984) Presynaptic regulation of neurotransmitter release in the brain. *Neuroscience 12*, 347-75

Cheung, W.Y. (1980) Calmodulin plays a pivotal role in cellular regulation. *Science 207*, 19-27

Cheung, W.Y. (1981) Discovery and recognition of calmodulin: a personal account. *J. Cyclic Nucleotide Res.* 7, 71-84

Chneiweiss, H., Prochiantz, A., Glowinski, J. and Premont, J. (1984) Biogenic amine-sensitive adenylate cyclase in primary cultures of neuronal or glial cells from mesencephalon. *Brain Res.* 302, 363-70

Cooper, J.R., Bloom, F.E. and Roth, R.H. (1986) Catecholamines I: General aspects. In: *The Biochemical basis of neuropharmacology*, 5th edn, pp. 203-58, Oxford University Press, New York and Oxford

Cooper, J.R. and Meyer, E.M. (1984) Possible mechanisms involved in the release and modulation of neuroactive agents. *Neurochem. Int.* 6, 419-33

Corda, M.G., Casu, M. and Biggio, G. (1979) Decrease of cyclic GMP in cerebellar cortex by intrastriatal (-)sulpiride. *Eur. J. Pharmacol.* 55, 327-30

Cottrell, G.A. and Powell, B. (1971) Formation of serotonin by isolated serotonin-containing neurones and by isolated non-amine-containing neurones. *J. Neurochem.* 18, 1695-7

Coyle, J.T. and Henry, D. (1973) Catecholamines in fetal and newborn rat brain. *J. Neurochem.* 21, 61-7

Cumming, R., Eccleston, D. and Steiner, A. (1977) Immunohistochemical localization of cyclic GMP in rat cerebellum. *J. Cyclic Nucleotide Res.* 3, 275-82

DeLorenzo, R.J. (1980) Role of calmodulin in neurotransmitter release and synaptic function. *Ann. NY Acad. Sci.* 356, 92-109

Dewhurst, S.A., Croker, S.G., Ikeda, K. and McCaman, R.E. (1972) Metabolism of biogenic amines in *Drosophila* nervous tissue. *Comp. Biochem. Physiol. 43B*, 975-81

Dolphin, A.C. and Greengard, P. (1981) Serotonin stimulates phosphorylation of Protein I in the facial motor nucleus of rat brain. *Nature (Lond.)* 289, 76-9

Donnellan, J.F., Alexander, K. and Chandeik, R. (1976) The isolation of cholinergic terminals from flesh fly heads. *Insect Biochem.* 6, 419-23

Drummond, A.H. (1985) Bidirectional control of cytosolic free calcium by thyrotropin-releasing hormone in pituitary cells. *Nature (Lond.)* 315, 752-5

Drummond, A.H., Bucher, F. and Levitan, I.B. (1980a) d-[^{3}H]Lysergic acid diethylamide binding to serotonin receptors in the molluscan nervous system. *J. Biol. Chem.* 255, 6679-86

Drummond, A.H., Bucher, F. and Levitan, I.B. (1980b) Distribution of serotonin and dopamine receptors in *Aplysia* tissues: analysis by [^{3}H]LSD-binding and adenylate cyclase stimulation. *Brain Res. 184*, 163-77

Drummond, A.H., Bucher, F. and Levitan, I.B. (1980c) Serotonin-induced hyperpolarization of an identified *Aplysia* neuron is mediated by cyclic AMP. *Proc. Natl Acad. Sci. USA 77*, 5013-17

Drummond, G.I. (1985) *Cyclic nucleotides in the nervous system.* Raven Press, New York

Dudai, Y. (1985) Analysis of receptors and binding sites in nervous tissue of insects. In: Breer, H. and Miller, T.A. (eds) *Neurochemical techniques in insect research*, pp. 79-101. Springer Verlag, Berlin

Dudai, Y. and Zvi, S. (1984a) High-affinity [³H]octopamine-binding sites in *Drosophila melanogaster*: interaction with ligands and relationship to octopamine receptors. *Comp. Biochem. Physiol. 77c*, 145-51

Dudai, Y. and Zvi, S. (1984b) [³H]-serotonin binds to two classes of sites in *Drosophila* head homogenates. *Comp. Biochem. Physiol. 77c*, 305-9

El-Fakahany, E. and Richelson, E. (1980) Regulation of muscarinic receptor-mediated cyclic GMP synthesis by cultured mouse neuroblastoma cells. *J. Neurochem. 35*, 941-8

El-Mestikawy, S., Glowinski, J. and Haman, M. (1983) Tyrosine hydroxylase activation in depolarized dopaminergic terminals — involvement of Ca^{2+}-dependent phosphorylation. *Nature (Lond.) 302*, 830-2

Elofsson, R., Laxmyr, L. Rosengren, E. and Hansson, C. (1982) Identification and quantitative measurements of biogenic amines and dopa in the CNS and haemolymph of the crayfish *Pacifastocus leniusculus* (Crustacea). *Comp. Biochem. Physiol. 71c*, 195-201

Emson, P.C., Burrows, M. and Fonnum, F. (1974) Levels of glutamate decarboxylase, choline acetyltransferase and acetylcholinesterase in identified motorneurons of the locust. *J. Neurobiol. 5*, 33-42

Enna, S.J. (1981) GABA receptors. *Trends Pharmacol. Sci. 2*, 62-6

Evans, P.D. (1978) Octopamine: a high-affinity uptake mechanism in the nervous system of the cockroach. *J. Neurochem. 30*, 1015-22

Evans, P.D. (1980) Biogenic amines in the insect nervous system. *Adv. Insect Physiol. 15*, 317-473

Evans, P.D. (1981) Multiple receptor types for octopamine in the locust *J. Physiol. (Lond.) 318*, pp. 99-122

Evans, P.D. (1982) Properties of modulatory octopamine receptors in the locust. In: *Neuropharmacology of insects.* Ciba Foundation Symposium, Vol. 88, pp. 48-62. Pitman, London

Evans, P.D., Davenport, A.P., Elias, M.S., Morton, D.B. and Triminer, B.A. (1985) Assays for biogenic amines in insect nervous tissue. In: Breer, H. and Miller, T.A. (eds) *Neurochemical techniques in insect research*, pp. 25-46. Springer-Verlag, Berlin

Evans, P.D., Kravitz, E.A. and Talamo, B.R. (1976) Octopamine release at two points along lobster nerve trunks. *J. Physiol. (Lond.) 262*, 71-89

Evans, P.D. and O'Shea, M. (1977) The identification of an octopaminergic neurone which modulates neuromuscular transmission in the locust. *Nature (Lond.) 270*, 257-9

Evans, P.D., Talamo, B.R. and Kravitz, E.A. (1975) Octopamine neurons: morphology, release of octopamine and possible physiological role. *Brain Res. 90*, 340-7

Evans, P.H. and Fox, P.M. (1975) Enzymatic N-acetylation of indolealkylamines by brain homogenates of the honey bee *Apis mellifera. J. Insect Physiol. 21*, 343-53

Evans, P.H., Soderlund, D.M. and Aldrich, J.R. (1980) *In vitro* N-acetylation of

166

biogenic amines by tissues of the European corn borer, *Ostrinia nubilalis* (Hüber). *Insect Biochem. 10*, 375-80

Falck, B. and Owman, C. (1965) A detailed methodological description of the fluorescence method for the cellular demonstration of biogenic amines. *Acta Univ. Lund. Sect. II No. 7*, 1-23

Fein, A., Payne, R., Carson, D.W., Berridge, M.J. and Irvine, R.F. (1984) Photoreceptor excitation and adaptation by inositol-1,4,5-trisphosphate. *Nature (Lond.) 311*, 157-60

Frazier, W.T., Kandel, E.R., Kupfermann, I., Waziri, R. and Coggeshall, R. (1967) Morphological and functional properties of identified neurons in the abdominal ganglion of *Aplysia californica. J. Neurophysiol. 30*, 1288-1351

Frontali, N. (1968) Histochemical localization of catecholamines in the brain of normal and drug-treated cockroaches. *J. Insect. Physiol. 14*, 881-6

Garcia, J.L., Haro, A. and Muncio, A.M. (1981) Regulation of adenylate cyclase from brain membranes of the insect *Ceratitis capitata. Comp. Biochem. Physiol. 70B*, 59-62

Gilman, A.G. and Schrier, B.K. (1971) Effect of catecholamines on the adenosine-3',5'-cyclic-monophosphate concentrations of clonal satellite cells of neurons. *Proc. Natl Acad. Sci. USA 68*, 2165-8

Gnegy, M. and Treisman, G. (1981) Effect of calmodulin on dopamine-sensitive adenylate cyclase activity in rat striatal membranes. *Molec. Pharmacol. 19*, 256-63

Goodhardt, M., Ferry, N., Geynet, P. and Hanoune, J. (1982) Hepatic α_1-adrenergic receptors show agonist-specific regulation by guanine nucleotides. Loss of nucleotide effect after adrenalectomy. *J. Biol. Chem. 257*, 11577-83

Grab, D.J., Carlin, R.K. and Siekevitz, P. (1981) Function of calmodulin in post synaptic densities II. Presence of a calmodulin-activatable protein kinase activity. *J. Cell Biol. 89*, 440-8

Greengard, P. (1978a) Phosphorylated proteins as physiological effectors. *Science 199*, 146-52

Greengard, P. (1978b) *Cyclic nucleotides, phosphorylated proteins and neuronal function.* Raven Press, New York

Haidamous, M., Kouyoumdjain, J.C., Briley, P.A. and Gonnard, P. (1980). *In vivo* effects of noradrenaline and noradrenergic receptor agonists and antagonists on rat cerebellar cyclic GMP levels. *Eur. J. Pharmacol. 63*, 287-94

Hamon, M., Bourgoin, S., Artaud, F. and Glowinski, J. (1979) The role of intra-neurol 5-HT and of tryptophan hydroxylase activation in the control of 5-HT synthesis in rat brain slices incubated in K⁺-enriched medium. *J. Neurochem. 33*, 1031-42

Harden, T.K. and McCarthy, K.D. (1982) Identification of the beta-andrenergic receptor subtype on astroglia purified from rat brain. *J. Pharmacol. Exp. Ther. 22*, 600-5

Harmar, A.J. and Horn, A.S. (1977) Octopamine-sensitive adenylate cyclase in cockroach brain: effects of agonists, antagonists and guanyl nucleotides. *Molec. Pharmacol. 13*, 512-20

Hayashi, S., Murdock, L.L. and Florey, E. (1977) Octopamine metabolism in invertebrates (*Locusta, Astacus, Helix*): evidence for N-acetylation in arthropod tissues. *Comp. Biochem. Physiol. 58c*, 183-91

Heideman, W., Wierman, B.M. and Storm, D.R. (1982) GTP is not required for calmodulin stimulation of bovine brain adenylate cyclase. *Proc. Natl Acad. Sci. USA 79*, 1462-5

Hildebrand, J.G., Barker, D.L., Herbert, E. and Kravitz, E.A. (1971) Screening for neurotransmitters: a rapid radiochemical procedure *J. Neurobiol.* 2, 231-46

Hiripi, L. and S-Rosa, K.C. (1973) Fluorimetric determination of 5-hydroxy-tryptamine and catecholamines in the central nervous system and heart of *Locusta migratoriodes. J. Insect Physiol.* 19, 1481-5

Hökfelt, T. (1966) The effect of reserpine on the intraneuronal vesicles of the rat vas deferens, *Experientia 22*, 56-7

Hollingworth, R.M. (1976) Chemistry, biological activity and uses of formamidine pesticides. *Environ. Hlth Persp. 14*, 57-69

Hornykiewicz, O. (1981) Importance of topographic neurochemistry studying neurotransmitter systems in human brain: critique and new data. In: Riederer, P. and Usdin, E. (eds) *Transmitter biochemistry of human brain tissue*, pp. 9-24. Macmillan, London

Houk, E.J. and Beck, S.D. (1978) Monoamine oxidase in the brain of European corn borer larvae, *Ostrinia nubilalis* (Hüber). *Insect Biochem. 8*, 231-6

Hoyle, G. and Barker, D.L. (1975) Synthesis of octopamine by insect dorsalmedian unpaired neurons. *J. Exp. Zool. 193*, 433-9

Hoyle, G., Dagan, D., Moberly, B. and Colquhoun, W. (1974) Dorsal unpaired median insect neurons make neurosecretory endings on skeletal muscle. *J. Exp. Zool. 187*, 159-65

Huttner, W.B. and Greengard, P. (1979) Multiple phosphorylation sites in protein 1 and their differential regulation by cAMP and calcium. *Proc. Natl Acad. Sci. USA 76*, 5402-6

Huttner, W.B., Schiebler, W., Greengard, P. and De Camilli, P. (1983) Synapsin I (protein I), a nerve terminal-specific phosphoprotein: III. Its association with synaptic vesicles studied in a highly-purified vesicle preparation. *J. Cell Biol. 96*, 1374-88

Hulme, E.C., Berrie, C.P., Birdsall, N.J. and Burgen, A.S.V. (1981) Interactions of muscarinic receptors with guanine nucleotides and adenylate cyclase. In: Birdsall, N.J.M. (ed.) *Drug receptors and their effectors*, pp. 23-34. Macmillan, London

Iversen, L.L. (1985) Brain cells converse with each other through chemistry. And their vocabulary of neurotransmitters is turning out to be extraordinarily diverse. *New Scientist 106*, 11-14

Kajikawa, N., Kaibuchi, K., Matsubara, T., Kikkawa, U., Takai, Y. and Nishizuka, Y. (1983) A possible role of protein kinase C in signal-induced lysosomal enzyme release. *Biochem. Biophys. Res. Commun. 116*, 743-50

Kandel, E.R. and Schwartz, J.H. (1982) Molecular biology of an elementary form of learning: modulation of transmitter release through cAMP-dependent protein kinase. *Science 218*, 433-43

Kant, G.J., Meyerhoff, J.L. and Lenox, R.H. (1980) *In vivo* effects of apomorphine and 4-(3-butoxy-4-methoxybenzyl)-2-imidazolidinone (RO 20-1724) on cyclic nucleotides in rat brain and pituitary. *Biochem. Pharmacol. 29*, 369-73

Karnusina, I., Suzuki, R., Padgett, W. and Daly, W. (1983) Degeneration of CAI neurons in hippocampus after ischemia in Mongolian gerbils: cyclic AMP-systems. *Brain Res. 268*, 87-94

Kaufman, S. (1974) Properties of the pterin-dependent aromatic amino acid hydroxylases. In: Wolstenholme, G.E.W. and Fitzsimons, D.W. (eds) *Aromatic amino acids in the brain*, Ciba Foundation Symposium 22, pp. 85-108. Elsevier, Amsterdam

Kelly, L.E. (1981) The regulation of protein phosphorylation in synaptosomal fractions from *Drosophila* heads: the role of cyclic adenosine monophosphate and calcium/calmodulin. *Comp. Biochem. Physiol. 68B*, 61-7

Kelly, P.T., Cotman, C.W. and Largen, M. (1979) Cyclic AMP-stimulated protein kinases at brain synaptic junctions. *J. Biol. Chem. 254*, 1564-75

Kennedy, M.B. (1978) Products of biogenic amine metabolism in the lobster: sulphate conjugates. *J. Neurochem. 30*, 315-20

Kennedy, M.B. and Greengard, P. (1981) Two calcium/calmodulin-dependent protein kinases, which are highly concentrated in brain, phosphorylate protein I at distinct sites. *Proc. Natl Acad. Sci. USA 78*, 1293-7

Kilpatrick, A.T. (1981) Monoamine-sensitive adenylate cyclase in the nervous system of the locust *Schistocerca gregaria*. Ph.D. Thesis, University of Glasgow

Kilpatrick, A.T., Vaughan, P.F.T. and Donnellan, J.F. (1980) Monoamine sensitive adenylate cyclase in *Schistocerca gregaria* nervous tissue. In: *Insect neurobiology and pesticide action*, pp. 341-5. Proc. Soc. Chem. Ind., London

Kilpatrick, A.T., Vaughan, P.F.T. and Donnellan, J.F. (1982) The effect of guanyl-nucleotides on the monoamine-sensitive adenylate cyclase of *Schistocerca gregaria* nervous tissue. *Insect Biochem. 12*, 393-7

Kingan, T.G. and Hildebrand, J.G. (1985) Screening and assays for neurotransmitters in the insect nervous system. In: Breer, H. and Miller, T.A. (eds) *Neurochemical techniques in insect research*, pp. 79-101. Springer-Verlag, Berlin

Klemm, N. (1976) Histochemistry of putative transmitter substances in the insect brain. *Progr. Neurobiol. 7*, 99-169

Klemm, N. and Schneider, L. (1975) Selective uptake of indolamine into nervous fibres in the brain of the desert locust *Schistocerca gregaria* Forskal (Insecta). A fluorescence and electron microscopic investigation. *Comp. Biochem. Physiol. 50c*, 177-82

Knight, D.E. and Baker, P.F. (1983) The phorbal ester TPA increases the affinity of exocytosis for calcium in 'leaky' adrenal medullary cells. *FEBS Lett. 160*, 98-100

Kojima, I., Lippes, H., Kojima, K. and Rasmussen, H. (1983) Aldosterone secretion: effect of phorbal ester and A23187. *Biochem. Biophys. Res. Commun. 116*, 555-62

Krebs, E.G, and Beavo, J. (1979) Phosphorylation–dephosphorylation of enzymes *Ann. Rev. Biochem. 48*, 923-59

Krueger, B.K., Forn, J. and Greengard, P. (1977) Depolarization induced phosphorylation of specific proteins, mediated by calcium ion influx, in rat brain synaptosomes. *J. Biol. Chem. 252*, 2764-73

Kuo, J.F. and Greengard, P. (1970) Stimulation of adenosine-3′,5′-monophosphate-dependent and guanosine-3′,5′-monophosphate-dependent protein kinases by some analogues of adenosine-3′,5′-monophosphate. *Biochem. Biophys. Res. Commun. 40*, 1032-8

Kuo, J.F., Wyatt, G.R. and Greengard, P. (1971) Cyclic nucleotide-dependent protein kinases. ix. Partial purification and some properties of guanosine 3′,5′-monophosphate-dependent and adenosine 3′,5′-monophosphate-dependent protein kinases from various tissues and species of arthropoda. *J. Biol. Chem. 246*, 7159-67

Kupferman, I. (1980) Role of cyclic nucleotides in excitable cells. *Ann. Rev. Physiol. 42*, 629-41

Lafon-Cazal, M. and Bockaert, J. (1984) Pharmacological characterization of dopamine-sensitive adenylate cyclase in the salivary glands of *Locusta migratoria* L. *Insect Biochem. 14*, 541-6

Laxmyr, L. (1984) Biogenic amines and dopa in the central nervous system of decapod crustaceans. *Comp. Biochem. Physiol. 77c*, 139-43

Lazar, M.A., Lockfeld, A.J., Truscott, R.J.W. and Barchas, J.D. (1982) Tyrosine hydroxylase from bovine striatum: catalytic properties of the phosphorylated and non phosphorylated forms of the purified enzyme. *J. Neurochem. 39*, 409-22

Lemos, J.R., Novak-Hofer, I. and Levitan, I.B. (1982) Serotonin alters the phosphorylation of specific proteins inside a single living nerve cell. *Nature (Lond.)* *298*, 65-6

Lent, C.M. (1982) Serotonin-containing neurones within the segmental nervous system of the leech. In: Osborne, N.N. (ed.) *Biology of serotonergic transmission*, pp. 431-56. John Wiley, New York

Levitan, I.B. and Barondes, S.H. (1974) Octopamine- and serotonin-stimulated phosphorylation of specific protein in the abdominal ganglia of *Aplysia californica*. *Proc. Natl Acad. Sci. USA 71*, 1145-8

Levitan, I.B., Marsden, C.J. and Barondes, S.H. (1974) Cyclic AMP and amine effects on phosphorylation of specific protein in abdominal ganglion of *Aplysia californica*; localization and kinetic analysis. *J. Neurobiol. 5*, 511-25

Lincoln, T.M. and Carlin, J.D. (1983) Characterization and biological role of the cGMP dependent protein kinase. In: Greengard, P. and Robison, G.A. (eds) *Advances in cyclic nucleotides research*, Vol. 15, pp. 139-92. Raven Press, New York

Livingston, M.S. and Tempel, B.L. (1983) Genetic dissection of monoamine neurotransmitter synthesis in *Drosophila*. *Nature (Lond.) 303*, 67-70

Lunt, G.G. (1985) Analysis of neurochemical data. In: Breer, H. and Miller, T.A. (eds) *Neurochemical techniques in insect research*, pp. 296-316. Springer-Verlag, Berlin

Lykouras, E., Eccleston, D. and Marshall, E.F. (1980) The effect of a 5HT agonist on cyclic guanosine monophosphate in rat cerebellum. *Biochem. Pharmacol. 29*, 827-8

McNall, S.J. and Mansour, T.E. (1985) Forskolin activation of serotonin-stimulated adenylate cyclase in the liver fluke *Fasciola hepatica*. *Biochem. Pharmacol. 34*, 1683-8

Malkinson, A.M. (1975) Effect of calcium on cyclic AMP-dependent and cyclic GMP-dependent endogenous protein phosphorylation in mouse brain cytosol. *Biochem. Biophys. Res. Commun. 67*, 752-9

Mancini, G. and Frontali, N. (1970) On the ultrastructural localization of catecholamines in the beta lobes (corpora pedunculata) of *Periplaneta americana*. *Z. Zellforsch. mikrosk. Anat. 103*, 341-50

Maranda, B. and Hodgetts, R. (1977) A characterization of dopamine acetyltransferase in *Drosophila melanogaster*. *Insect Biochem. 7*, 33-43

Maxwell, D.J. (1978) Fine structure of axons associated with the salivary apparatus of the cockroach, *Nauphoeta cinerea*. *Tissue and Cell 10*, 699-706

Maxwell, G.D., Tait, J.F. and Hildebrand, J.G. (1978) Regional synthesis of neurotransmitter candidates in the CNS of the moth, *Manduca sexta*. *Comp. Biochem. Physiol. 61c*, 109-19

Mir, A.K. and Vaughan, P.F.T. (1981a) Biosynthesis of N-acetyldopamine and N-acetyloctopamine by *Schistocerca gregaria* nervous tissue. *J. Neurochem. 36*, 441-6

Mir, A.K. and Vaughan, P.F.T. (1981b) The conversion of N-acetyltyramine to N-acetyldopamine by *Schistocerca gregaria* thoracic ganglia. *Insect Biochem. 11*, 571-7

Molinoff, P.B. and Axelrod, J. (1972) Distribution and turnover of octopamine in tissues. *J. Neurochem. 19*, 157-63

Morton, D.B. (1984) Pharmacology of the octopamine-stimulated adenylate cyclase of the locust and tick CNS. *Comp. Biochem. Physiol, 78c*, 153-8

Murdock, L.L. (1971) Catecholamines in arthropods: a review. *Comp. Gen. Pharmacol. 2*, 254-74

170

Murdock, L.L. and Hollingworth, R.M. (1980) Octopamine-like actions of formamidines in firefly light organs. In: *Insect neurobiology and pesticide action*, pp. 341-5. Proc. Soc. Chem. Ind., London

Murdock, L.L. and Omar, D. (1981) N-acetyldopamine in insect nervous tissue. *Insect Biochem. 11*, 161-6

Murdock, L.L., Wirtz, R.A. and Kohler, G. (1973) 3,4-dihydroxyphenylalanine (DOPA) decarboxylase activity in the arthropod nervous system. *Biochem. J. 132*, 681-8

Nässel, D.R. and Laxmyr, L. (1983) Quantitative determination of biogenic amines and DOPA in the CNS of adult and larval blow flies, *Calliphora erythrocephala. Comp. Biochem. Physiol. 75c*, 259-65

Nathanson, J.A. (1977) Cyclic nucleotides and nervous system function. *Physiol. Rev. 57*, 157-256

Nathanson, J.A. and Greengard, P. (1973) Octopamine-sensitive adenylate cyclase: evidence for a biological role of octopamine in nervous tissue. *Science 180*, 308-10

Nathanson, J.A. and Greengard, P. (1974) Serotonin-sensitive adenylate cyclase in neural tissue and its similarity to the serotonin receptor: a possible site of action of lysergic acid diethylamide. *Proc. Natl Acad. Sci USA 71*, 797-801

Nelson, D.L. and Molinoff, P.B. (1976) Distribution and properties of adrenergic storage vesicles in nerve terminals. *J. Pharmacol. Exp. Ther. 196*, 346-59

Nestler, E.J. and Greengard, P. (1982a) Distribution of protein I and regulation of its state of phosphorylation in the rabbit superior cervical ganglion. *J. Neurosci. 2*, 1011-23

Nestler, E.J. and Greengard, P. (1982) Nerve impulses increase the phosphorylation state of protein I in rabbit superior cervical ganglion. *Nature (Lond.) 296*, 452-4

Nestler, E.J. and Greengard, P. (1983) Protein phosphorylation in the brain. *Nature (Lond.). 305*, 583-8

Nishizuka, Y. (1983) Calcium, phospholipid turnover and transmembrane signalling. *Phil. Trans. R. Soc. B 302*, 101-12

Nishizuka, Y. (1984) The role of protein kinase C in cell surface signal transduction and tumour promotion. *Nature (Lond.) 308*, 693-8

Novak-Hofer, I. and Levitan, I.B. (1983) Ca^{2+}/calmodulin regulated protein phosphorylation in the *Aplysia* nervous system. *J. Neurosci. 3*, 473-81

Oertel, D., Linberg, K.A. and Case, J.F. (1975) Ultrastructure of the larval firefly light organ as related to control of light emission. *Cell Tiss. Res. 164*, 27-44

Omar, D., Murdock, L.L. and Hollingworth, R.M. (1982) Actions of pharmacological agents on 5-hydroxytryptamine and dopamine in the cockroach nervous system *Periplaneta americana* L. *Comp. Biochem. Physiol. 73c*, 423-9

Osborne, N.N. (1972) The *in vivo* synthesis of serotonin in an identified serotonin-containing neuron of *Helix pomatia. Int. J. Neurosci. 3*, 215-19

Osborne, N.N. (1973) Tryptophan metabolism in characterised neurons of *Helix. Br. J. Pharmacol. 48*, 546-9

Osborne, N.N. (1977) Adenosine-3'5'-monophosphate in snail (*Helix pomatia*) nervous system: analysis of dopamine receptors. *Experientia 33*, 917-19

Osborne, N.N., Hiripi, L. and Neuhoff, V. (1975a) The *in vitro* uptake of biologic amines by snail (*Helix pomatia*) nervous tissue. *Biochem. Pharmacol. 24*, 2141-8

Osborne, N.N. and Neuhoff, V. (1974) Formation of serotonin in insect (*Periplaneta americana*) nervous tissue. *Brain Res. 74*, 366-9

Osborne, N.N., Priggemeier, E. and Neuhoff, V. (1975b) Dopamine metabolism in characterised neurones of *Planorbis corneus. Brain Res. 90*, 261-71

171

Pau, R.N. and Kelly, C. (1975) The hydroxylation of tyrosine by an enzyme from the third-instar larvae of the blowfly *Calliphora erythrocephala*. *Biochem. J.* *147*, 565-73

Pellmar, T.C. (1981) Transmitter-induced calcium current. *Fed. Proc.* *40*, 2631-6

Pellmar, T.C. and Carpenter, D.O. (1980) Serotonin induces a voltage-sensitive calcium current in neurons of *Aplysia californica*. *J. Neurophysiol.* *44*, 423-39

Peroutka, S.J. and Snyder, S.H. (1979) Multiple serotonin receptors; differential binding of [3H]5-hydroxytryptamine, [3H]lysergic acid diethylamide and [3H]spiroperidol. *Molec. Pharmacol.* *16*, 687-99

Pollock, R.J., Kapatos, G. and Kaufman, S. (1981) Effect of cyclic AMP-dependent protein phosphorylating conditions on the pH-dependent activity of tyrosine hydroxylase from beef and rat striata. *J. Neurochem.* *37*, 855-60

Powell, B. and Cottrell, G.A. (1974) Dopamine in an identified neuron of *Planorbis corneus*. *J. Neurochem.* *22*, 605-6

Ram, J.L. and Ehrlich, Y.H. (1978) Cyclic GMP-stimulated phosphorylation of membrane-bound proteins from nerve roots of *Aplysia californica*. *J. Neurochem.* *30*, 487-91

Robertson, H.A. (1974) The innervation of the salivary gland of the moth *Manduca sexta*. *Cell Tiss. Res.* *148*, 237-45

Robertson, H.A. (1976) Octopamine, dopamine and noradrenaline content of the brain of the locust, *Schistocerca gregaria*. *Experientia 32*, 552-4

Robertson, H.A. and Juorio, A.V. (1976) Octopamine and some related non-catecholic amines in invertebrate nervous systems. *Int. Rev. Neurobiol.* *19*, 173-224·

Robertson, H.A. and Osborne, N.N. (1979) Putative neurotransmitters in the annelid (*Lumbricus terrestris*) central nervous system: presence of 5-hydroxy-tryptamine and octopamine-stimulated adenylate cyclases. *Comp. Biochem. Physiol.* *64c*, 7-14

Robinson, N.L., Cox., P.M. and Greengard, P. (1982) Glutamate regulates adenylate and guanylate cyclase activities in an isolated membrane preparation from insect muscle. *Nature* (Lond.) *296*, 354-6

Ross, E.M. and Gilman, A.G. (1980) Biochemical properties of hormone-sensitive adenylate cyclase. *Ann. Rev. Biochem.* *49*, 533-64

Rotondo, D. (1984) The effect of octopamine, cyclic adenosine 3',5'-monophosphate and calcium on protein phosphorylation in *Schistocerca gregaria* central nervous system. Ph.D Thesis, University of Glasgow

Rotondo, D., Vaughan, P.F.T. and Donnellan J.F. (1987a) Octopamine and cyclic AMP stimulate protein phosphorylation in the CNS of *Schistocerca gregaria*. *Insect Biochem.* *17*, 283-90

Rotondo, D., Vaughan, P.F.T. and Donnellan, J.F. (1987b) A study of cyclic AMP-dependent protein phosphorylation in *Schistocerca gregaria* CNS: a comparison to that in mammalian CNS. *Comp. Biochem. Physiol.* in press

Routtenberg, A., Morgan, D., Conway, R.G., Schmidt, M.S. and Ghetti, B. (1981) Human brain protein phosphorylation *in vitro* cyclic AMP stimulation of electro-phoretically separated substrates. *Brain Res.* *222*, 323-33

Rutschke, E. and Thomas, H. (1975) Histochemical and ultrastructural investi-gations on occurrence of catecholamines in the deutocerebrum of the cockroach, *Periplaneta americana* L. *Zool. Jb. Anat.* *94*, 474-98

Saavedra, J.M., Brownstein, M.J., Carpenter, D.O. and Axelrod, J. (1974) Octo-pamine: presence in single neurons of *Aplysia* suggests neurotransmitter function. *Science 185*, 364-5

Saller, C.G. and Salaman, A.I. (1984) Dopamine synthesis in synaptosomes:

relation of autoreceptor functioning to pH, membrane depolarisation and intra-synaptosomal dopamine content. *J. Neurochem. 43*, 675-88

Salter, R.S., Krinks, M.H., Klee, C.B. and Neer, E.J. (1981) Calmodulin activates the isolated catalytic unit of brain adenylate cyclase. *J. Biol. Chem. 256*, 9830-3

Sano, K., Takai, Y., Yamanishi, J. and Nishizuka, Y. (1983) A role of calcium-activated phospholipid dependent protein kinase in human platelet activation. *J. Biol. Chem. 258*, 2010-13

Schoeffeniels, E. and Dandrifosse, G. (1980) Protein phosphorylation and sodium conductance in nerve membrane. *Proc. Natl Acad. Sci USA 77*, 812-16

Seaman, K. and Daly, J.W. (1982) Calmodulin stimulation of adenylate cyclase in rat brain membranes does not require GTP. *Life Sciences 30*, 1457-64

Sekeris, C.E. and Karlson, P. (1966) Biosynthesis of catecholamines in insects. *Pharmacol. Rev. 18*, 89-94

Sieghart, W., Forn, J. and Greengard, P. (1979) Ca^{2+} and AMP regulate phosphor-ylation of some two membrane-associated proteins specific to nerve tissue. *Proc. Natl Acad. Sci. USA 76*, 2475-9

Singh, S.J.P., Orchard, I. and Loughton, B.G. (1981) Octopamine actions of formamidines on hormone release in the locust *Locusta migratoria. Pest. Biochem. Physiol. 16*, 249-55

Sloley, B.D. and Owen, M.D. (1982) The effects of reserpine on amine concen-trations in the nervous system of the cockroach (*Periplaneta americana*). *Insect Biochem. 12*,469-76

Stadel, J.M., DeLean, A. and Lefkowitz, R.J. (1982) Molecular mechanisms of coupling in hormone receptor-adenylate cyclase systems. *Adv. Enzymol. 53*, 1-43

Stefano, G.B. and Aiello, E. (1975) Histofluorescent localization of serotonin and dopamine in the nervous system and gill of *Mytilus edulis* (Bivalvia). *Biol. Bull. 148*, 141-56

Stuart, A.E., Hudspeth, A.J. and Hall, Z.W. (1974) Vital staining of specific monoamine-containing cells in the leech nervous system. *Cell Tiss. Res. 153*, 55-61

Taylor, D.P., Dyer, K.A, and Newburgh, R.W. (1978) Cyclic nucleotides in neuronal and glial-enriched fractions from the nerve cord of *Manduca sexta. J. Insect. Physiol. 22*, 1303-4

Thomas, A.P., Alexander, J. and Williamson, J.R. (1984) Relationship between inositol polyphosphate production and the increase of cytosolic free Ca^{2+} induced by vasopressin in isolated hepatocytes. *J. Biol. Chem. 259*, 5574-84

Treistman, S.W. and Levitan, I.B. (1986) Alteration of electrical activity in mollus-can neurones by cyclic nucleotides and peptide factors. *Nature (Lond.) 261*, 62-4

Ueda, T. and Greengard, P. (1977) Adenosine 3′,5′-monophosphate-regulated phosphorylation system of neuronal membranes. 1. Solubilization, purification and some properties of an endogenous phosphoprotein. *J. Biol. Chem. 252*, 5155-63

Ueda, T., Maeno, H. and Greengard, P. (1973) Regulation of endogenous phospho-rylation of specific proteins in synaptic membrane fractions in rat brain by adeno-sine 3′,5′-monophosphate. *J. Biol. Chem. 248*, 8295-305

Ungerstedt, U. (1971) Stereotaxic mapping of the monoamine pathways in the rat brain. *Acta Physiol. Scand. Suppl. 367*, 1-48

Uzzan, A. and Dudai, Y. (1982) Aminergic receptors in *Drosophila melanogaster*: responsiveness of adenylate cyclase to putative neurotransmitters. *J. Neurochem, 38*, 1542-50

Vandenberg, C.A. and Montal, M. (1984) Light-regulated biochemical events in invertebrate photoreceptors. 2. Light-regulated phosphorylation of rhodopsin and phosphoinositides in squid photoreceptor membranes. *Biochemistry 23*, 2347-52

Vaughan, P.F.T. and Neuhoff, V. (1976) The metabolism of tyrosine, tyramine and 3,4-dihydroxyphenylalanine by thoracic and cerebral ganglia of the locust (*Schistocerca gregaria*). *Brain Res. 117*, 175-80

Walaas, S.I., Aswad, D.W. and Greengard, P. (1983c) A dopamine- and cyclic AMP-regulated phosphoprotein enriched in dopamine-innervated brain regions. *Nature (Lond.) 301*, 69-71

Walaas, S.I. and Greengard, P. (1984) DARPP-32, a dopamine- and adenosine 3',5'-monophosphate-regulated phosphoprotein enriched in dopamine-innervated brain regions. 1. Regional and cellular distribution in the rat brain. *J. Neurosci. 4*, 84-98

Walaas, S.I., Nairn, A.C. and Greengard, P. (1983a) Regional distribution of calcium- and cyclic adenosine-3',5'-monophosphate-regulated protein phosphorylation systems in mammalian brain I. Particulate systems. *J. Neurosci. 3*, 291-301

Walaas, S.I., Nairn, A.C. and Greengard, P. (1983b) Regional distribution of calcium- and cyclic adenosine-3',5'-monophosphate-regulated protein phosphorylation systems in mammalian brain II. Soluble systems. *J. Neurosci. 3*, 302-11

Wallace, B.G. (1976) The biosynthesis of octopamine–characterization of lobster tyramine β-hydroxylase. *J. Neurochem. 26*, 761-70

Walter, U. and Greengard, P. (1981) Cyclic AMP-dependent and cyclic GMP-dependent protein kinases of nervous tissue. *Curr. Top. Cell Reg. 19*, 219-56

Walter, U., Kanof, P., Schulman, H. and Greengard, P. (1978) Adenosine 3',5'-monophosphate receptor proteins in mammalian brain. *J. Biol. Chem. 253*, 6275-80

Walter, U., Miller, P., Wilson, F., Menkes, D. and Greengard, P. (1980) Immunological distinction between guanosine 3',5'-monophosphate-dependent and adenosine 3',5'-monophosphate-dependent protein kinases. *J. Biol. Chem. 255*, 3757-62

Weinreich, D., McCaman, M.W., McCaman, R.E. and Vaughn, J.E. (1973) Chemical, enzymatic and ultrastructural characterization of 5-hydroxytryptamine containing neurones from the ganglia of *Aplysia californica* and *Tritonia diomedia*. *J. Neurochem. 20*, 969-76

Welsh, J.H. (1972) Catecholamines in the invertebrates. In: Blaschko, H. and Muscholl, E. (eds) *Catecholamines, Vol. 33 Handb. Exp. Pharmakol.* pp. 79-109. Springer-Verlag, Berlin

Whitehead, D.L. (1969) New evidence for the control mechanism of sclerotization in insects. *Nature (Lond.) 224*, 721-3

Wood, J.G., Wallace, R., Whittaker, J. and Cheung, W.Y. (1980) Immunocytochemical localization of calmodulin in regions of rodent brain. *Ann. NY Acad. Sci. 356*, 75-82

Wu, C.F., Berneking, J.M. and Barker, D.L. (1983) Acetylcholine synthesis and accumulation in the CNS of *Drosophila* larvae: analysis of shibire[ts] a mutant with a temperature-sensitive block in synaptic transmission. *J. Neurochem. 40*, 1386-96

Zawalich, W., Brown, C. and Rasmussen, H. (1983) Insulin secretion: combined effects of phorbol esters and A23187. *Biochem. Biophys. Res. Commun. 117*, 448-55

5

Invertebrate Neuropeptides

Nicholas Platt and Stuart E. Reynolds

INTRODUCTION

Peptides are now recognised as by far the largest and most structurally diverse class of neuroregulatory substances (Snyder, 1980; Iversen, 1983; Krieger, 1983). This is true not only for vertebrates (Krieger, 1983, listed 38 known brain peptides) but also for invertebrates (see Table 5.1). The realisation that peptides play an important role within the brain has come only recently, being largely a result of improved techniques for physically handling and chemically manipulating the tiny quantities of peptides characteristically present in nervous tissue. Of course, peptides have been studied for much longer than this in their role as neurohormones. Indeed it is true to say that we still know most about those neuropeptides (vaso-pressin is a vertebrate example) that have neurohormonal functions, many (if not all) of which also have modulatory or transmitter roles within the brain. However, it is now clear that neuropeptides also exist (e.g. substance P) that have no known function as circulating regulatory agents.

Why study invertebrate neuropeptides? In terms of basic science the reward for using invertebrate preparations is the small size (i.e. fewer neurones) of the invertebrate nervous system, and its simpler organisation. Frequently invertebrate neurones are large in size and uniquely identifiable from preparation to preparation, offering experimental opportunities to both physiologist and biochemist that are rarely, if ever, paralleled in vertebrate preparations. These opportunities are particularly important because knowledge of neuropeptide function within the vertebrate brain has lagged behind appreciation of the chemical diversity of neuropeptides, their widespread occurrence, and even the structure of their genes and details of their biosynthesis. This is a direct result of the complexity of the vertebrate CNS and the inaccessibility of peptidergic neurones within it. As has been the case in the past, study of particularly favourable experimental preparations will be of paramount importance in elucidating physiological principles. In the case of brain peptides we are confident that inverte-

175

Table 5.1: Fully and partially characterised invertebrate neuropeptides

Peptide	Sequence	Reference
Coelenterates		
Hydra head activator	pQPPGGSKVILF	Schaller and Bodenmuller (1981)
Arthropods: Crustaceans		
Red pigment concentrating hormone (RPCH)	pQLNFSPGW$_{NH_2}$	Fernlund and Josefsson (1972)
Distal retinal pigment hormone (DRPH) *(Pandalus)*	NSGMINSILGIPRVTMEA$_{NH_2}$	Fernlund (1986)
DRPH *(Uca)*	NSGLINSILGLPLVMNDA$_{NH_2}$	Rao *et al.* (1985)
Crustacean cardioactive peptide (CCAP)	PFCNAFTGC$_{NH_2}$	Stangier *et al.* (1986)
Arthropods: Insects		
Proctolin	RYLPT	Starratt and Brown (1975)
(Adipokinetic hormone AKH-I)	pQLNFTPNWGT$_{NH_2}$	Stone *et al.* (1976)
AKH-II *(Locusta)*	pQLNFSTGW$_{NH_2}$	Siegert *et al.* (1985)
AKH-II *(Schistocerca)*	pQLNFSAGW$_{NH_2}$	Gäde *et al.* (1986)
AKH-*Carausius*-II	pQLTFTPNWGT$_{NH_2}$	Gäde (1986)
MI (HGH-I, CC-I neurohormone D)	pQVNFSPNW$_{NH_2}$	Witten *et al.* (1984)
MII (HGH-II, CC-II)	pQLTFTPNW$_{NH_2}$	Scarborough *et al.* (1984)
AKH: *Manduca* *(Heliothis)*	pQLTFTSSWG$_{NH_2}$	Ziegler *et al.* (1985) Jaffe *et al.* (1986)
Leucopyrokinin (LPK)	pQTSFTPRL$_{NH_2}$	Holman *et al.* (1986a)
Leucokinin I	DPAFNSWG$_{NH_2}$	Holman *et al.* (1986b)
Leucokinin II	DPGFSSWG$_{NH_2}$	Holman *et al.* (1986b)
Leucokinin III	DQGFNSWG$_{NH_2}$	Holman *et al.* (1986c)
Leucokinin IV	DASFHSWG$_{NH_2}$	Holman *et al.* (1986c)
Leucomyosupressin	pQDVDHVFLRF$_{NH_2}$	Holman *et al.* (1986d)
Leucosulfakinin-I	EEFEDY (SO$_3$H) GHMRF$_{NH_2}$	Nachman *et al.* (1986a)
Leucosulfakinin-II	pQSDDY (SO$_3$H) GHMRF$_{NH_2}$	Nachman *et al.* (1986b)
Bombyx 4K PTTH-I[a]	GVVDECCFRPCTLDVLLSY-	Nagasawa *et al.* (1984)
4K PTTH-II[a]	GIVDECCLRPCSVDVLLSY-	Nagasawa *et al.* (1984)
4K PTTH-III[a]	GVVDECCLZPCT?DVVATY-	Nagasawa *et al.* (1984)
Bombyx eclosion hormone[a]	SPAIASSYDAMEI-	Nagasawa *et al.* (1985)
Bombyx MRCH-I[a]	LSEDMPATPADQEMYQ-	Matsumoto *et al.* (1985)
MRCH-II[a]	??EDMPATPADQEMYQ-	Matsumoto *et al.* (1986)
MRCH-III[a]	PLSEDMPATPADQEMY-	Matsumoto *et al.* (1985)
Molluscs: Bivalves		
FMRFamide	FMRF$_{NH_2}$	Price and Greenberg (1977)
Molluscs: Gastropods		
Helix FMRFamide-like peptide	pQDPFLRF$_{NH2}$	Price *et al.* (1983)
Met-enkephalin	YGGFM	Leung and Stefano (1984)
Leu-enkephalin	YGGFL	Leung and Stefano (1984)

Met-enkephalin Arg[6]Phe[7]	YGGFMRF	Stefano and Leung (1984)
Met-enkephalin Arg[6]Phe[7] amide	YGGFMRF$_{NH_2}$	Kiehling et al. (1984)
Aplysia small cardioactive peptide A (SCP$_A$)[b]	ARPGYLAFPRM	Mahon et al. (1985)
SCP$_B$	MNYLAFPRM	Morris et al. (1982)
Aplysia egg-laying hormone (ELH)	ISINQDLKAITDMLLTDQIRGRQ RYLADLRQRLLDK$_{NH_2}$	Chiu et al. (1979)
Aplysia bag-cell peptides:		
α-BCP[b]	APRLRFYSL	Rothman et al. (1983)
α-BCP[b]	RLRFH	Scheller et al. (1983)
α-BCP[b]	RLRFD	Scheller et al. (1983)
acidic peptide	SSGVSLLTSNLDEEQRELL-KAISNLLD	Scheller et al. (1983)
Aplysia atrial gland peptides:		
peptide A	AVKLSSDGNYPFDLSKEDGAQ PYFMTPRLRFYPI$_{NH_2}$	Heller et al. (1980)
peptide B	AVKSSSYGKYPFDLSKEDGAQ PYFMTPRLRFYPI$_{NH_2}$	Heller et al. (1980)
Egg-releasing hormone I	IVSLFKAITDMLLTEQIYA NYFSTPRLRFYPI	Schlesinger et al. (1981)
Lymnaea CDCH[c]	LSITNDLRAIADSYLYDQHWLRE RQEENLRRRFLEL$_{NH_2}$	Ebberink et al. (1985)
Aplysia R3-R8, R14 peptides:		
peptide I[b]	AWSEEVFDDTDVGDGL	Nambu et al. (1983)
peptide II[b]	EAEEPSAFMTRL	Nambu et al. (1983)
peptide III[b]	YGGGHLSDA	Nambu et al. (1983)

[a] Partial sequence (N-terminal).
[b] Inferred from cDNA nucleotide sequence.
[c] C-terminal amidation probable.

brates will continue to provide many of the model systems that illustrate the principles of peptide action. An extremely lucid review of some of the progress that has been made already is given by O'Shea and Schaffer (1985). Unfortunately, progress in studying invertebrate neuropeptide physiology has undoubtedly been slowed by lack of knowledge of their chemistry. As discussed below, this limitation is one that is rapidly being overcome thanks to technical advances.

The study of invertebrate biochemistry and physiology is often justified in terms of the potential benefits that might accrue in controlling organisms that directly or indirectly cause disease in man or his livestock, or damage to his crops. Much of the impetus to the study of 'conventional' neurotransmitters in insects has come from such considerations, as readers of other chapters in this book will realise. The rationale is that more complete knowledge of neurotransmitter systems (many of which are already the targets of existing pesticides) will allow the design of improved chemical pesticides. The benefits to be gained from studying neuropeptides in insects and other pests are perhaps less obvious. Peptides, it is true, regulate many aspects of behaviour and physiology, and disruption of peptidergic control systems by the massive and untimely release of neurohormones have even been postulated to be the proximate cause of death in the chain of events that is initiated by the conventional neurotoxicant insecticide poisoning of insects (Maddrell and Casida, 1971; Maddrell and Reynolds, 1972; Singh and Orchard, 1982). However, peptides are expensive to manufacture (even with recombinant DNA techniques) and are likely to be inadequate practical insecticides because of poor penetration and environmental instability. Nevertheless their potential for disrupting normal function within the insect could be exploited by employing unconventional means of delivery. Genes for insect neuropeptides could, for example, be inserted into recombinant insect pathogenic viruses, bacteria or fungi. These organisms would provide specificity in targeting, and in turn would gain increased pathogenicity from the peptide genes (Kirschbaum, 1985). There are of course environmental questions to be addressed before such recombinant organisms could be used in the field.

IDENTITY OF INVERTEBRATE NEUROPEPTIDES

Peptide sequencing by classical methods

The structural characterisation of neuropeptides (i.e. the determination of amino acid sequences) remains an important focus for peptide research, since studies of physiology, mode of action and pharmacology are greatly facilitated when chemical structures are known and synthetic peptides can be prepared. This information is still largely to be gained by the traditional methodology of isolation and purification (usually using a bioassay or radioimmunoassay (RIA) to follow physiological or biochemical activity through the purification protocol), followed by sequence determination usually using some variant of the Edman degradation procedure (Walsh *et al.*, 1981).

Although rather smaller than the number of vertebrate neuropeptides of known sequence, the list of peptides from invertebrate sources that have been fully or even partially characterised is growing rapidly. Table 5.1 lists the 48 examples known to us. This list is growing so rapidly at the time of writing that it is unlikely to be complete now, and will certainly be rapidly out of date. Within the invertebrates, studies have so far concentrated on arthropods and molluscs almost to the exclusion of other phyla. Only one neuropeptide has been characterised from an animal outside these two taxa, from the sea anemone *Anthopleura*. As it seems extremely unlikely that neuropeptides will not be found in the nervous systems of members of other invertebrate phyla, the extension of neuropeptide research to these

178

groups promises the discovery of new peptides. Since the number of invertebrate species far outstrips the number of vertebrate species, it may be presumed that eventually the invertebrate list will be much longer than the vertebrate one.

At first sight, the task of sequencing an invertebrate peptide might appear no more demanding than for a vertebrate example. The size and organisation of the invertebrate nervous system can, however, present problems. Typically, the invertebrate CNS is several orders of magnitude smaller than that of a vertebrate. This can mean that the peptide of interest exists in only a small number of neurones in each animal. As an extreme example of this, the insect peptide prothoracicotrophic hormone (PTTH), a neurohormone that is periodically secreted into the blood to initiate moulting, is believed to be the product of only two cells in the brain of the lepidopteran *Manduca sexta* (Agui *et al.*, 1979). Hence workers interested in characterising invertebrate peptides are usually faced with accumulating a large quantity of tissue (which can be both time-consuming and expensive), and then isolating minute quantities of peptide from a large bulk of irrelevant material. Fortunately, recent improvements in technique have radically improved yields of peptides during isolations and reduced the quantities that are required for sequencing. In particular, high performance liquid chromatography (HPLC) and sensitive automatic microsequencers are now routinely used to separate and subsequently determine the primary structures of peptides (Hunkapiller *et al.*, 1984). Even so, the problems are daunting. Again using the example of PTTH, this time from the commercial silkworm, *Bombyx mori*, three heterogeneous species with PTTH bioactivity have recently been isolated and partially sequenced using gas-phase sequencing techniques (Nagasawa *et al.*, 1984). The starting material was some 648 000 adult male heads. This is actually a small number, compared with the 10 million heads anecdotally reported to have been used by the same group of workers during previous unsuccessful attempts when only classical techniques were available (Ishizaki and Susuki, 1984).

This particular example illustrates a number of instructive points. First, a sensitive and reliable bioassay is an absolute prerequisite for success in isolating novel neuropeptides. The developmental response of debrained silkmoth pupae to injected PTTH used in the above work is tedious, but adequate to the task. Some previous attempts to purify PTTH came unstuck when inadequately specific assays were used (see Williams, 1967). Secondly, the small size of invertebrate nervous systems frequently means that irrelevant material may have to be included in the original extraction, because it is impractical to separate out peptide-rich regions of the CNS. In the *Bombyx* example, the use of whole heads rather than dissected nervous tissue undoubtedly caused problems during purification, which ultimately required no fewer than 15 separation steps. This must have considerably reduced the final yield of peptide. Thirdly, where the end result of peptide isolation and purification is not one but several related peptides, it is not certain that these are all present in any one animal. The use of large numbers of animals from genetically undefined populations may result in the simultaneous isolation of polymorphic forms of a single gene product. (It is, of course, possible to check this retrospectively.)

Other recent successes in the characterisation of invertebrate neuropeptides using modern techniques include the sequencing of the *Aplysia* cardioactive peptide SCP$_B$ (Morris *et al.*, 1982) and the cockroach myotropic peptides MI and MII (O'Shea *et al.*, 1974; Witten *et al.*, 1984) using fast-atom bombardment mass spectrometry (FAB–MS) instead of the classical Edman approach. FAB-MS is less sensitive than gas-phase Edman sequencing, but offers some important advantages, notably that sequence determination is not impaired when the peptide's N-terminal residue is blocked, as is the case in many neuropeptides (Rinehart, 1982). The signal-to-noise ratio in the FAB mass spectrum is considerably improved when tandem mass spectrometry (MSMS) is combined with the use of FAB as the ionisation technique (Eckart *et al.*, 1986). This was recently used to elucidate the sequence of *Manduca sexta* apidokinetic hormone (AKH) (Ziegler *et al.*, 1985). MSMS is a very expensive investment and is still less sensitive than the gas-phase Edman technique, so that its use is likely to be restricted to difficult cases, e.g. where the N-terminal residue is blocked. Even where FAB–MS is not used to attribute an entire sequence, it can be very useful in establishing an exact molecular weight (e.g. Jaffe *et al.*, 1986) or in attributing partial sequences (e.g. Gäde *et al.*, 1986).

The achievements of those workers who succeeded in completely characterising RPCH (Fernlund and Josefsson, 1972), proctolin (Starratt and Brown, 1975), AKH (Stone *et al.*, 1976), DRPH (Fernlund, 1976), Phe-Met-Arg-Phe-amide (FMRFamide) (Price and Greenberg, 1977) and *Hydra* head-activating factor (Schaller and Bodenmuller, 1981) without the use of HPLC or modern microsequencing techniques can only be admired.

Quantity is not always so serious a problem. Neuropeptides may be present in low concentration in the nervous system as a whole, but regions have been recognised that contain relatively large amounts of peptides. Two arthropod neurohaemal organs, the corpora cardiaca (CC) of insects and the sinus glands (SG) of Crustacea, are notable among such structures. The CC of the cockroach *Periplaneta americana* contain about 140 pmol of the myotropic octapeptides periplanetin I and II (same as MI and MII) (Scarborough *et al.*, 1984). The same gland in the locust *Schistocerca gregaria* stores up to 1 500 pmol of the lipid-mobilising neurohormones AKH I and II (Siegert and Mordue, 1986). The SG of the shore crab *Carcinus maenas* contain 150 pmol of the hyperglycaemic peptide, hyperglycaemic hormone (HGH) (Jaros and Keller, 1979). There are obvious advantages in utilising these rich sources of peptides, and they can be expected to be exploited in the future for the characterisation of the other peptides they are believed to contain. On the other hand it must be admitted that most invertebrate neuropeptides are not available in such convenient packaging, nor in such quantity.

In addition to problems of scale, there are others that make the handling and separation of peptides difficult. Low recoveries of biological activity have often been recorded. This can sometimes be explained by the inherent chemical properties of the peptides concerned. Peptides are liable to proteolytic degradation by enzymes released during tissue homogenisation, a problem that can sometimes be overcome by heat-treating extracts, provided that the peptide of interest is not thermolabile. Extraction in acid or organic solvent is helpful in this regard. Peptides that have sulphur-containing amino acid residues (methionine, cysteine) are prone to oxidation (for example, DRPH — Fernlund, 1976). They can be protected by the use of mild reducing agents such as thiodiglycol. Many small peptides are very hydrophobic, and losses may occur particularly in late stages of purification when the peptide sticks to the walls of the tube in which it is held. Scarborough *et al.* (1984) reported such losses when working with periplanetins I and II. Larger peptides are particularly demanding. These are often unstable during isolation, and subsequently require controlled cleavage before sequencing. Even partial sequences may, however, still be very useful, when used as part of an alternative strategy involving recombinant DNA methods (see below).

Molecular genetic approaches to peptide sequencing

Essentially, the molecular genetic approach determines the structure of the peptide by isolating the gene that encodes it and inferring the amino acid sequence from that of the nucleotides. One approach to the problem of finding the relevant gene is the strategy of differential screening, which has been successfully employed to determine the sequences of a number of neuropeptides from the CNS of the sea hare *Aplysia californica* (reviewed by Scheller *et al.*, 1984; Kaldany *et al.*, 1985). The procedures used are illustrated here by the work of Nambu *et al.* (1983) on a family of cardioactive neuropeptides expressed in an identified group of nerve cells (R3–R14) in the abdominal ganglion (see Figure 5.1).

First poly A^+ mRNA was isolated from individually dissected cells R2 (a control), and from R14 and the R3–R8 group. This was facilitated, of course, by the large size of *Aplysia* neuronal cell bodies. [^{32}P]cDNA was prepared from this mRNA using reverse transcriptase with an oligo dT primer, and used as a probe for a cDNA library constructed in λgt 10 from total abdominal ganglion mRNA. Clones that were specific for cells of

181

Figure 5.1: Organisation of the gene coding for the *Aplysia* R3–R14 neuropeptides. Genomic DNA blot analysis shows that the R3-R14 peptides are probably encoded by a single gene. (A) shows the organisation of the neuropeptide precursor as inferred from the nucleotide sequence of cloned cDNA prepared from R3–R14-specific mRNA. (B) shows the sequence of this cDNA, together with the amino acid sequence of the precursor. The coding region is derived from the only in-phase reading frame that matches the known molecular weight of the *in vitro* translation product. The exact cleavage position of the leader signal sequence is not known, but the indicated position between amino acid residues 20 and 21 is plausible and consistent with the known size of the precursor. The cleavage sites that generate peptides I, II and III are pairs of dibasic residues. (C) is a diagrammatic representation of the *Aplysia* abdominal ganglion, showing the positions of the R3–R8 and R14 neurones that express this gene.
(A) and (B) based on Nambu *et al.* (1983); (C) based on Scheller *et al.* (1984)

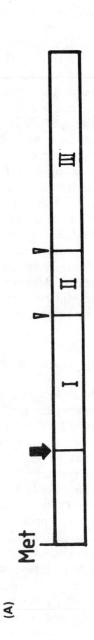

(A)

GCGCTCACTCACTCAGTCACAAACACACGCACTATCTCTCACTAACAGACTCAC

(B)

TCACCCCAAGCAGAGTGACTTTCAATTCTTTCGCTCGTCGTGAGAGCGTTGGCCGTGAGTCGGTAACAAAGTGAGATCGACTTTCTCATCCAGACATCAACTACCCAAGCAGCC
55

```
 1
met gln val leu his leu cys leu ala val ser ilu ala val ala leu leu ser gln ala ala trp ser glu glu val phe asp asp thr
ATG CAA GTC CTC CAC CTG TGT CTA GCG GTG TCC ATC GCT GTG GCC CTC CTG TCC CAG GCT GCG TGG TCA GAA GAG GTG TTT GAT GAC ACA
174

31
asp val gly asp glu leu thr asn ala leu glu ser val leu thr asp phe lys asp lys arg glu ala glu glu pro ser ala phe met
GAC GTC GGT GAT GAG CTG ACC AAC GCC TTG GAG TCA GTT CTG ACA GAT TTC AAA GAC AAA CGG GAA GCA GAA CCA TCA GCC TTC ATG
264

61
thr arg leu arg gln val ala gln met his val trp arg ala val asn his asp arg asn his gly ser gly arg his gly
ACC CGC CTG CGC AGA CAA GTT GCG CAG ATG CAC GTC TGG CGT GCC GTC AAC CAC GAC CGT AAC CAC CAC GGC TCC GGC CGT CAC GGT
354

91
arg phe leu ilu arg asn arg tyr arg arg tyr gly gly his leu ser asp ala ***
CGC TTC CTC ATT CGT AAC CGA TAC CGA TAC GGT GGC GGT CAT CTC TCC GAC GCG TAA ATTGATCCAAAACAACTACATCGACGATGACTTCAACACAAC
444
```

AATAACACAACAACAACAACAACAACACGCAGAACAACAACTACAACAACAATACAACACATAACCCCACCCCCCCCCCCCCCACCCCACCCCTCACCCCCG
544

CAAAACAATCACCACAACTACACAGAGTATAAACGGGTCATAAGGGTTGAGAGACTAAAACTCGACAAAGTGTTCGATCTCCCCCCCCCCCCCCCCACCCCTCCCAACTCTCTCCCAAT
663

AAAAAATCAACAGCAAAGACTCTGAATGGGCTTAAAAGAACAACAAAGAAAAAAAACAGTGATGATGAAAAGTGTTCGACATCAACTGTCCAGATTGCTTGTGTAATGTACAAAATAGCTCGTGACT
782

GTTTGCACTAAAATTAAATAAGCTAAAATGTTAAAACATGCAAAGATAAGATATTAATTGATAATAAAAGCACTTCAGTTATTCAGTAATTCAAAATGTAATGTGATTGTCACCGAATGTTAAA
901

CCAGGATCGGGTAAAGGGAGGCTTGGGTAGGGGTGAGCCCGTTAGATTAGCTATCACATAATATAATAGATAGTAAATTATAAGATTAGCACCATTTCCTGTGCCGGTTTCCAGTCGAAT
1020

TTATCCAGTTCATTGTGTGTCGTGATGCGTCACAAATAAAATAAAATACTTGTCTTTCCAAAAAAAAAAAAAAAAAAAAA
1139

Figure 5.1 continued

(C)

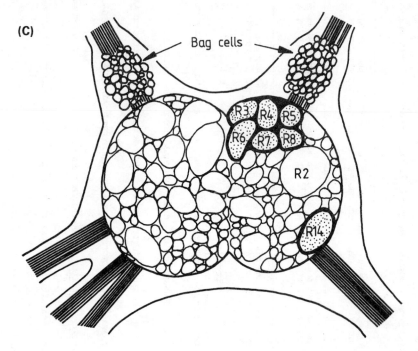

interest were identified as those that hybridised to probes from R3–R8 and R14, but not R2. An 800-bp restriction fragment from a clone specific to R14 was now used to probe an abdominal ganglion total mRNA cDNA library, and thus to identify sequences that code for the R3–R14 neuropeptides. Sequencing these clones allowed the prediction of a 108-amino acid protein, presumably the polyprotein precursor of the neuropeptide(s) synthesised by these cells. The details of this precursor's subsequent processing have not yet been worked out, but three cleavage sites have been identified within the precursor sequence, so that three smaller peptides may be produced in addition to the signal sequence.

Where antibodies to the peptide of interest already exist, an alternative strategy may be employed. In this approach (which has been used by P. Taghert (personal communication) to study genes coding for the insect tanning hormone, bursicon), mRNA extracted from the CNS is used to make cDNA which is cloned into a vector (e.g. λgt 11) that allows efficient expression of the cloned sequences. The recombinants produced are screened using an immunoblot procedure that recognises the peptide of interest. Once recognised, relevant cDNA sequences can be used as probes to find the original genes in a cDNA or genomic library as before.

Still another approach can be used where the amino acid sequence of the peptide is already at least partially known. In this case, a cocktail of synthetic oligonucleotides all of which code for the relevant peptide (the genetic code is redundant) are either used directly as probes to screen a cDNA or genomic library, or are used as primers for reverse transcriptase. In the latter approach, the resulting cDNA will be highly enriched for sequences of interest. The procedure automatically selects the correct nucleotide sequence from among the cocktail, as this is the sequence that hybridises best. This cloning strategy has been successfully used by Schaffer *et al.* (1984) to clone and sequence cDNA to the messenger for the decapeptide adipokinetic hormone I (AKH-I) in *Locusta*.

Last, where cDNA clones are available for peptide genes known to be related to an unknown neuropeptide (e.g. where an invertebrate neuro-peptide is recognised by an antiserum directed against a vertebrate peptide — see below), it is possible to screen a cDNA library prepared from invertebrate CNS mRNA using the 'foreign' peptide cDNA as a probe. This probe will recognise homologous DNA sequences by hybridising with them. The closer the sequence relationship between the unknown gene and the probe, the higher the stringency (temperature) that can be used in the screen. This strategy of homology hybridisation has been much used to isolate invertebrate genes coding for vertebrate proteins (for example the *Drosophila* homologue of the oncogene c-*erb*B, which is discussed below in the section on receptors) and *vice versa*, (e.g. homoeotic genes, first sequenced from *Drosophila*, but now recognised in a wide range of animals). However, to our knowledge it has not so far successfully been used to isolate neuropeptide genes.

There can be no doubt that recombinant DNA methods such as these will not only allow the sequencing of invertebrate neuropeptides that have proved difficult by classical methods, but will also predict the existence of bioactive neuropeptides not previously known to exist (see below).

Invertebrate neuropeptide families

The elucidation of the primary structures of numbers of neuropeptides has revealed that vertebrate peptides exist in groups or families, the members of which share many common residues in homologous sequences. Although fewer invertebrate neuropeptides have so far been characterised (Table 5.1), it is already clear that some of these are members of such families. Table 5.2 shows two of these groups and their structural homologies. A third invertebrate neuropeptide family is based on the *Aplysia* egg-laying hormone (ELH). This is discussed in the section on neuropeptide genes (below).

The first example consists of six peptides related to the insect lipid-

185

Table 5.2: Two invertebrate peptide families

A. Adipokinetic hormone family[a]

LPK	pGlu-Thr-Ser-Phe-Thr-Pro-Arg-Leu$_{NH_2}$
MI	pGlu-Val-Asn-Phe-Ser-Pro-Asn-Trp$_{NH_2}$
MII	pGlu-Leu-Thr-Phe-Thr-Pro-Asn-Trp$_{NH_2}$
Carausius AKH	pGlu-Leu-Thr-Phe-Thr-Pro-Asn-Trp-Gly-Thr$_{NH_2}$
AKH-I	pGlu-Leu-Asn-Phe-Thr-Pro-Asn-Trp-Gly-Thr$_{NH_2}$
RPCH	pGlu-Leu-Asn-Phe-Ser-Pro-Gly-Trp$_{NH_2}$
AKH-II (L)	pGlu-Leu-Asn-Phe-Ser-Thr-Gly-Trp$_{NH_2}$
AKH-II (S)	pGlu-Leu-Asn-Phe-Ser-Ala-Gly-Trp$_{NH_2}$
Manduca AKH	pGlu-Leu-Thr-Phe-Thr-Ser-Ser-Trp-Gly$_{NH_2}$
Glucagon (N-terminal sequence	His-Ser-Glu-Gly-Thr-Phe-Thr-Ser-Asp-Tyr-Ser-Lys-

B. FMRFamide family[a]

Leu-enkephalin	Tyr-Gly-Gly-Phe-Leu
Met-enkephalin	Tyr-Gly-Gly-Phe-Met
Met-enkephalin Arg[6]Phe[7]	Tyr-Gly-Gly-Phe-Met-Arg-Phe
Met-enkephalin Arg[6]Phe[7] amide	Tyr-Gly-Gly-Phe-Met-Arg-Phe$_{NH_2}$
FMRFamide	Phe-Met-Arg-Phe$_{NH2}$
FLRFamide	Phe-Leu-Arg-Phe$_{NH_2}$
Helix FMRFamide-like peptide	pGlu-Asp-Pro-Phe-Leu-Arg-Phe$_{NH_2}$
Leucomyosupressin	pGlu-Asp-Val-Asp-His-Val-Phe-Leu-Arg-Phe$_{NH_2}$
LPLRFamide (avian)	Leu-Pro-Leu-Arg-Phe$_{NH_2}$
Antinociceptive octadecapeptide[b] (mammalian)	-Ser-Leu-Ala-Ala-Pro-Gln-Arg-Phe$_{NH_2}$
Antinociceptive octapeptide (mammalian)	Phe-Leu-Phe-Gln-Pro-Gln-Arg-Phe$_{NH_2}$

[a] Peptides are arranged to maximise correspondence of sequence where possible.
[b] Complete sequence is AGEGLSSPFWSLAAPQRF$_{NH_2}$

mobilising hormone, AKH-I. All are amino-terminally blocked by pGlu and are amidated at the carboxy-terminal. All of the peptides lack charged groups and are therefore relatively hydrophobic. These peptides were all purified from arthropod sources, four from orthopteran insects, one from two species of Lepidoptera, and the last (RPCH) from a prawn, but it is not yet clear if this family of peptides is restricted to this phylum. Immuno-cytochemical evidence suggests that AKH or AKH-like peptides may be present in other invertebrates, as well as in other insects (Schoonveld *et al.*, 1985; Witten and O'Shea, 1985). It seems certain that further work will reveal additional members of this family. Similarities between the receptors that correspond to these neuropeptides are indicated by the ability of the various peptides to cross-react in bioassays. For example, AKH-I, which elevates insect haemolymph lipid levels (Stone *et al.*, 1976) can induce erythrophore concentration in the crab *Uca pugilator*, and in the reciprocal test, RPCH (the endogenous crustacean colour change hormone) can mobilise lipids in the locust (Mordue and Stone, 1977). AKH-I and RPCH, as well as MI and MII, can cause acceleration of the cockroach heart *in vitro* (Scarborough *et al.*, 1984). These results, in combination with structure–activity studies of the peptides on individual preparations, indicate that the conserved sequences are necessary for biological activity. The physiological significance of the occurrence of more than one member of the family in a single animal (e.g. AKH-I and AKH-II in locusts, MI and MII in cockroaches) is not yet clear. It has been suggested (Loughton and Orchard, 1981) that AKH-I and AKH-II may have different functions in locusts.

A second family is based on the cardioactive tetrapeptide, FMRFamide. All the molluscan peptides in this family have an amidated C-terminal sequence that is identical with or closely related to FMRFamide, but some are extended N-terminally (see Table 5.2). A structure–activity study of FMRFamide using synthetic analogues (Painter *et al.*, 1982) found that the intact carboxy terminal and the basic -Arg- at the penultimate position are both required for receptor activation in molluscs. These are the regions that are conserved in the other molluscan peptides in the family, and are presumably also essential for activity. Like the AKH family, there are good indications that the FMRFamide family is more extensive than is indicated in Table 5.2. Different anti-FMRFamide sera have detected immunoreactivity in several invertebrates including insects (Schot *et al.*, 1981; Myers and Evans, 1985) but as yet the nature of these peptides is mostly unknown. Recently, Holman *et al.* (1968d) have characterised an insect neuropeptide, leucomyosuppressin, that inhibits activity in visceral muscle; this peptide's C-terminal sequence is clearly related to the FMRFamide family. It has been suggested that the FMRFamide family may be related to that of the vertebrate enkephalins (see below).

Other invertebrate peptide families almost certainly exist. As seen in Table 5.1, two pigment dispersing hormones from Crustacea have been sequenced, and bioassay data from other species (including chelicerates and insects) show that other peptides with similar structures exist (Rao *et al.*, 1985). Other known families are based on the leucokinins (Holman *et al.*, 1986b,c), and on the gastropod egg-laying hormones (see below).

Structural relationships between invertebrate and vertebrate peptides

We have just discussed some of the evidence that supports the existence of families of invertebrate neuropeptides. Clearly, in the knowledge that many neuropeptides are utilised as regulators by the vertebrate CNS, a question to be addressed is whether there are homologies between characterised invertebrate and vertebrate peptides. This is interesting because (a) it may suggest the evolutionary origins of individual neuropeptides, and (b) it may be useful in predicting the existence of undiscovered peptides in either vertebrates or invertebrates. In the following short account of the evidence for such relationships we shall consider peptides as being of 'vertebrate type' or 'invertebrate type', according to the source from which they were first purified. Hence, Leu- and Met-enkephalin are 'vertebrate' peptides, and the *Hydra* head-activating peptide is an 'invertebrate' peptide, even though they have now both been isolated and sequenced from both invertebrate and vertebrate sources (Bodenmuller and Schaller, 1981; Leung and Stefano, 1984).

Studies on the possible existence of vertebrate-type peptides in invertebrates have been facilitated by the availability of antisera against known vertebrate peptides. Although immunocytochemical studies employing these antisera have been popular (see below), only a few workers have gone on to investigate the chemical nature of the immunoreactivity demonstrated histologically.

In fact, only three vertebrate-type peptides have so far been purified and completely sequenced from an invertebrate source using this strategy. These are Leu- and Met-enkephalin (Leung and Stefano, 1984) and Met-enkephalin-Arg^6-Phe^7 (Stefano and Leung, 1984) all from the marine mussel, *Mytilus edulis*. Purification in these cases was followed by using a conspecific opiate binding assay.

A number of other invertebrate peptides of vertebrate type (insulin, glucagon, gastrin/CCK, α-endorphin) have been extensively purified and characterised, although sequences have not yet been determined. In most cases the evidence so far available indicates that the invertebrate peptides are extremely similar to their vertebrate counterparts (Thorpe and Duve, 1984).

Insulin is a good example. Insulin-like peptides have been purified from the blowfly *Calliphora vomitoria* (Duve *et al.*, 1979), and from the tobacco hornworm *Manduca sexta* (Kramer *et al.*, 1982). In both cases the chromatographic properties of the insect peptides were extremely similar to those of vertebrate insulins, and in the case of the *Manduca* peptide the amino acid composition differed only slightly from porcine insulin. The *Manduca* insulin-like peptide showed a parallel binding curve to porcine insulin in an RIA, but the *Calliphora* peptide showed non-parallelism compared with bovine insulin, indicating that some differences must exist. However, the *Calliphora* insulin-like peptide was nevertheless able to displace specifically bound ^{125}I-labelled insulin from rat liver insulin receptors, and to increase [^{3}H]glucose transport into isolated rat epididymal adipocytes, indicating that it must resemble the vertebrate peptide very closely at its active site. It will be very interesting, when full sequence data are available, to see how closely these invertebrate insulins resemble their vertebrate counterparts. Perhaps it will not be surprising to find close homology:

188

classical neurotransmitters are, after all, ubiquitous.

Recently, the partial sequencing of two other insect neuropeptides, the moult-initiating prothoracicotropic hormone (PTTH), and the melanisation and reddish colouration hormone, MRCH, unexpectedly revealed close structural homologies with insulin. The amino-terminal sequences of three peptides with PTTH activity from *Bombyx* heads (Table 5.3) are each homologous to human insulin A-chain, and to an insulin-like growth factor (Nagasawa *et al.*, 1984). The remainder of the PTTH molecule (still unsequenced) must differ considerably from insulin, however, as its molecular weight is quite different (PTTH is 4.4 kDa, insulin A-chain is 2.4 kDa, insulin A + B chains are 6.0 kDa). Mammalian insulins have no PTTH activity (H. Ishizaki, personal communication). The N-terminal sequence of MRCH similarly shows close homology with insulin-like growth factor II (IGF-II) (Matsumoto *et al.*, 1985).

Table 5.3: Insulin-like invertebrate peptides

A. Bombyx prothoracicotropic hormone (PTTH)	
	42 60
IGF-I	-Gly-Ile-Val-Asp-Glu-Cys-Cys-Phe-Arg-Ser-Cys-Asp-Leu-Arg-Arg-Leu-Glu-Met-Tyr-
	1 19
4K PTTH-II[a]	H-Gly-Ile-Val-Asp-Glu-Cys-Cys-Leu-Arg-Pro-Cys-Ser-Val-Asp-Val-Leu-Leu-Ser-Tyr-
	1 19
Human insulin A-chain	H-Gly-Ile-Val-Glu-Gln-Cys-Cys-Thr-Ser-Ile-Cys-Ser-Leu-Tyr-Gln-Leu-Glu-Asn-Tyr-
B. Bombyx melanisation and reddish colouration hormone (MRCH)	
	16
MRCH-III[b]	H-Pro-Leu-Ser-Glu-Asp-Met-Pro-Ala-Thr-Pro-Ala-Asp-Gln-Glu-Met-Tyr-
	54 67
IGF-II	-Ala-Leu-Leu-Glu-Thr-Tyr-Cys-Ala-Thr-Pro-Ala-Lys-Ser-Glu-OH

[a] Three 4K PTTHs have been sequenced from *Bombyx*. All are highly homologous.
[b] Three MRCHs have been sequenced from *Bombyx*. All are highly homologous.

These sets of data not only confirm the ancient status and conservation of the insulin sequence but in addition suggest the possible existence of an ancestral peptide that evolved to give (at least) PTTH, MRCH and insulin. It is perhaps of value to point out that these structural similarities may have consequences for cytochemical studies employing anti-insulin antibodies. One might expect that antisera specific to insulin A-chain would cross-react with PTTH, but that B-chain antibodies would not. However, El-Salhy *et al.* (1983) report that in *Manduca* the brain contains no cells that are immunoreactive to anti-A-chain sera, but cells immunoreactive to anti-B-chain sera are present. This could simply indicate that the A-chain antiserum used was unsuitable for the task, or that PTTH in *Manduca* is markedly different from that in *Bombyx*. The last is certainly a possibility since Bollenbacher *et al.* (1984) report quite different molecular weights (29 kDa and 7 kDa) for the two forms of PTTH that they isolated from *Manduca* brains.

It is not yet clear whether the structural homologies found for these peptides also extend to their functions. The insulin-like peptide of *Calliphora* (which probably originates in the medial neurosecretory cells (MNC) of the brain — Duve *et al.*, 1979) is certainly capable of acting in a classical insulin-like fashion to reduce blood sugars when injected into flies, although this was only demonstrable in insects made experimentally hypertrehalosaemic by extirpation of the MNC (Duve *et al.*, 1979). There is as yet no unambiguous proof that insect insulin-like peptides act to regulate blood sugar levels under normal conditions, although the circumstantial evidence is compelling, at least for flies. It is possible that insulin-like peptides might act primarily as growth factors in insects, rather than as regulators of blood sugar levels (Seecoff and Dewhurst, 1974; Thompson *et al.*, 1985).

Interestingly, a homology can be made between the N-terminus of glucagon, a peptide that causes elevated blood sugar levels in vertebrates, and the AKH-family peptides MI and MII from cockroaches, and *Manduca* AKH. The homology is not a particularly strong one (Table 5.2). Although MI and MII were originally isolated on the basis of their ability to stimulate skeletal muscle preparations (O'Shea *et al.*, 1984) or to cause cardioacceleration (Scarborough *et al.*, 1984), both peptides are also able to cause hypertrehalosaemia on injection into cockroaches. As yet, however, the normal functions of MI and II in cockroaches are not known. *Manduca* AKH causes mobilisation of carbohydrate reserves in caterpillars by activating fat-body phosphorylase (Siegert and Ziegler, 1983).

Another unexpected homology that has recently emerged from a study of myoactive neuropeptides in the cockroach *Leucophaea* is between two sulphated insect peptides (leucosulfakinin I and II) and mammalian gastrin/CCK (Nachman *et al.*, 1986a,b).

Parallel to attempts to identify vertebrate-type peptides in invertebrates there has been a handful of studies investigating the possible occurrence of invertebrate neuropeptides in vertebrates. The most extensively examined example of this is the *Hydra* head-activating peptide. Bodenmuller and Schaller (1981) found peptides in rat intestine and in bovine and human hypothalamus consistent with the known properties of the *Hydra* peptide. The extracted vertebrate peptides had the same biological activity (in *Hydra*), the same retention time on HPLC, and the same amino acid composition and carboxy- and amino-terminal groups. Finally, sequencing of the rat intestinal peptide showed it to be identical with the *Hydra* peptide, confirming that it had been conserved through evolution from coelenterates to mammals. In *Hydra* the peptide is thought to play a role in regulating morphogenesis. The demonstration that the *Hydra* peptide is

190

present during mammalian embryogenesis may even suggest conservation of function. The ability of the head activator peptide to stimulate contraction of the rat uterus, and a weak structural homology with the bradykinins in the sequence X-Pro-Pro-Gly-X-Ser, may perhaps represent another relationship (Schaller *et al.*, 1984).

The presence of invertebrate peptides in vertebrate nervous systems has been implied by some immunocytochemical reports. FMRFamide is a good example. Dockray *et al.* (1983) have purified one of the vertebrate peptides that cross-reacted with antisera raised against molluscan FMRF-amide. This was characterised as a pentapeptide, LPLRFamide (see Table 5.2). Subsequently, two other endogenous mammalian neuropeptides which cross-react with anti-FMRFamide sera have been sequenced (Yang *et al.*, 1985). They are AGEGLSSPFWSLAAPQRFamide and FLFQPQRFamide. These peptides are structurally unique among vertebrate peptides. Their sequence similarity to the molluscan peptide is limited, and the evolutionary relationship to FMRFamide and to each other is not yet clear. That the peptides not only have an immunological relationship but also share some biological activity (e.g. Tang *et al.*, 1984) probably reflects the importance of the shared -RFamide C-terminus. This work emphasises the benefits of complete chemical characterisation of the immunoreactive material stained in histochemical procedures. LPLRF-amide may function as a regulator of the circulation. It has potent pressor effects in rats (Barnard and Dockray, 1984). The other two peptides both appear to have actions opposite to those of opiates in nociception (Yang *et al.*, 1985).

There is evidence of a possible evolutionary relationship between FMRFamide and the enkephalins — a hypothesis that is largely based on the sequence of a possible evolutionary intermediate peptide, YGGFMRF. This peptide was first found in vertebrate adrenal, and subsequently CNS (Stern *et al.*, 1979). It has since been found in a bivalve mollusc, the mussel *Mytilus* (Stefano and Leung, 1984). Even more intriguingly, the C-terminal amidated form YGGFMRFamide has been isolated from the cephalopod mollusc, *Octopus* (Kiehling *et al.*, 1984). As can easily be seen (Table 5.2), this peptide contains within its structure the sequences of both Met-enkephalin and FMRFamide. The implication is that YGGFMRF-amide may be ancestral to both of the smaller peptides. If this is so, then the split had already occurred by the time that ancestral molluscs and vertebrates diverged, since molluscs not only possess YGGFMRF, but also both YGGFM and FMRFamide. Pharmacological studies (e.g. Greenberg *et al.*, 1983) indicate that there are separate receptor classes for all three peptide types.

More speculatively, Sullivan and Newcomb (1982) have suggested the possibility of a familial relationship between the insect neuropeptide proctolin (RYLPT) and the vertebrate vasoactive intestinal peptide (VIP),

191

which has a proctolin-like sequence at residues 21–25 (KYLNS). The resemblance is not particularly convincing. Antibodies to another insect peptide, AKH I, reveal an antigen in rat brain and adrenal (Sasek *et al.*, 1985). Preliminary data on the chemical nature of the rat peptide(s) responsible reveal, however, that they probably do not resemble the insect peptide very closely (Schueler *et al.*, 1986).

NEUROPEPTIDE GENES

The vertebrate paradigm

Molecular genetic techniques have permitted the detailed analysis of the primary structures of a large number of vertebrate neuropeptide-encoding genes. Those so far studied resemble structurally the majority of eukaryotic genes (Douglass *et al.*, 1984). In many cases the nucleotide sequence of the gene includes not only regions that encode the mature mRNAs (exons), but also intervening non-coding sequences (introns). Both are initially transcribed into pre-mRNAs which are subsequently processed (by excision of introns and the addition of a poly A tail) into mature mRNAs. These are translated to yield a prepropeptide that is larger than the final peptide product(s). The N-terminus of this polypeptide is usually Met followed by a short hydrophobic 'signal sequence' that enables the packaging of the molecule within membrane-bound vesicles (Blobel, 1980). Once so packaged, the signal sequence is cleaved off to leave a propeptide still larger than the peptide that is to be secreted. This propeptide is now subject to further enzymic cleavage into smaller peptides and also to other kinds of post-translational modification.

An extremely interesting result of the molecular genetic analysis of vertebrate neuropeptide genes is the finding that many such genes encode more than one biologically active peptide. As an illustration we will briefly consider the pro-opiomelanocortin (POMC) gene (Nakanishi *et al.*, 1979). POMC is a 240 amino acid propeptide synthesised predominantly in the pituitary. Nucleotide sequencing of the POMC gene and amino acid sequencing of various fragments of the protein have revealed that the original mRNA transcript can be processed so as to yield an almost bewildering array of biologically active peptide products. These include adrenocorticotrophic hormone (ACTH); corticotrophin-like intermediate lobe peptide (CLIP); α- and β-endorphins; α-, β- and γ-melanocyte stimulating hormones (MSH); and α-, β- and γ-lipotrophins (LPH). Processing actually differs in various species and in different tissues so that (for instance) the predominant secretory products in porcine anterior pituitary are ACTH, β- and γ-LPH, β-endorphin and a 16 KD N-terminal glycoprotein, whereas, in bovine intermediate lobe, ACTH is further divided into α-MSH and CLIP, with β-LPH also being cleaved to give β-MSH, β-endorphin, and the N-terminal peptide. The β-endorphin is then further cleaved to smaller peptides. Depending on the location, some of these peptides may be N-terminally acetylated, which either enhances activity (α-MSH) or inactivates them (β-endorphin) (Smyth et al., 1981).

The POMC gene is perhaps a particularly complex example, but the principle of encoding a number of peptide products within a single gene is also seen in (for example) genes for β-neoendorphin/dynorphin (Kakidani et al., 1982) and tachykinins (substance P/substance K) (Nawa et al., 1983). A particularly interesting point that has emerged from the study of the gene for the enkephalins is that a single gene may encompass many copies of the sequence for a single peptide secretory product. Proenkephalin contains no fewer than six copies of Met-enkephalin as well as one of Leu-enkephalin (Noda et al., 1982; Gubler et al., 1982).

One of the structural features revealed by study of the POMC gene and others is that the biologically active peptide fragments within the propeptide sequence are almost always flanked by pairs of basic amino acid residues. Lys-Arg is the commonest of such sequences although other combinations are found. Another common structural feature is that C-terminal amidation is signalled by a Gly residue on the N-terminal side of the basic pair (Loh et al,, 1984).

The realisation that neuropeptide genes may encode more than one secretory product has led to predictions, based on nucleotide sequences, of novel biologically active peptide species. For instance, calcitonin-gene-related peptide (CGRP) was predicted to result from alternative processing of the primary mRNA transcript of the rat calcitonin gene (Amara et al., 1982), was subsequently detected immunologically in rat brain (Rosenfeld et al., 1983), and was ultimately purified and sequenced from human medullary thyroid carcinoma (Morris et al., 1984). The peptide has potent vasodilatory effects and may be involved in neural control of vascular tone and blood flow (Brain et al., 1985). In another example, a 56 amino acid peptide that is a part of the precursor for gonadotrophin releasing hormone (GnRH) has been found to be a potent inhibitor of prolactin secretion, and to stimulate gonadotrophin release. This peptide may be involved in the normal regulation of reproductive hormone titres (Nikolics et al., 1985).

Aplysia egg-laying hormone (ELH) and related genes

By comparison, molecular genetic studies of invertebrate neuropeptide genes are still in their infancy, perhaps with the exception of the work by Scheller and his associates on neuropeptides from the CNS of the gastropod mollusc *Aplysia californica*. This animal has several characteristics advantageous to studies on neuropeptides. The large size of the individually identifiable neurones (up to 1 mm in diameter) facilitates not only neurophysiological investigations but also biochemical and molecular genetic analyses, since individual cells contain relatively large quantities of peptides and specific mRNA. A second major attraction is that the abdominal ganglion has been very strongly implicated in the regulation of specific (though simple) behaviours, thereby offering an excellent opportunity to put the knowledge of neuropeptide genes so obtained to use in studying the function of peptidergic neuroregulators during natural integrative activity.

The work of the Stanford group on *Aplysia* neuropeptides has used almost exclusively the differential screening strategy (described above) to identify cDNA or genomic clones of interest. Some of this work has recently been reviewed (Scheller *et al.*, 1984; Kaldany *et al.*, 1985). The case of the *Aplysia* egg-laying hormone (ELH) is used here to illustrate the kind of information which it is possible to obtain.

ELH is a 36 amino acid peptide, purified and sequenced from the bag cells of the abdominal ganglion (the location of these cells is indicated in Figure 5.1C), that induces egg-laying and related behavioural activity when injected (Chiu *et al.*, 1979). Two other peptides (A and B) from a non-neural source, the atrial gland (a part of the reproductive system), are also capable of causing egg-laying behaviour on injection. Peptides A and B probably act on the abdominal ganglion to cause ELH release. The sequences of peptides A and B are very similar to each other, but are dissimilar to that of ELH (Table 5.4) (Heller *et al.*, 1980).

A total of nine clones, related by the possession of sequences homologous to ELH, was isolated from an *Aplysia* genomic library by a differential screening strategy (Scheller *et al.*, 1982). Some of the genes are probably alleles, and Mahon and Scheller (1983) have concluded that there are probably five ELH family genes per genome. Although these genes

Table 5.4: ELH gene family peptides

A. Aplysia ELH and Lymnaea CDCH are homologous					
	1	5	10	15	20
ELH	Ile-Ser-Ile-Asn-Gln-Asp-Leu-Lys-Ala-Ile-Thr-Asp-Met-Leu-Leu-Thr-Asp-Gln-Ile-Arg-				
CDCH	Leu-Ser-Ile-Thr-Asn-Asp-Leu-Arg-Ala-Ile-Ala-Asp-Ser-Tyr-Leu-Tyr-Asp-Gln-His-Trp-				
	21	25	30	35	
ELH	-Gly-Arg-Gln-Arg-Tyr-Leu-Ala-Asp-Leu-Arg-Gln-Arg-Leu-Leu-Asp-Lys$_{NH_2}$				
CDCH	-Leu-Arg-Glu-Arg-Gln-Glu-Glu-Asn-Leu-Arg-Arg-Arg-Phe-Leu-Glu-Leu$_{NH_2}$				

B. A and B peptides from the atrial gland are homologous with the α-bag cell peptide (α-BCP)					
	1	5	10	15	20
A	Ala-Val-Lys-Leu-Ser-Ser-Asp-Gly-Asn-Tyr-Pro-Phe-Asp-Leu-Ser-Lys-Glu-Asp-Gly-Ala-				
B	Ala-Val-Lys-Ser-Ser-Ser-Tyr-Gly-Lys-Tyr-Pro-Phe-Asp-Leu-Ser-Lys-Glu-Asp-Gly-Ala-				
	21	25	30	34	
A	-Gln-Pro-Tyr-Phe-Met-Thr-Pro-Arg-Leu-Arg-Phe-Tyr-Pro-Ile$_{NH_2}$				
B	-Gln-Pro-Tyr-Phe-Met-Thr-Pro-Arg-Leu-Arg-Phe-Tyr-Pro-Ile$_{NH_2}$				
α-BCP	Ala-Pro-Arg-Leu-Arg-Phe-Tyr-Ser-Leu				

share 90 per cent sequence homology, nevertheless they do not all direct synthesis of ELH. The ELH family then comprises three classes of genes: two genes coding for ELH itself, a single gene coding for peptide A and at least two coding for peptide B. ELH genes are expressed principally in the bag cells, whereas A and B genes are expressed in the atrial gland, in accordance with the known sites of synthesis of these peptides.

A clue to the relationship between these genes is given by the finding that the polyprotein precursors synthesised by bag cells and atrial gland cells are immunologically related (Scheller *et al.*, 1983). Restriction enzyme analysis and nucleotide sequencing of three representative genes (coding for ELH, peptide A and peptide B) show how these closely related genes nevertheless direct the synthesis of different products (Table 5.4 and Figure 5.2).

The ELH gene encodes a 271 amino acid precursor that begins with a typical signal sequence (Met followed by a succession of hydrophobic residues). In addition to this signal peptide, the sequences of as many as 11 other peptides may be discerned within the polyprotein precursor on the basis of potential enzymic cleavage sites. There is evidence that at least four of these are actually secreted by the bag cells. These are the α- and β-bag cell peptides (BCPs), ELH itself, and the C-terminal acidic peptide. α-BCP has been isolated and sequenced, being found to have the predicted amino acid sequence; β-BCP has at least the correct amino acid composition (Rothman *et al.*, 1983). As would be expected if they arose by processing of a single precursor, α-BCP, ELH and acidic peptide are found in approximately equimolar amounts in bag cell extracts. Before the existence of the BCPs was predicted from nucleotide sequence analysis, it had been realised that injecting ELH alone could not duplicate all of the behavioural effects of activating bag cell secretory activity (Mayeri *et al.*, 1979a,b). At least some of these missing actions can be duplicated by giving α- or β-BCP (Rothman *et al.*, 1983). Thus, as in the case of some of the vertebrate neuropeptides mentioned above, analysis of a neuropeptide gene has led to the prediction of novel bioactive peptides and the discovery of their physiological roles.

The genes that code for peptides A and B differ from the ELH gene mainly in the absence of a 240 bp sequence that occurs towards the N-terminal end of the polyprotein precursor (Table 5.4). The removal of this sequence leads to the loss of a number of potential proteolytic cleavage sites, so that the sequence that in the ELH gene codes for α-BCP is in both A and B genes extended N-terminally to include an extra 25 amino acids. Thus, the C-terminal amino acid sequences of A and B are both highly homologous with that of α-BCP, differing in only the first and last two residues (Table 5.4).

It is the possession by the ELH of this insert that explains why it does not direct the synthesis of a peptide resembling A or B. There are also

195

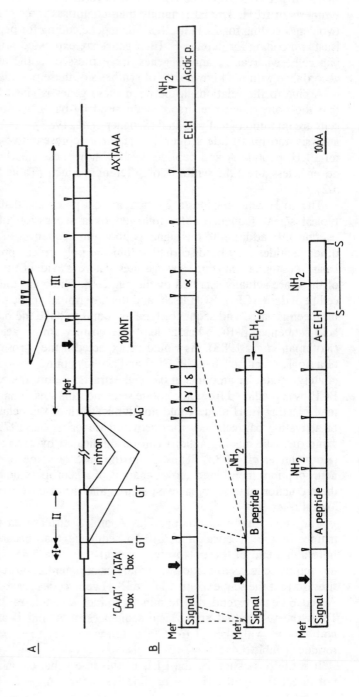

Figure 5.2: ELH gene family organisation. (A) shows the relative arrangement of ELH exons I, II and III, their possible RNA splicing pathways and the putative CAAT, TATA and polyadenylation consensus sequences. Exon III encodes the known peptide products of ELH family genes, as shown in (B). The presumed signal sequences run between the initiator methionine (Met) and the broad arrow. Additional known or potential proteolytic processing sites occur at single arginine or multiple basic residues, indicated by thin arrows. The positions of cysteine residues are indicated (S). Note the inferred disulphide bridge between the A-gene ELH-like peptide and that gene's C-terminal peptide. Based on Mahon *et al.* (1985).

differences in the exon III nucleotide sequences of the three genes that affect the expression of ELH peptides. Analysis of cDNAs made from atrial gland mRNA (Mahon *et al.*, 1985) has shown a single deletion at nucleotide 699 followed by a transition of nucleotide 700 from a C to a T. This causes a frame shift, giving rise to an additional cysteine codon and a stop codon further down stream. Mahon *et al.* hypothesise that the cysteine residue forms a disulphide bridge with another cysteine within the ELH region. This clearly would radically alter the character of the expressed peptide (Figure 5.2). The B-peptide gene differs from the ELH gene principally by the deletion of a single base in the third position of the sixth ELH codon. This causes a frame shift so that the next triplet is a nonsense codon that terminates translation. Thus only a fragment of the ELH region is expressed. Nevertheless, the rest of the sequence of ELH is clearly recognisable in the untranslated part of the B gene.

These features of the ELH gene family may suggest how these genes evolved. The A/B genes would appear to be ancestral. The 240-bp insert at the 5′ end of the transcribed portion of the ELH gene has been suggested to be the evolutionary remnant of an insertion by transposition, since there are (imperfect) repeats at both ends of the insert (Scheller *et al.*, 1983). There is considerable homology between the amino acid sequences of three of the small, basic peptides (α-, β- and γ-BCP) that are coded within the ELH gene (Table 5.4), suggesting that these sections of the gene may have arisen by duplication. All three peptides share the sequence Arg-Leu-Arg-Phe. The members of the ELH gene family may all be described as having three principal structural subdivisions: an A/B peptide region, an ELH region, and an acidic peptide region (Figure 5.2). Scheller *et al.* (1983) point out that there are homologies between the sequences of these three products that are considerably greater than would be predicted by chance, suggesting that these three regions of the gene may share a common ancestor. The existence of multiple copies of each subclass of the ELH gene family within the *Aplysia* genome is also presumably the result of gene duplication.

Interestingly, further work which characterised cDNA clones that hybridised to an ELH-1 probe (Mahon *et al.*, 1985a) has revealed that all mRNAs that code for either ELH or A/B peptides carry one or two untranslated 5′ sequences (exons I and II) that were not present in the genomic clones originally isolated by Scheller *et al.* (1982). When restriction fragments from one of those cDNA clones were used to screen genomic DNA, genomic clones were isolated that included exons I and II, exon III, and in one case, all three exons. Exons I and II are contiguous, with no intervening sequence, but exon III is physically distant, at a minimum distance of 5.3 kb from exons I/II. The organisation of the ELH gene is illustrated in Figure 5.2.

Sequencing of the new genomic clones that included exons I and II revealed that GT donor sequences are present at the 3′ ends of both exons I and II, while an AG acceptor site is present at the 5′ end of exon III in the ELH, A and B genes. This is consistent with the observed arrangement of I, II, III; I, III; and II, III in the cDNA clones isolated.

Consensus sequences for the proper initiation of translation ('TATA' and 'CAAT' boxes) are present upstream of the exon I sequence in all the genomic clones examined. Up to this point homology is strong, but it diminishes markedly upstream of the 'TATA' box. Mahon *et al.* (1985a) suggest that this region may be concerned with tissue-specific or developmental regulation of transcription.

Still more recently, it has been confirmed that the ovulation hormone, or caudodorsal cell hormone (CDCH), of the aquatic pulmonate snail *Lymnaea stagnalis* is (as suspected) homologous with *Aplysia* ELH (Table 5.4). Ebberink *et al.* (1985) isolated two closely related forms of CDCH and were able to attribute all but one amino acid residue of each of the two sequences on the basis of gas-phase sequencing. The missing residue (at position 13) was deduced from the sequence of a cDNA clone isolated and sequenced by Vreugdenhil *et al.* (1985) The two forms that were isolated as peptides differ at only two amino acid positions of 36. A number of other peaks of bioactivity were noted but not sequenced. The molecular genetic work indicates that there is a CDCH gene family in *Lymnaea* just as there is an ELH gene family in *Aplysia.* The homology at the amino acid level between ELH and CDCH is considerable (44%) considering the 400 MY that are estimated to have passed since the opisthobranch and fulmonate lines of the Mollusca diverged.

Other *Aplysia* neuropeptide genes

Other *Aplysia* genes studied by the Stanford group are those coding for a group of peptides specific to the R3–R14 group of neurones in the abdominal ganglion (Nambu *et al.*, 1983); a peptide synthesised by the abdominal neurone L11 (Taussig *et al.*, 1984); two myoactive peptides (SCP_A and SCP_B) contained in the buccal neurones B1 and B2 (Mahon *et al.*, 1985b); and FMRFamide, which is present in many neurones throughout the *Aplysia* nervous system (Schaefer *et al.*, 1985).

There is not space here to review all of this work in detail, but the case of FMRFamide is particularly interesting. A differential screening strategy resulted in the isolation of three distinct cDNA clones. Restriction endonuclease mapping shows that these are extensively homologous but not identical. Southern analysis of genomic DNA from four individuals revealed only a single gene, however, so that it must be inferred that the different cDNAs arise from multiple mRNA transcripts of the same gene. The existence of multiple transcripts was confirmed by RNA blotting using a cDNA clone as a probe. It remains possible that other genes containing the FMRF sequence exist within the genome, but if this is the case, then they would differ substantially from the gene isolated by Schaefer *et al* (1985).

The FMRFamide gene is remarkable in that it contains no less than 22 copies of the FMRF sequence (Figure 5.3). The initiator ATG sequence in fact corresponds to the Met residue of the first FMRF sequence. The FMRF-1 cDNA clone corresponds to a message that could yield as many as 19 separate copies of the FMRF peptide. Nearly all of the FMRF sequences are flanked by amino-terminal Lys-Arg and carboxy-terminal Gly-Lys pairs. This last is usual since it implies that the C-terminus of FMRFamide is defined by an endopeptidase that recognises a single Lys residue.

Presumably the multiple transcripts of the FMRFamide gene arise as a result of different ways of splicing the primary transcript. Interestingly, the nucleotide sequence to the amino acid pair Arg-Phe (present in very many copies within the gene) can give rise to RNA splicing sites for both donors (GT) and acceptors (AG). A sequence difference between two of the FMRFamide cDNA clones that were isolated is clearly attributable to such an excision.

Another interesting feature of the FMRFamide gene is that it codes for a polypeptide precursor that apparently contains no classical hydrophobic signal sequence. Although some other secretory proteins are known that lack such a sequence, the FMRFamide precursor is the first neuropeptide precursor of this kind.

The existence within the FMRFamide gene's sequence of large numbers of copies of the active peptide sequence is strongly reminiscent of the pro-enkephalin gene (Gubler *et al.*, 1982), although perhaps a much more extreme case. This is particularly interesting, given the possible existence of an evolutionary relationship between Met-enkephalin and FMRFamide (see above). However, it must be pointed out that there is no hint of any such link in the sequence of the genes for either peptide. Schaefer *et al.* (1985) comment that the existence of multiple copies of the FMRF sequence within the FMRFamide precursor may have implications for the quantal content of FMRFamide-containing vesicles. Additionally, it is possible that efficient translocation of a polypeptide precursor across membranes may require a product of a minimum size. In this case, economy would be promoted by incorporating many copies of the required peptide product into the primary gene product.

The sequence of the FMRFamide gene clearly reveals that it arose from repeated duplication of a 48-bp unit that encodes the basic FMRF and an acidic spacer peptide. The overall electroneutrality of this unit may be necessary for efficient packing of vesicles.

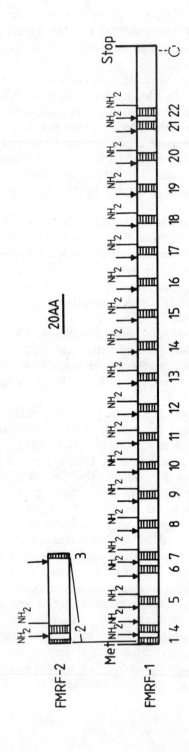

Figure 5.3: Organisation of the precursor polypeptide encoded by an *Aplysia* FMRFamide gene. The precursor is inferred from the nucleotide sequence of the cDNA clones FMRF-1 and FMRF-2. Horizontal lines indicate (numbered) FMRF amino acid sequences, arrows indicate proteolytic cleavage sites at dibasic residues, lines indicate cleavage at single basic residues, NH_2 indicates potential amidation sites (-Gly-Lys-). The dotted circle indicates a potential C-terminal glycosylation site. Note that the precursor lacks a classical signal sequence. Southern blotting shows that the FMRF-1 clone corresponds to a single homologous gene. However, RNA blots show that this gene gives rise to several distinct transcripts of varying sizes. Alternate splicing at the many potential donor and acceptor sites encoded by the -Arg-Phe- sequence is a likely explanation for this (see text). After Schaefer *et al.* (1985)

QUANTIFICATION OF INVERTEBRATE NEUROPEPTIDES

Bioassay is probably the simplest and most direct means of determining the quantity of peptide in a sample, provided authentic peptide is available for calibration of the assay response. A good bioassay should ideally detect picomole or subpicomole quantities of peptide and respond quantitatively over two or three orders of magnitude. O'Shea and colleagues, for their studies of the insect pentapeptide proctolin, employed as an assay a spontaneous rhythm of contraction that occurs in the extensor tibiae muscle of the locust leg. This preparation is capable of responding with an accelerated rhythm to as little as 10 fmol of proctolin (Bishop *et al.*, 1981) which can be applied in volumes as small as 1 µl, allowing measurement of the proctolin content of individual neurones (O'Shea and Adams, 1981). It is, of course, necessary in bioassays to verify that bioactivity is due to the peptide under study, and not another substance to which the bioassay might also be sensitive. Co-migration of bioactivity and authentic peptide on HPLC (which offers both high resolution and recovery) is a suitable way of verifying this and has been used by a number of authors working with proctolin in various invertebrates. The cockroach *Periplaneta americana* (Bishop *et al.*, 1981), the locust *Schistocerca nitens* (Keshishian and O'Shea, 1985), the lobster (Schwarz *et al.*, 1984), and the crayfish (Bishop *et al.*, 1984) all have authentic proctolin. By contrast, in the king crab, *Limulus polyphemus* (a chelicerate), endogenous proctolin-like bioactivity may not all be due to proctolin itself (Watson *et al.*, 1983).

Bioassays have the disadvantages that they are frequently time-consuming, do not usually permit parallel estimations, and may lack precision. Therefore, researchers have where possible increasingly turned to other procedures for quantifying neuropeptides.

Chemical methods rely on separation of the peptide of interest from other substances (e.g. by HPLC) followed by measurement of UV light absorption, or some other physical attribute of the molecule. Usually this will require substantial pre-purification. However, O'Shea *et al.* (1984) were able to quantify the AKH-like neuropeptides MI and MII from the corpora cardiaca of the cockroach *Periplaneta americana* by measuring UV absorbance at known retention times on reversed-phase HPLC of tissue extracts after only minimal (Sep-Pak) clean-up. It was only possible to do this because cockroach corpora cardiaca contain relatively large amounts of MI and MII. Siegert and Mordue (1986) have done the same for AKH-I and II in the locusts *Locusta migratoria* and *Schistocerca gregaria*. For most invertebrate neuropeptides this would not be a practical method of quantification. The use of electrochemical detection (instead of UV absorption) may offer increased detector selectivity, so that the presence of co-eluting contaminants may be less important. This method has been successfully applied to vertebrate neurophysins (Bennett *et al.*, 1985) but so far not (to our knowledge) to invertebrate peptides.

Immunoassays are undoubtedly now the most important method of peptide quantification. Of these radioimmunoassay (RIA) is by far the most popular. Peptide radiolabelled to high specific activity is essential to a sensitive RIA. Some tritiated tracers have been successful (e.g. ^{3}H-enkephalins), but labelling by iodination of tyrosine residues using ^{125}I is more common. This has the advantages of higher activity and ease of counting, but has the disadvantages of shorter label life and (more seriously) the potential disruption of chemical conformation and biological activity by the introduction of a large substituent atom into the molecule. Iodination is a particularly drastic solution to this problem where very small peptides are the subjects (e.g. the pentapeptide proctolin contains a Tyr residue that can be iodinated. The resulting (^{125}I-Tyr)-proctolin is the radiolabelled ligand used in the proctolin RIA devised by Kingan and Titmus (1983). However, (I-Tyr)-proctolin is not biologically active. Also many small peptides do not contain a suitable Tyr residue that can be iodinated. In these cases, ^{125}I can be introduced by substituting a Tyr residue into the sequence of the peptide. For instance (Tyr[1])-AKH I has been prepared by Schooneveld et al. (1983). Again this derivative of the native peptide is not biologically active (W. Mordue, personal communication). RIAs have been reported for three invertebrate peptides: crustacean hyperglycaemic hormone, CHH (Jaros and Keller, 1979); FMRFamide (Dockray and Williams, 1983); and proctolin (Kingan and Titmus, 1983).

A criticism that can be levelled against RIA determinations is that the identity of the molecules in the sample that are bound by the antibody is often not known. In other words, all or part of the quantity estimated could be due to the presence of substances that are not identical with the original antigen, but which possess similar antigenic determinants. The use of antibodies of recognised specificity, preferably directed against a distinguishable region of the peptide, can be useful in this regard. A region-specific RIA can be most useful when distinction between two structurally related peptides (e.g. Leu- and Met-enkephalin) is demanded. Although non-parallelism of the binding curves for authentic and unknown peptides is useful in excluding obvious cases of cross-reactivity, parallel curves do not provide a guarantee of identity. This can only be established by full chemical characterisation of the unknown.

Cross-reactivity in RIA is of course sometimes useful. For example, RIA has confirmed the previously suspected species variability in crustacean CHH, peptides from different decapods differing in their ability to displace Carcinus CHH from the antibody (Keller, 1983). Cross-reactivity can also be useful in revealing the existence of novel peptides related to the original antigen, as is illustrated by the work of Thorpe and Duve (described above) on vertebrate-type peptides in blowfly brain. Similarly, the existence of the vertebrate peptide LPLRFamide was predicted on the basis of its cross-reactivity in an FMRFamide RIA (Dockray *et al.*, 1983).

Other immunological assay techniques are currently less important than RIA, but may become more popular in the future. Enzyme-linked immunoabsorbent assays (ELISA) are important because they eliminate the need for radiolabelled peptide. This may be particularly important where it is not possible to iodinate a peptide without substantial modification of its structure. An ELISA has been employed to measure circadian changes in circulating levels of melanophore dispersing hormone (MDH) in the blood of a crab (Quackenbush and Fingerman, 1985).

Radioreceptor assays use the same principles as RIA, but use a purified receptor preparation instead of antibody to bind the peptide. Although descriptions of radioreceptor assays for several vertebrate peptides have been published, none has been reported for an 'invertebrate-type' peptide. However, Leung and Stefano (1984) used an opiate binding assay to quantitate Leu- and Met-enkephalin during their isolation from the mussel, *Mytilus edulis.* The same receptor assay also revealed the existence of YGGFMRF (Stefano and Leung, 1984). Radioreceptor assays may have the advantage over RIA that they may give a more accurate indication of biological activity. On the other hand it will usually be less convenient to use isolated receptors than to use an antiserum in a competitive binding assay.

LOCALISATION OF INVERTEBRATE NEUROPEPTIDES

The peptidergic phenotype

Neuropeptides, in invertebrates as in vertebrates, are synthesised in neurones that contain numerous large, membrane-bound vesicles with contents that appear electron dense in conventional electron microscopic (EM) images. These cells are well endowed with rough endoplasmic reticulum and have a prominent Golgi apparatus. Such cells, which correspond to the traditional description of 'neurosecretory' cells (Maddrell and Nordmann, 1979), are not rare in the nervous systems of either vertebrates or invertebrates. Snyder (1980) estimated that 50–60 per cent of synapses in the vertebrate CNS might be peptidergic. It seems likely that a similarly high proportion of invertebrate neurones are peptidergic. Many of these cells secrete their peptide products into the blood as neurohormones. The populations of neurosecretory cells that constitute the arthropod neuroendocrine system have been the subject of numerous reports (see reviews by Maddrell, 1974; Raabe, 1982; Gupta, 1983), with perhaps rather fewer for molluscs (see Joosse and Geraerts, 1983) and other invertebrate phyla.

Neuroendocrine cells often have axons that project to neurohaemal structures, such as the insect corpora cardiaca and perivisceral organs (PVOs) and the crustacean sinus glands. These organs are the sites of storage of neurosecretory products prior to release — a point we made earlier when we considered their exploitation for peptide purification.

Invertebrate neurones are often uniquely identifiable. The nervous systems of gastropod molluscs are particularly suitable for this approach on account of the large size of many neuronal cell bodies. Peptidergic-type cells in *Aplysia* can frequently be identified visually in fresh preparations by their white or light blue appearance under suitable incident illumination. This colouration (due to Tyndall light scattering caused by the presence of large numbers of neurosecretory vesicles), contrasts sharply with the orange colour of other neuronal cell bodies, which is due to carotenoid pigments. The ease with which peptide-containing cells can be identified in *Aplysia* was an early stimulus to the study of their function. The bag cells were discovered by Kupfermann in 1967 to be the source of a factor that induces egg-laying behaviour, and the giant, bursting cell R15 was shown by Kupfermann and Weiss in 1976 to be the source of a peptide that regulates water balance. The synaptic connections and activity of these and other identified peptidergic cells in *Aplysia* have been readily established by electrophysiological investigations (Kandel, 1976). Similar studies have revealed the peptidergic functions of identified neurosecretory cells in other snails, particularly *Lymnaea* (Joosse, 1984).

Although the Tyndall effect can sometimes be used to locate peptidergic cells in arthropod nervous systems, their more complex anatomy and smaller cell sizes have made for slower progress. Some cells have been identified by microdissection and bioassay, as in the molluscan examples. Agui *et al.* (1979) were able to pinpoint the site of PTTH synthesis in the brain of the lepidopteran, *Manduca sexta*, as a single cell in each of the paired lateral neurosecretory cells of the protocerebrum. In other cases, identification of peptidergic cells has brought a battery of techniques to bear on the problem. Often, the cell bodies of neurosecretory cells have been located by backfilling with dye from the relevant neurohaemal organ, usually using cobalt chloride. This may establish the cell as a member of a known group of cells with a peptidergic phenotype. Details of the cell's projection and dendritic anatomy may allow it to be individually identified. The contents of the individually microdissected cells, or small groups of cells, can be subjected to microbioassay or microchemical determination. In this way, the identities were established of cells in each of the abdominal ganglia of a moth *Manduca* that contain the (as yet unsequenced) peptide neurohormone, bursicon (Taghert and Truman, 1982). Three pairs of median neurones in the same ganglia appear to synthesise and secrete two small cardioactive peptides (Tublitz and Truman, 1985). The same approach can be used with peptidergic cells that are not neuroendocrine, but which project to peripheral targets. O'Shea and Bishop (1982) showed that individually dissected cockroach motor neurones (D_s) contain material that is indistinguishable from proctolin on HPLC.

Immunocytochemistry

Antibodies have been extensively used not only to quantify neuropeptides, but also to determine the distribution and identity of peptidergic cells. Immunocytochemistry has been one of the dominant themes of invertebrate peptide research, and the field continues to expand as new immunoreactivities are reported. This very probably reflects the ease with which tissues can be screened using this method. An extensive recent review is that of Greenberg and Price (1983).

Immunocytochemical mapping has revealed that cells of a particular phenotype are normally relatively rare, and that the distribution patterns of cells reactive for individual peptides vary widely. Immunoreactive cells may be restricted to a single structure (e.g. CHH-reactive cells are apparently found only in the crab X-organ/sinus gland complex — Jaros and Keller, 1979), or may be widespread throughout the entire CNS (e.g. cells that are stained by an anti-proctolin serum are found in most ganglia of cockroaches and grasshoppers — Bishop and O'Shea, 1982; Keshishian and O'Shea, 1985). Observations on proctolin illustrate the important point that neurones that are diverse in structure, location and (presumably) function may utilise the same neuropeptide. Proctolin-containing neurones in the cockroach include cells whose axons project peripherally to skeletal muscles (O'Shea and Bishop, 1982), and to visceral muscles (Eckert *et al.*, 1981), and also neurones whose projections do not leave the CNS (i.e. interneurones — Bishop and O'Shea, 1982). As many as 5–10 per cent of all motor neurones in cockroaches and locusts may be proctolingeric (O'Shea and Schaffer, 1985).

A large amount of work has been directed towards identification within invertebrate nervous systems of cells immunoreactive to antibodies generated against vertebrate-type peptides. Greenberg and Price (1983) attempt a voluminous list of the antibodies used and the animals screened. Here we will give just two examples of this kind of work to illustrate typical results.

Table 5.5 summarises the findings of El Salhy *et al.* (1983) who tested a variety of antisera on the brain of the moth *Manduca sexta*, and of Schot *et al.* (1981) who worked with the pond snail, *Lymnaea stagnalis*. The impression to be gained from a review of such studies is that almost all the antisera used have discovered immunoreactive material somewhere in an invertebrate nervous system. A tentative conclusion from this must be that vertebrate-type peptides are phylogenetically ancient, and have been conserved (at least in their antigenic determinants) over long periods of

Table 5.5: Immunological evidence for vertebrate-type neuropeptides in invertebrates: two examples of immunohistochemical studies

A. *Vertebrate-type neuropeptides found in the brain of the moth,* Manduca sexta (El Salhy *et al.*, 1983)

Found	Not found
Insulin B-chain	Insulin A-chain
Somatostatin	Insulin C-peptide
Glucagon	Nerve growth factor (NGF)
Bovine pancreatic polypeptide (BPP)	Epidermal growth factor (EGF)
Secretin	Neuropeptide YY
Vasoactive intestinal peptide (VIP)	Neurotensin
Glucose-dependent insulinotropic peptide (GIP)	Bombesin
Gastrin (C-terminal specific)	Motilin
Enkephalin (Met or Leu[a])	ACTH
α-Endorphin	
β-Endorphin	
Substance P	
Calcitonin	

B. *Vertebrate-type neuropeptides found in the brain of the pond snail,* Lymnaea stagnalis, (Schot *et al.*, 1981)

Found	Not found
Vasopressin	β-Endorphin
Vasotocin	Cholecystokinin (CCK)
Oxytocin	Neurophysin I and II
α-Melanocyte stimulating hormone (α-MSH)	
Met-enkephalin	
Somatostatin	
Glucagon	
Insulin	
Glucose-dependent insulinotropic peptide (GIP)	
Vasoactive intestinal peptide (VIP)	
Gastrin	
Secretin	
Pancreatic polypeptide	
Substance P	
Calcitonin	

[a] The antiserum used did not differentiate Met- and Leu-enk.

evolutionary time. However, as we noted above when considering the specificity of RIA, immunological evidence gives little idea of the chemical nature of the substance found by the antibody, other than that it is recognised in at least one region as conformationally similar to the original antigen. There are problems of spurious cross-reactivity and non-specific binding. The abolition of immunostaining by preincubation of the antibody with authentic antigen can eliminate non-specific binding but cannot exclude the possibility of cross-reactivity (as is often implied). To make best use of immunocytochemical investigations, they really need to be supplemented by chemical characterisation of the immunoreactive peptides. This is particularly the case for vertebrate-type peptides present in invertebrates, where doubts must exist about their chemical nature (see above).

It is nevertheless possible to gain useful information from immunological studies using antibodies raised against purified antigens of unknown chemical nature. Copenhaver and Truman (1986) have raised antibodies against purified eclosion hormone (EH), on 8.5 kDa neuropeptide from the moth *Manduca sexta*. This was used to locate the site of EH synthesis in a bilaterally paired group of 5–6 neurones in the lateral group of brain neurosecretory cells.

The use of monoclonal antibodies (MABs) is sometimes supposed to offer an advantage in immunohistochemistry by assuring specificity. Actually MABs do not necessarily provide greater specificity than polyclonals (Lane and Koprowski, 1982). Additionally many MABs prove unsuitable for use in histochemistry (Haaijman *et al.*, 1984). Perhaps the major advantage offered by the use of hybridoma technology lies in the standardisation of the primary antibody used in staining. Additionally, *in vitro* immunisation may allow MABs to be prepared from vanishingly small amounts of immunogen (Reading, 1982).

Another potential use of MABs is in generating immunocytochemical staining reagents that distinguish populations and subpopulations of neurones within the CNS. Such 'libraries' of MABs can be generated by immunising with a homogenate of whole CNS, and using an immunocyto-chemical screen to detect clones of hybridoma cells that produce interesting antibodies. The unknown antigens recognised by such antibodies may include neuropeptides (see Valentino *et al.*, 1985). MABs against *Manduca* cardioactive peptides have been prepared by Taghert *et al.* (1984) by immunising mice with an unpurified homogenate of abdominal neurohaemal organs, which are a rich source of these peptides. The MABs were used to locate populations of neurones containing the peptides, and also to delete biological activity (Tublitz *et al.*, 1985).

In situ hybridisation

This technique relies on the selective hybridisation in tissue sections of a gene-specific radiolabelled DNA probe (usually a restriction fragment from an appropriate cDNA clone but sometimes a synthetic oligonucleotide) with cytoplasmic mRNA. The location of the hybridised probe is subsequently revealed by autoradiography. This technique reveals not the presence of the peptide itself, but the genetic instructions for making it. It is potentially valuable in revealing neurones that transcribe peptide genes but do not translate the message, and those cells that continually release the peptide product so that the quantity of stored peptide is small. Additionally, of course, it is useful in examining the expression of genes for which the sequence of the protein product is unknown, or where antibodies to the gene product are not available. This technique has, in the present context of invertebrate neuropeptides, so far only been applied to the case of *Aplysia*, since it is only in this case that the necessary DNA probes are available.

In situ hybridisation was used in conjunction with immunocytochemistry (using antibodies directed against synthetic peptides) to map ELH gene expression in the *Aplysia* nervous system (McAllister *et al.*, 1983). All ganglia except the pedal ganglion proved to contain neurones expressing ELH genes. Some of these cells could be identified as interneurones, implying that ELH gene-related peptides have transmitter or modulatory functions within the CNS. The same methods were used to trace the embryonic origins and developmental fates of some of these ELH cells. The bag cells arise early as an extension of the larval body wall, later migrating to their final positions in the adult CNS. Since ELH gene transcripts are detected early in development, before these cells are fully differentiated as neurones, they must be committed to their eventual fate while still in the body wall. By contrast, the atrial gland (which also expresses ELH family genes) arises from part of the developing reproductive system, and cellular migration is not involved.

In situ hybridisation, like immunocytochemistry, can suffer from the problem of lack of specificity. The probes used in the study by McAllister *et al.* (1983) were unable to distinguish between the various members of the ELH gene family. Thus when it was found that certain neurones in the head ganglia were labelled, it could not be determined whether these cells expressed ELH, A, B or some combination of these. Shymala *et al.* (1986) addressed this problem by cloning cDNA from head ganglion message. Sequence information revealed that both ELH and (to a lesser extent) peptide A genes are transcribed in the head ganglia. As yet the function of the neuropeptide gene products in the head ganglion is unknown.

Colocalisation

Evidence has accumulated that, contrary to previous belief, neurones may contain more than one transmitter substance. Frequently a peptide has been shown to be localised with a 'classical' transmitter. For example, substance P and serotonin have been shown to occur together in the same vertebrate CNS (medulla oblongata) neurones (Chan-Palay *et al.*, 1978) and in the same vesicles isolated from the dorsal horn of the spinal cord (Pelletier, 1981). This exciting field has been reviewed by Hokfelt *et al.* (1984).

There are a number of reports of colocalisation of invertebrate neuropeptides. For example, proctolin and serotonin have been shown to coexist in specific cells of the lobster (Siwicki and Kravitz, 1984). An α-endorphin-like immunoreactive substance has been localised to cells in the suboesophageal ganglion of the silkworm, *Bombyx*, that also contains dopamine (Takeda *et al.*, 1986). The finding of peptide and monoamine transmitters in the same neurone obviously poses questions about neural transmission. Do the two transmitters coexist in the same vesicles or are there separate populations? Are both released when the cell fires an action potential? Is the proportion of each transmitter released always the same? Do the transmitters mediate the same or different responses in their post-synaptic targets? Answers to such questions, which are of general import-ance in neurobiology, are likely to be provided largely by the study of invertebrate preparations, in which neurones can be uniquely identified. For example, Adams and O'Shea (1983) have investigated the role of proctolin in an identified motor neurone (D_s) in the cockroach metathor-acic ganglion. They showed that the release of the classical excitatory trans-mitter at the neuromuscular junction (thought to be glutamate) is accom-panied by the co-release of proctolin. The response of the muscle to nerve stimulation is biphasic. The glutamate-like transmitter is responsible for a transient contraction, whereas proctolin produces a sustained, catch-like contraction that is only evident after a number of neural stimuli have been received. Preparations like these are well suited to increasing our under-standing of the functional significance of colocalisation and cotransmission.

In addition to cases where neuropeptides coexist with amines, there are neurones that synthesise and secrete more than one neuropeptide. We have already seen (above) that the bag cells of the *Aplysia* abdominal ganglion contain several small peptide products which are all derived from the post-translational processing of a single gene product (Rothman *et al.*, 1983). The finding that the pentapeptide proctolin apparently co-exists with the much larger peptide tanning hormone, bursicon, in certain identified neurones of the cockroach has led to the suggestion (Adams and Phelps, 1983) that these two peptides may both be derivatives of a single, larger, polyprotein precursor. Since the regional distribution within the CNS of the two peptides differs, differential processing must occur if the two are indeed encoded by a single gene.

MODE OF ACTION OF INVERTEBRATE NEUROPEPTIDES

Receptors

Biologically active peptides are generally accepted to exert their effect by interacting with receptors on the plasma membranes of target cells. Pharmacological and biochemical studies of receptors for 'classical' neuro-transmitters have shown that in most cases receptors for a single transmitter exist as multiple types within the same animal — for instance ACh receptors may be of nicotinic or muscarinic type depending on the conformational requirements of the receptor (see elsewhere in this volume). Similar pharmacological studies of vertebrate neuropeptides have revealed that multiple receptor types exist for peptides too. For example, receptors for endogenous opiates are considered to fall into at least four categories, according to their relative affinities for a range of different agonists (Zukin and Zukin, 1984). It is not known whether these multiple receptor types are normally required to distinguish among a range of endogenous neuropeptides *in vivo*, or whether each receptor type is normally exposed only to a single peptide species.

Very little is known about receptors to invertebrate neuropeptides: probably the best documented study concerns opioid receptors in the marine mussel, *Mytilus edulis*. Nervous tissue from this bivalve contains high-affinity stereospecific binding sites for a number of opioid ligands, including D-Ala^2 N-Me-Phe^4, Met (O)-ol enkephalin; etorphine; and levallorphan (Kream *et al.*, 1980); naloxone (Stefano *et al.*, 1980); and D-Ala^2, Met^5 enkephalinamide (DAMA) (Stefano and Leung, 1984). The pharmacology of binding resembles that of the δ-type opioid receptor of mammalian brain. *Mytilus* ganglia have been shown to contain at least three neuropeptides that are able to bind to these receptors: Met-enkephalin, Leu-enkephalin (Leung and Stefano, 1984) and Met-enkephalin-Arg^6-Phe^7 (Stefano and Leung, 1984).

The brain of the cockroach *Leucophaea maderae* has been shown to possess high binding sites for D-Ala^2, Leu^5 enkephalinamide (DALA) (Stefano and Scharrer, 1981). Interestingly there were 30 per cent more binding sites in the brains of adult females than in adult males, but there was no difference between the brains of larval (i.e. sexually immature) insects. The sites appear to be kappa-like. They may have some role in regulating locomotory activity (Ford *et al.*, 1986). There are also opioid sites in the midgut of the same insect (Stefano *et al.*, 1982), in this case demonstrated by the binding of DAMA.

This work correlates nicely with that of Gros *et al.* (1978) which showed the presence of Met- and Leu-enkephalin-like immunoreactive material in the brain of *Locusta*. This material was separable from Met-enkephalin on gel filtration and was therefore different from it. Remy and Dubois (1981) have mapped the enkephalin-immunoreactive cells in the locust brain and suggest that they may play a role in regulating the release of other neurosecretory peptides from the neurohaemal corpora cardiaca.

Some recent work has focused on proteins in *Drosophila* that may be receptors for endogenous epidermal growth factor (EGF) and insulin-like peptides. Petruzzelli *et al.* (1985) used affinity chromatography to isolate from adult fruit flies a protein that binds insulin, and shares with its vertebrate homologue many characteristics (e.g. α- and β-subunits of about the right size). In parallel, Thompson *et al.* (1985) have identified membrane proteins in two *Drosophila* cell lines that are recognised by antisera against mammalian EGF and insulin receptors. One of these proteins is recognised by both antisera, and is also able to bind ^{125}I-insulin, although ^{125}I-EGF was not bound. The relationship between the affinity-purified receptor from adult flies, and the immunoreactive proteins from the cell lines is not yet clear. However, one of the other membrane proteins studied by Thompson *et al.* (1985) may be identical with the product of a gene cloned by Livneh *et al.* (1985) who isolated a *Drosophila* cDNA that has extensive sequence homology with the vertebrate oncogene C-erbB. The *Drosophila* gene has 41 per cent homology at the amino acid level with human EGF receptor.

The only other studies that relate to invertebrate peptide receptors are those providing pharmacological information about the structural requirements of receptors. Examples include the insect peptides proctolin (Starratt and Brown, 1979; Sullivan and Newcomb, 1982), and AKH-I (Stone *et al.*, 1978; Goldsworthy *et al.*, 1986), and a number of studies for the molluscan peptide FMRFamide (e.g. Painter *et al.*, 1982; Muneoka and Saitoh, 1986).

Second messengers

Adenosine 3′,5′-cyclic monophosphate (cAMP) is now well established as an intracellular second messenger in the actions of many activating signals (Schramm and Selinger, 1984). There is evidence that the actions of a number of invertebrate neuropeptides involve adenylate cyclase activation. The techniques used to show this include: mimicking the normal response to peptide by administering cAMP (or more usually a more permeant and longer-lasting derivative such as dibutyryl cAMP or 8-(4-chlorophenyl-thio)-cyclic AMP; the use of the diterpene compound forskolin to increase endogenous cAMP levels by activating adenylate cyclase; potentiating the action of a subthreshold dose of peptide by administering an antagonist of cAMP phosphodiesterase, such as 3-isobutyl-1-methylxanthine (IBMX); showing that stimulation by peptide results in a dose-dependent increase in adenylate cyclase activity; and demonstrating that peptide-activated tissues contain increased quantities of cAMP. More often than not, any one investigation will include only one or two of these experimental approaches. Unfortunately, this sometimes leaves the conclusions drawn open to some doubt.

Enhanced cAMP levels have been found in many invertebrate tissues following exposure to a biologically active peptide. Examples selected from many include *Aplysia* (Lloyd *et al.*, 1984) tail sensory neurones (Ocorr and and Byrne, 1985), and heart (Lloyd *et al.*, 1985), all on exposure to the

211

molluscan small cardioactive peptide B (SCP$_B$); the heart of the clam *Mercenaria mercenaris* when perfused with FMRFamide (Higgins *et al.*, 1978); and the prothoracic glands of the caterpillar *Manduca sexta* when incubated with PTTH (Smith *et al.*, 1984). However, the results of other experiments performed in similar ways show that cAMP is not always involved in the actions of invertebrate peptides. Thus, although proctolin may exert its effects at the cockroach hindgut through cAMP (Jennings *et al.*, 1983), its action in the lobster claw opener muscle is achieved without accumulation of cAMP (Kravitz *et al.*, 1985). Similarly, although FMRF-amide stimulates cAMP production in cardiac muscle of *Mercenaria* (Higgins *et al.*, 1978), the peptide has no effect on cAMP levels in the anterior byssus retractor muscle (ABRM) of the mussel *Mytilus edulis* (Painter *et al.*, 1982) in which it induces a long-lasting catch contraction. The obvious conclusion is that the possibility of cAMP mediation of any particular peptidergic response requires individual examination.

Responses to biologically active peptides may alternatively involve mediation by guanosine $3',5'$ cyclic monophosphate (cGMP). The role of cGMP as a second messenger is much less well understood than is the case for cAMP, except perhaps in vertebrate photoreceptors (Kaupp and Koch, 1986), where light acts through an enzyme to activate a cGMP phospho-diesterase and thus decrease cGMP levels. Relationships between cGMP levels and the actions of the crustacean hyperglycaemic hormone, CHH (Sedlmeier and Keller, 1981) and the insect eclosion hormone (Truman *et al.*, 1979) have been demonstrated. In the latter case, the absence of a response to the peptide in Ca^{2+}-free medium implies that activation of guanylate cyclase may in turn be dependent upon an increase in the intra-cellular concentration of calcium ions (Truman, 1980). This illustrates that more than a single intracellular messenger may be required to couple the response of the cell to the peptidergic stimulus. Levitan *et al.* (1979) have proposed that the characteristic response of the *Aplysia* abdominal ganglion neurone R15 to an unknown neuropeptide isolated from another snail, *Helix*, may involve elevation of both cAMP and cGMP levels.

Calcium ions are important regulators of cellular function, and are frequently involved in cellular activation as second messengers (Campbell, 1983). In the locust, proctolin may act on the muscles of the hindgut by increasing intracellular Ca^{2+} levels (Dunbar and Huddart, 1982). PTTH stimulation of steroid synthesis in *Manduca* prothoracic glands is depend-ent on the presence of external calcium. Smith *et al.* (1985) have suggested that the peptide may act to cause Ca^{2+} influx, which in turn activates a Ca^{2+}-sensitive adenylate cyclase; the resulting elevation of cAMP levels is responsible for increased ecdysteroid synthesis. This rather complicated scheme illustrates how several cell regulatory systems may interact with each other.

It has recently been realised that a wide variety of cellular responses to various activating stimuli may be mediated by catabolism of inositol lipids to produce, among other products, inositol trisphosphate (IP_3) and diacyl-glycerol (DG). It has been proposed that IP_3 and DG may be important second messengers, IP_3 acting to release calcium from intracellular stores, and DG to activate protein kinase C (Berridge and Irvine, 1984; Nishizuka, 1984). It seems exceedingly likely that at least some invertebrate neuropeptides will be found to act via this second messenger system. As far as we know there is at the time of writing no direct evidence for this proposition. However, the response of blowfly salivary gland to the amine serotonin has been shown to involve increased phosphoinositide metabolism (Berridge et al., 1984). Thus the elements of this potential second messenger system are known to be available in invertebrates, and rapid progress may be expected in this area.

Ion channels in cell membranes

If neuropeptides function as neurotransmitters and/or neuromodulators in the invertebrate CNS, they might be expected to influence the electrical activity of postsynaptic cells by opening or closing membrane channels. Invertebrate examples of changes in neuronal activity following peptide application are numerous (reviewed by Haynes, 1980; O'Shea and Schaffer, 1985), but rather few systems have been investigated in detail so as to reveal the effects of peptides on individual membrane ionic conductances. Probably the best example is the work of Cottrell (1982) and Cottrell et al. (1984), who have studied the responses to FMRFamide of several identified neurones in the CNS of the garden snail Helix aspersa using voltage-clamp methods. Peptide application alters membrane conductance in different ways according to the neurone studied. Some cells are depolarised by increased membrane permeability to sodium ions, while others respond with hyperpolarisation caused by increased potassium permeability. These responses are in accord with both excitatory and inhibitory transmitter-like actions for FMRFamide. A voltage-dependent inward current is seen in identified cell C1 when FMRFamide is applied. This appears to result from the closure of a Ca^{2+}-dependent potassium channel, and perhaps indicates a modulatory role for the peptide at this neurone.

Another molluscan neuropeptide, Aplysia ELH, also exerts a variety of effects on different target cells, including both excitatory and inhibitory effects (Brownell and Mayeri, 1979). In this case the very long-lasting nature of these changes has led Mayeri and Rothman (1982) to term these actions 'non-synaptic'. By contrast the effects on neuronal activity in the abdominal ganglion of another Aplysia neuropeptide, α-BCP, are of short duration (Rothman et al., 1983) indicating that its actions may be more local than those of ELH. Neuropeptides may act to modulate transmitter release from target neurones by altering membrane ion channels. The *Aplysia* peptides SCP_A and SCP_B facilitate neurotransmitter release by the siphon sensory neurones as a result of a cAMP-dependent closure of the $S-K^+$ channel that broadens the presynaptic action potential (Abrams *et al.*, 1984). Better understanding of the ionic mechanisms of neuronal responses to neuropeptides will undoubtedly further explain their functions as neuroregulators.

INACTIVATION OF INVERTEBRATE NEUROPEPTIDES

Little is known about how the actions of peptide neuroregulators are terminated in either vertebrate or invertebrate nervous systems. Enzymic cleavage and active re-uptake are two possible mechanisms for neuropeptide inactivation. Paradigms for these processes are the hydrolysis of acetylcholine by acetylcholinesterase in cholinergic synapses, and the active re-uptake of catecholamines by specific neuronal and glial active-transport mechanisms. There is at present no evidence for active re-uptake, so that extracellular enzymatic inactivation is favoured by most workers.

Neuropeptides are obviously susceptible to enzymic degradation by hydrolysis of their peptide bonds, through the activities of peptidases or proteases. Though there is considerable *in vitro* evidence that suitable enzymes can be prepared from vertebrate brain, there is only limited evidence that they are involved in the termination of neuropeptide action *in vivo* (see reviews by Dua *et al.*, 1985; White *et al.*, 1985).

An important question is that of enzyme specificity. Despite the demonstration of peptidases that cleave at identified sites (e.g. the enkephalin inactivating enzymes reviewed by Schwartz, 1983), it is not known if specific peptidases exist for each neuropeptide, or whether a single enzyme normally serves to degrade many peptides. The number of vertebrate brain peptidases so far examined is small. Most interest centres on those that are located in the cell membrane. 'Enkephalinase' (enkephalin dipeptidyl-carboxypeptidase) is a promising example, being membrane bound and unevenly distributed within the brain, occurring in areas rich in opioid receptors. Inhibitors of the enzyme enhance enkephalin activity (Schwartz, 1983). Like many other candidate neuropeptidases, however, enkephalinase is not unique to the nervous system, being apparently identical to an enzyme (endopeptidase-24.11) previously isolated from kidney and intestine (Kenny, 1986). It may be that it is the location of these enzymes that defines their function, rather than their specificity.

It has been suggested that peptides with terminal modifications (i.e. C-terminal amidation, or N-terminal pGlu) may be less susceptible to enzymic degradation, and therefore particularly suitable to exert long-term

214

effects within the CNS, or to act as circulating neurohormones (O'Shea and Schaffer, 1985). This may be particularly important in invertebrates, which often have rather sluggish circulations. It is notable that a high proportion of the neuropeptides listed in Table 5.1 have C-terminal amide groups. The AKH family of peptides (several of which are known to act as neuro-hormones) are in fact blocked at both ends, which may be supposed to render them resistant to the actions of most carboxypeptidases and amino-peptidases. We have, however, found that both *Manduca* AKH and locust AKH-I are degraded by an enzyme present in *Manduca* haemolymph (Fox and Reynolds, 1986).

So far, studies of the inactivation of invertebrate neuropeptides have been limited. Proctolin may apparently be inactivated by two pathways in the cockroach *Periplaneta americana*, being cleaved at either Pro-Thr by a carboxypeptidase, or at Tyr-Leu by an endopeptidase (Quistad *et al.*, 1984; Starratt and Steele, 1984; Steele and Starratt, 1985). These studies were, however, concerned with haemolymph metabolism. It is not clear that proctolin ever circulates as a neurohormone in the blood, and so it may be more pertinent to look for proctolin-inactivating enzymes at known release sites. R.E. Isaac (University of Leeds — personal communication) has found that membrane preparations from both nervous system and muscle of *Schistocerca gregaria* will also cleave proctolin. In this case the major route at high substrate concentration involves an aminopeptidase, but at low substrate concentration an endopeptidase that cleaves at the Tyr-Leu bond is operative. The presence of local inactivating enzymes in muscle supports the view (O'Shea and Schaffer, 1985) that proctolin is not a neurohormone but a locally increased regulator.

The blood of *Astacus* and *Limulus* both contain several enzymes that degrade enkephalins (Coletti-Previero *et al.*, 1985). Since it is not known whether peptides do in fact circulate in these arthropods, the relevance of this finding is not clear.

As is true for most of the areas of study reviewed in this chapter, progress in our understanding of invertebrate neuropeptide inactivation will be to a large extent dependent on the elucidation of the sequences of more peptides.

REFERENCES

Abrams, T.W., Castellucci, V.F., Camardo, J.S., Kandel, E.R. and Lloyd, P.E. (1984) Two endogenous neuropeptides modulate the gill and siphon withdrawal reflex in *Aplysia* by presynaptic facilitation involving cAMP-dependent closure of a serotonin-sensitive potassium channel. *Proc. Natl Acad. Sci. USA 81*, 7956-60

Adams, M.E. and O'Shea, M. (1983) Peptide cotransmitter at a neuromuscular junction. *Science, Wash. 221*, 286-9

Adams, M.E. and Phelps, M.N. (1983) Colocalization of bursicon bioactivity with proctolin in identified neurons. *Soc. Neurosci. Abstr. 9*, 313

Agui, N., Granger, N.A., Gilbert, L.I. and Bollenbacher, W.E. (1979) Cellular localization of the insect prothoracicotropic hormone: *in vitro* assay of a single neurosecretory cell. *Proc. Natl Acad. Sci. USA 76*, 5694-8

Amara, S.G., Jonas, V., Rosenfeld, M.G., Ong, E.S. and Evans, R.H. (1982) Alternative RNA processing in calcitonin gene expression generates mRNAs encoding different polypeptide products. *Nature (Lond.) 298*, 240-4

Barnard, C.S. and Dockray, G.J. (1984) Increases in arterial blood pressure in the rat in response to a new vertebrate neuropeptide, LPLRFamide, and a related molluscan peptide, FMRFamide. *Reg. Peptides 8*, 209-15

Bennett, G.W., Brazell, M.P. and Marsden, C.A. (1981) Electrochemistry of neuropeptides: a possible method for assay and *in vivo* detection. *Life Sci. 29*, 1001-7

Berridge, M.J., Buchan, P.B. and Heslop, J.P. (1984) Relationship of polyphosphoinositide metabolism to the hormonal activation of the insect salivary gland by 5-hydroxytryptamine. *Molec. Cell. Endocrinol. 36*, 37-42

Berridge, M.J. and Irvine, R.F. (1984) Inositol triphosphate, a novel second messenger in cellular signal transduction. *Nature (Lond.) 312*, 315-21

Bishop, C.A. and O'Shea, M. (1982) Neuropeptide proctolin (H-Arg-Tyr-Leu-Thr-OH): immunocytochemical mapping of neurons in the central nervous system of the cockroach. *J. Comp. Neurol. 207*, 223-38

Bishop, C.A., O'Shea, M. and Miller, R.J. (1981) Neuropeptide proctolin (H-Arg-Tyr-Leu-Pro-Thr-OH): immunological detection and neuronal localization in the insect central nervous system. *Proc. Natl Acad. Sci. USA 78*, 5899-6002

Bishop, C.A., Wine, J.J. and O'Shea, M. (1984) Neuropeptide proctolin in postural motorneurons of the crayfish. *J. Neurosci. 4*, 2001-9

Blobel, G. (1980) Intracellular protein topogenesis. *Proc. Natl Acad. Sci. USA 77*, 1496-1500

Bodenmuller, H. and Schaller, H.C. (1981) Conserved amino acid sequence of a neuropeptide, the head activator, from coelenterates to humans. *Nature (Lond.) 293*, 579-80

Bollenbacher, W.E., Katahira, E.J., O'Brien, M., Gilbert, L.I., Thomas, M.K., Agui, N. and Baumhover, A.H. (1984) Insect prothoracicotropic hormone: evidence for two molecular forms. *Science, Wash. 224*, 1243-5

Brain, S.D., Williams, T.J., Tippins, J.R., Morriss, H.R. and McIntyre, I. (1985) Calcitonin gene-related peptide is a potent vasodilator. *Nature (Lond.) 313*, 54-6

Brownell, P. and Mayeri, E. (1979) Prolonged inhibition of neurons by neuroendocrine cells in *Aplysia. Science, Wash. 204*, 417-20

Campbell, A.K. (1983) *Intracellular calcium: its universal role as regulator.* Wiley, Chichester

Chan-Palay, V., Jonsson, G. and Palay, S.L. (1978) Serotonin and substance P coexist in neurons of the rat's central nervous system. *Proc. Natl Acad. Sci. USA 75*, 1582-6

Chiu, A.Y., Hunkapiller, M.W., Heller, E., Stuart, D.K., Hood, L.E. and Strumwasser, F. (1979) Neuropeptide egg-laying hormone of *Aplysia*: purification and primary structure. *Proc. Natl Acad. Sci. USA 76*, 6656-60

Coletti-Previero, M-A., Mattras, H., Zwilling, R. and Previero, A. (1985) Enkephalin-degrading activity in arthropod hemolymph. *Neuropeptides 6*, 405-15

Copenhaver, P.F. and Truman, J.W. (1986) Identification of the cerebral neurosecretory cells that contain eclosion hormone in the moth *Manduca sexta. J. Neurosci. 6*, 1738-47

Cottrell, G.A. (1982) FMRFamide neuropeptides simultaneously increase and

216

decrease potassium currents in an identified neuron. *Nature (Lond.) 296*, 87-9

Cottrell, G.A., Davies, N.W. and Green, K.A. (1984) Multiple actions of a molluscan cardio-excitatory neuropeptide and related peptides on identified *Helix* neurons. *J. Physiol. (Lond.) 356*, 315-34

Dockray, G.J., Reeve, J.R., Shively, J., Gayson, R.J. and Barnard, C.S. (1983) A novel active pentapeptide from chicken brain identified by antibodies to FMRFamide. *Nature (Lond.) 305*, 328-30

Dockray, G.J. and Williams, R.G. (1983) Phenylalanylmethionylarginylphenylalaninamide-like immunoreactivity in rat brain: development of a radioimmunoassay and its application in studies of distribution and chromatographic properties. *Brain Res. 266*, 295-303

Douglass, J., Civelli, O. and Herbert, E. (1984) Polyprotein gene expression: generation of diversity of neuroendocrine peptides. *Ann. Rev. Biochem. 53*, 665-750

Dua, A.K., Pinsky, C. and LaBella, F.S. (1985) Peptidases that terminate the action of enkephalins. Consideration of physiological importance for amino-, carboxy-, and pseudoenkephalinase. *Life Sci. 37*, 985-92

Dunbar, S.J. and Huddart, H. (1982) Calcium movements in insect visceral muscle. *Comp. Biochem. Physiol. 71A*, 425-37

Duve, H., Thorpe, A. and Lazarus, N.R. (1979) Isolation of material displaying insulin-like immunological and biological activity from the brain of the blowfly, *Calliphora vomitoria. Biochem. J. 184*, 221-7

Ebberink, R.H.M., van Loenhout, H., Geraerts, W.P.M. and Joosse, J. (1985) Purification and amino acid sequence of the ovulation hormone of *Lymnaea stagnalis. Proc. Natl Acad. Sci. USA 82*, 7767-71

Eckart, K., Schwartz, H., Chorev, M. and Gilon, C. (1986) Sequence determination of N-terminal and C-terminal blocked peptides containing N-alkylated amino acids and structure determination of these amino acid constituents by using fast-atom bombardment/tandem mass spectrometry. *Eur. J. Biochem. 157*, 209-16

Eckert, M., Agricola, H. and Penzlin, H. (1981) Immunocytochemical identification of proctolin-like immunoreactivity in the terminal ganglion and hindgut of the cockroach *Periplaneta americana* (L). *Cell Tiss. Res. 217*, 633-45

El-Salhy, M., Falkmer, S., Kramer, K.J. and Spiers, R.D. (1983) Immunohistochemical investigations of neuropeptides in the brain, corpora cardiaca and corpora allata of an adult lepidopteran insect, *Manduca sexta* (L). *Cell Tiss. Res. 232*, 295-317

Fernlund, P. (1976) Structure of a light-adapting hormone from the shrimp, *Pandalus borealis. Biochim. Biophys. Acta 439*, 17-25

Fernlund, P. and Josefsson, L. (1972) Crustacean color-change hormone: amino acid sequence and chemical synthesis. *Science, Wash. 177*, 173-5

Ford, R., Jackson, D.M., Tetrault, L., Torres, J.C., Assanah, P., Harper, J., Leung, M.K. and Stefano, G.B. (1986) A behavioural role for enkephalins in regulating locomotor activity in the insect *Leucophaea maderae*: evidence for high affinity kappa-like opioid binding sites. *Comp. Biochem. Physiol. 55C*, 61-6

Fox, A.M. and Reynolds, S.E. (1986) Enzymatic degradation of an insect neuropeptide by haemolymph. *Bull. Soc. Zool. France III*, 36

Gäde, G. (1986) Relative hypertrehalosaemic activities of naturally occurring neuropeptides from the AKH/RPCH family. *Z. Naturforsch. 41C*, 315-20

Gäde, G., Goldsworthy, G., Schaffer, M.H., Cook, J.C. and Rinehart, K.L. (1986) Sequence analysis of adipokinetic hormones II from corpora cardiaca of *Schistocerca nitens, Schistocerca gregaria*, and *Locusta migratoria* by fast atom bombardment mass spectrometry. *Biochem. Biophys. Res. Commun. 134*, 723-30

Goldsworthy, G.J., Mallison, K., Wheeler, C.H. and Gäde, G. (1986) Relative adipokinetic activities of members of the adipokinetic hormone/red pigment concentrating hormone family. *J. Insect. Physiol. 32*, 433-8

Greenberg, M.J., Painter, S.D., Doble, K.E., Nagle, G.T., Price, D.A. and Lehman, H.K. (1983) The molluscan neurosecretory peptide FMRFamide: comparative pharmacology and relationships to enkephalins. *Fed. Proc, 42*, 82-6

Greenberg, H.J. and Price, D.A. (1983) Invertebrate neuropeptides: native and naturalized. *Ann. Rev. Physiol. 45*, 271-88

Gros, C., Lafon-Cazal, M. and Dray, F. (1978) Presence de substances immuno-réactivement apparentées aux encephalines chez un insecte, *Locusta migratoria. C. R. Acad. Sci. Paris 287*, 647-50

Gubler, U., Seeburg, P., Hoffman, B.J., Gage, L.P. and Udenfriend, S. (1982) Molecular cloning establishes pro-enkephalin as precursor of enkephalin-containing peptides. *Nature (Lond.) 295*, 206-8

Gupta, A.P. (ed). (1983) *Neurohormonal organs of arthropods.* C. Thomas, Spring-field, Ill.

Haaijman, J.J., Deen, C., Krose, C.J.M., Zijlstra, J.J., Coolen, J. and Radl, J. (1984) Monoclonal antibodies in immunocytology; a jungle of pitfalls. *Immunology Today 5*, 56-8

Haynes, L.W. (1980) Peptide neuroregulators in invertebrates. *Progr. Neurobiol. 15*, 205-45

Heller, E., Kaczmarek, L.K., Hunkapiller, M.W., Hood, L.E. and Strumwasser, F. (1980) Purification and primary structure of two neuroactive peptides that cause bag cell after discharge and egg-laying in *Aplysia. Proc. Natl Acad. Sci. USA 77*, 2328-32

Higgins, W.J., Price, D.A. and Greenberg, M.J. (1978) FMRFamide increases the adenylate cyclase activity and cyclic AMP level of molluscan heart. *Eur. J. Pharmacol. 48*, 425-30

Hökfelt, T., Johansson, O. and Goldstein, M. (1984) Chemical anatomy of the brain. *Science, Wash. 225*, 1326-34

Holman, G.M., Cook, B.J. and Nachman, R.J. (1986a) Primary structure and synthesis of a blocked myotropic neuropeptide isolated from the cockroach *Leucophaea maderae. Comp. Biochem. Physiol. 85C*, 219-24

Holman, G.M., Cook, B.J. and Nachman, R.J. (1986b) Isolation, primary structure and synthesis of two neuropeptides from *Leucophaea maderae*: members of a new family of cephalomyotropins. *Comp. Biochem. Physiol, 84C*, 205-11

Holman, G.M., Cook, B.J. and Nachman, R.J. (1986c) Primary structure and synthesis of two additional neuropeptides from *Leucophaea maderae*: members of a new family of cephalomyotropins. *Comp. Biochem. Physiol. 84C*, 271-6

Holman, G.M., Cook, B.J. and Nachman, R.J. (1986d) Isolation, primary structure and synthesis of leucomyosuppressin, an insect neuropeptide that inhibits spontaneous contractions of the cockroach hindgut. *Comp. Biochem. Physiol. 850*, 329-33

Hunkapiller, M., Kent, S., Caruthers, M., Dreyer, W., Firca, J., Giffin, C., Horvath, S., Hunkapiller, T. and Hood, L. (1984) A microchemical facility for the analysis and synthesis of genes and proteins. *Nature (Lond.) 310*, 105-11

Ishizaki, H. and Suzuki, A. (1984) The prothoracicotropic hormone of *Bombyx mori.* In: Hoffmann, J. and Parchet, M. (eds) *Biosynthesis, metabolism and mode of action of invertebrate hormones*, pp. 63-77. Springer-Verlag, Berlin

Iversen, L.L. (1983) Neuropeptides — what next? *Trends Neurosci. 6*, 293-94

Jaffe, H., Raina, A.K., Riley, C.T., Fraser, B.A., Holman, G.M., Wagner, R.M., Ridgeway, R.L. and Hayes, D.K. (1986) Isolation and primary structure of a peptide from the corpora cardiaca of *Heliothis zea* with adipokinetic activity. *Biochem. Biophys. Res. Commun. 135*, 622-8

Jaros, P.P. and Keller, R. (1979) Radioimmunoassay of an invertebrate peptide hormone — the crustacean hyperglycaemic hormone. *Experientia 35*, 1252-3

Jennings, K.R., Steele, R.W. and Starratt, A.N. (1983) Cyclic AMP actions on proctolin- and neurally-induced contractions of the cockroach hindgut. *Comp. Biochem. Physiol. 74C*, 69-74

Joosse, J. (1984) Recent progress in the endocrinology of molluscs. In: Hoffmann, J. and Parchet, M. (eds) *Biosynthesis, metabolism and mode of action of invertebrate hormones*, pp. 19-35. Springer-Verlag, Berlin

Joosse, J. and Geraerts, W.P.M. (1983) In: Salenddin, A.S.M. and Wilbur, K.M. (eds) *The Mollusca*, Vol IV. *Physiology*, Part I, pp. 317-406. Academic Press, London.

Kakidani, H., Furutani, Y., Takahashi, H., Noda, M., Morimoto, Y., Hirose, T., Asai, M., Inayama, S., Nakanishi, S. and Numa, S. (1982) Cloning and sequence analysis of cDNA for porcine β-neo-endorphin/dynorphin precursor. *Nature (Lond.) 298*, 245-9

Kaldany, R.J., Namton, J.R. and Scheller, R.H. (1985) Neuropeptides in identified *Aplysia* neurons. *Ann. Rev. Neurosci. 8*, 431-55

Kandel, E.R. (1976) *Cellular basis of behaviour. An introduction to behavioural neurobiology*. W.H. Freeman, San Francisco

Kaupp, U.B. and Koch, K.W. (1986) Mechanism of photoreception in vertebrate vision. *Trends Biochem.* Sci. *11*, 43-7

Keller, R. (1983) Biochemistry and specificity of the neurohaemal hormones in crustacea. In: Gupta, A.P. (ed.) *Neurohaemal organs of arthropods*, pp. 118-48. C. Thomas, Springfield, Ill.

Kenny, J. (1986) Cell surface peptidases are neither peptide- nor organ-specific. *Trends Biochem. Sci. 11*, 40-2

Keshishian, H. and O'Shea, M. (1985) The distribution of a peptide neurotransmitter in the postembryonic grasshopper central nervous system. *J. Neurosci. 5*, 992-1004

Kiehling, C., Martin, R., Geis, R., Bickel, U. and Voigt, K.H. (1984) Cardioexcitatory and opioid activity in extracts from nerve-tissue of *Octopus vulgaris. Gen. Comp. Endocrinol. 53*, 467-8

Kingan, T.G. and Titmus, M. (1983) Radioimmunologic detection of proctolin in arthropods. *Comp. Biochem. Physiol. C 74*, 75-8

Kirschbaum, J.B. (1985) Potential implication of genetic engineering and other biotechnologies to insect control. *Ann. Rev. Entomol. 30*, 51-70

Kramer, K.J., Childs, C.N., Spiers, R.D. and Jacobs, R.M. (1982) Purification of insulin-like peptides from insect haemolymph and royal jelly. *Insect Biochem. 12*, 91-8

Kream, R.M., Zukin, R.S. and Stefano, G.B. (1980) Demonstration of two classes of opiate binding sites in the nervous tissue of the marine mollusc *Mytilus edulis.* Positive homotrophic cooperativity of lower affinity binding sites. *J. Biol. Chem. 225*, 9218-24

Krieger, D.T. (1983) Brain peptides: what, where and why? *Science, Wash. 222*, 975-85

Kupfermann, I. (1967) Stimulation of egg-laying: possible neuroendocrine function of bag cells of abdominal ganglion of *Aplysia californica. Nature (Lond.) 216*, 814-15

Kupfermann, I. and Weiss, K.R. (1976) Water regulation by a presumptive hormone contained in identified neurosecretory cell R15 of *Aplysia. J. Gen. Physiol. 67*, 113-23

Lane, D. and Koprowski, H. (1982) Molecular recognition and the future of monoclonal antibodies. *Nature (Lond.) 296*, 200-2

Leung, M.K. and Stefano, G.B. (1984) Isolation and identification of enkephalins in pedal ganglia of *Mytilus edulis* (Mollusca). *Proc. Natl Acad. Sci. USA 81*, 955-8

Levitan, I.B., Harmar, A.J. and Adams, W.B. (1979) Synaptic and hormonal modulation of a neuronal oscillator — search for molecular mechanisms. *J. exp. Biol. 81*, 131-51

Livneh, E., Glazer, L., Segal, D., Schlessinger, J. and Shilsh, B-Z. (1985) *Drosophila* EGF receptor gene homolog — conservation of both hormone-binding and kinase domains. *Cell 40*, 599-607

Lloyd, P.E., Kupfermann, I. and Weiss, K.R. (1984) Evidence for parallel actions of a molluscan peptide (SCP_B) and serotonin in mediating arousal in *Aplysia*. *Proc. Natl Acad. Sci. USA 81*, 2934-7

Lloyd, P.E., Kupfermann, I. and Weiss, K.R. (1985) Two endogenous neuropeptides (SCP_A and SCP_B) produce a cAMP-mediated stimulation of cardiac activity in *Aplysia*. *J. Comp. Physiol. A 156*, 659-7

Loh, Y.P., Brownstein, M.J. and Gainer, H. (1984) Proteolysis in neuropeptide processing and other neural functions. *Ann. Rev. Neurosci. 7*, 189-222

Loughton, B.G. and Orchard, I. (1981) The nature of the hyperglycaemic factor from the glandular lobe of the corpus cardiacum of *Locusta migratoria*. *J. Insect Physiol. 27*, 383-5

McAllister, L.B., Scheller, R.H., Kandel, E.R. and Axel, R. (1983) *In situ* hybridisation to study the origin and fate of identified neurons. *Science, Wash. 222*, 800-8

Maddrell, S.H.P. (1974) Neurosecretion. In: Treherne, J.E. (ed.) *Insect neurobiology*, pp. 307-57, North-Holland, Amsterdam

Maddrell, S.H.P. and Casida, J.E. (1971) Mechanism of insecticide-induced diuresis in *Rhodnius*. *Nature (Lond.) 231*, 55-6

Maddrell, S.H.P. and Nordmann, J.J. (1979) *Neurosecretion*. Blackie, Glasgow.

Maddrell, S.H.P. and Reynolds, S.E. (1972) Release of hormones in insects after poisoning with insecticides. *Nature (Lond.) 236*, 404-6

Mahon, A.C. and Scheller, R.H. (1983) The molecular basis of a neuroendocrine fixed action pattern: egg laying in *Aplysia*. *Cold Spring Harbor Symp. Quant. Biol. 48*, 405-12

Mahon, A.C., Nambu, J.R., Taussig, R., Shyamala, M., Roach, A. and Scheller, R.H. (1985) Structure and expression of the egg-laying hormone gene family in *Aplysia*. *J. Neurosci. 5*, 1872-80

Mahon, A.C., Lloyd, P.E., Weiss, K.R., Kupfermann, I. and Scheller, R.H. (1985b) The small cardioactive peptides A and B of *Aplysia* are derived from a common precursor molecule. *Proc. Natl Acad. Sci. USA 82*, 3925-9

Matsumoto, S., Isogai, A. and Suzuki, A. (1985) N-terminal amino acid sequence of an insect neurohormone, melanization and reddish coloration hormone (MRCH): heterogeneity and sequence homology with human insulin-like growth factor II. *FEBS Lett. 189*, 115-18

Matsumoto, S., Isogai, A. and Suzuki, A. (1986) Isolation and amino terminal sequence of melanization and reddish coloration hormone (MRCH) from the silkworm, *Bombyx mori*. *Insect Biochem. 16*, 775-9

Mayeri, E. and Rothman, B.S. (1982) Nonsynaptic peptidergic neurotransmission in the abdominal ganglion of *Aplysia*. In: Farner, D.S. and Lederis, K. (eds) *Neurosecretion: molecules, cells, systems*, pp. 305-15. Plenum, New York

Mayeri, E., Brownell, P. and Branton, W.D. (1979a) Multiple, prolonged actions of neuroendocrine bag cells on neurons in *Aplysia*. I. Effects on bursting pacemaker neurons. *J. Neurophysiol. 42*, 1165-84

Mayeri, E., Brownell, P. and Branton, W.D. (1979b) Multiple, prolonged actions of neuroendocrine bag cells on neurons in *Aplysia*. II. Effects on beating pacemaker and silent neurons. *J. Neurophysiol. 42*, 1185-97

Mordue, W. and Stone, J.V. (1977) Relative potencies of locust adipokinetic hormone and prawn red-pigment concentrating hormone in insect and crustacean systems. *Gen. Comp. Endocrinol. 33*, 103-8

Morris, H.R., Panico, M., Karplus, A., Lloyd, P.E. and Riniker, B. (1982) Elucidation by FAB-MS of the structure of a new cardioactive peptide from *Aplysia. Nature (Lond.) 300*, 643-5

Morris, H.R., Panico, M., Etienne, T., Tippins, J., Girgis, S.I. and MacIntyre, I. (1984) Isolation and characterization of human calcitonin gene-related peptide. *Nature (Lond.) 308*, 746-8

Muneoka, Y. and Saitoh, H. (1986) Pharmacology of FMRFamide in *Mytilus* catch muscle. *Comp. Biochem. Physiol. 85C*, 201-14

Myers, C.M. and Evans, P.D. (1985) The distribution of bovine pancreatic polypeptide FMRFamide-like immunoreactivity in the ventral nervous system of the locust. *J. Comp. Neurol. 234*, 1-16

Nachman, R.J., Holman, G.M., Haddon, W.F. and Ling, N. (1986a) Leucosulfakinin, a sulfated insect neuropeptide with homology to gastrin and cholecystokinin. *Science, Wash. 234*, 71-3

Nachman, R.J., Holman, G.M., Cook, B.J., Haddon, W.F. and Ling, N. (1986b) Leucosulfakinin — II, a blocked sulfated insect neuropeptide with homology to cholecystokinin and gastrin. *Biochem. Biophys. Res. Commun. 140*, 357-66

Nagasawa, H., Kataoka, H., Isogai, A., Tamura, S., Suzuki, A., Ishizaki, H., Mizoguchi, A., Fujiwara, Y. and Suzuki, A. (1984) Amino-terminal amino acid sequence of the silkworm prothoracicotropic hormone: homology with insulin. *Science, Wash. 226*, 1344-45

Nagasawa, H., Kamito, T., Takahashi, S., Kogai, A., Fugo, H. and Suzuki, A. (1985) Eclosion hormone of the silkworm, *Bombyx mori*; purification and determination of the N-terminal amino acid sequence. *Insect Biochem. 15*, 573-8

Nakanishi, S., Inoue, A., Kita, T., Nakamu-a, M., Chang, A.C.Y., Cohen, S.N. and Numa, S. (1979) Nucleotide sequence of cloned cDNA for bovine corticotropin-β-lipotropin precursor. *Nature (Lond.) 278*, 423-7

Nambu, J.R., Taussig, R., Mahon, A.C. and Scheller, R.H. (1983) Gene isolation with cDNA probes from identified *Aplysia* neurons: neuropeptide modulators of cardiovascular physiology. *Cell 35*, 47-56

Nawa, H., Hirose, T., Takashima, H., Inayama, S. and Nakanishi, S. (1983) Nucleotide sequences of cloned cDNAs for two types of bovine brain substance P precursor. *Nature (Lond.) 306*, 32-6

Nikolics, K., Mason, A.J., Szonyi, E., Ramachandran, J. and Seebury, P.H. (1985) A prolactin-inhibiting factor within the precursor for human gonadotropin-releasing hormone. *Nature (Lond.) 316*, 511-17

Nishizuka, Y. (1984) Turnover of inositol phospholipids and signal transduction. *Science, Wash. 225*, 1365-70

Noda, M., Furutani, Y., Takahashi, H., Toyosato, M., Hirose, T., Inayama, S., Nakanishi, S. and Numa, S. (1982) Cloning and sequence analysis of cDNA for bovine adrenal preproenkephalin. *Nature (Lond.) 295*, 202-6

Ocorr, K.A. and Byrne, J.H. (1985) Membrane responses and changes in cAMP levels in *Aplysia* sensory neurons produced by serotonin, tryptamine, FMRF-amide and small cardioactive peptide$_B$ (SCP$_B$). *Neurosci. Lett. 55*, 113-18

O'Shea, M. and Adams, M.E. (1981) Pentapeptide (proctolin) associated with an identified neuron. *Science, Wash. 213*, 567-9

O'Shea, M. and Bishop, C.A. (1982) Neuropeptide proctolin associated with an identified skeletal motoneuron. *J. Neurosci. 2*, 1242-51

O'Shea, M. and Schaffer, M. (1985) Neuropeptide function: the invertebrate

221

connection. *Ann. Rev. Neurosci. 8*, 171-98

O'Shea, M., Witten, J. and Schaffer, M. (1984) Isolation and characterization of two myactive neuropeptides: further evidence of an invertebrate peptide family. *J. Neurosci. 4*, 521-9

Painter, S.D., Morley, J.S. and Price, D.A. (1982) Structure–activity relations of the molluscan neuropeptide FMRFamide on some molluscan muscles. *Life Sci. 31*, 2471-8

Pelletier, G., Steinbusch, H.W.M. and Verhofstad, A.A.J. (1981) Immunoreactive substance P and serotonin present in the same dense-core vesicles. *Nature (Lond.) 293*, 71-2

Petruzzelli, I., Herrera, R., Garcia, R. and Posen, D.M. (1985) In: Feramisco, J., Ozanne, B. and Stiles, C. (eds) *Growth factors and transformation: cancer cells*, vol. 3, pp. 115-21. Cold Spring Harbor Laboratory, Cold Spring Harbor, New York

Price, D.A., Cottrell, G.A., Doble, K.E., Greenberg, M.J., Jorenby, W., Lehman, H.K. and Riehm, J.P. (1985) A novel FMRFamide-related peptide in *Helix* pQDPFLRFamide. *Biol. Bull. Mar. Biol. Lab. Woods Hole 169*, 256-66

Price, D.A. and Greenberg, M.J. (1977) Structure of a molluscan cardioexcitatory neuropeptide. *Science, Wash. 197*, 670-1

Quackenbush, L.S. and Fingerman, M. (1985) Enzyme-linked immunosorbent assay of black pigment dispersing hormone from the fiddler crab, *Uca pugilator*. *Gen. Comp. Endocrinol., 57*, 438-44

Quistad, G.B., Adams, M.E., Scarborough, R.M., Carney, R.L. and Schooley, D.A. (1984) Metabolism of proctolin, a pentapeptide neurotransmitter in insects. *Life Sci. 34*, 569-76

Raabe, M. (1982) *Insect neurohormones*. Plenum Press, New York

Rao, K.R., Riehm, J.P., Zahnour, C.A., Kleinholz, L.H., Tarr, G.E., Johnson, L., Norton, S., Landau, M., Semmes, O.J., Sattelberg, R.M., Jorenby, W.H. and Hintz, M.F. (1985) Characterization of a pigment dispersing hormone in eyestalks of the fiddler crab, *Uca pugilator*. *Proc. Natl Acad. Sci. USA 82*, 5319-22

Reading, C.L. (1982) Theory and methods for immunisation in culture and monoclonal antibody production. *J. Immunol. Meth. 53*, 261-91

Remy, C. and Dubois, M.P. (1981) Immunohistological evidence of methionine enkephalin-like material in the brain of the migratory locust. *Cell Tiss. Res. 218*, 271-8

Rinehart, K.L. (1982) Fast atom bombardment mass spectrometry. *Science, Wash. 218*, 254-60

Rosenfeld, M.G., Mermod, J-J., Amara, S.G., Swanson, L.W., Sawchenko, P.E., Rivier, J., Vale, W.W. and Evans, R.M. (1983) Production of a novel neuropeptide encoded by the calcitonin gene via tissue-specific RNA processing. *Nature (Lond.) 304*, 129-35

Rothman, B.S., Mayeri, E., Brown, R.O., Yuan, P-M. and Shively, J.E. (1983) Primary structure and neuronal effects of α-bag cell peptide, a second candidate neurotransmitter encoded by a single gene in bag cell neurons of *Aplysia*. *Proc. Natl Acad. Sci. USA 80*, 5753-7

Sasek, C.A., Schueler, P.A., Herman, W.S. and Elde, R.P. (1985) An antiserum to locust adipokinetic hormone reveals a novel peptidergic system in the rat central nervous system. *Brain Res. 343*, 172-5

Scarborough, R.M., Jamieson, G.C., Kalish, F., Kramer, S.J., McEnroe, G.A., Miller, C.A. and Schooley, D.A. (1984) Isolation and primary structure of two peptides with cardioacceleratory and hyperglycaemic activity from the corpora cardiaca of *Periplaneta americana*. *Proc. Natl Acad. Sci. USA 81*, 5575-9

Schaefer, M., Picciotto, M.R., Kreimer, T., Kaldany, R-R., Taussig, R. and Scheller,

R.H. (1985) *Aplysia* neurons express a gene encoding multiple FMRFamide neuropeptides *Cell 41*, 457-67

Schaffer, M.H., Noyes, B.E. and O'Shea, M. (1984) Molecule biological studies of the sequenced insect neuropeptides. *Soc. Neurosci. Abstr. 10*, 152

Schaller, H.C. and Bodenmuller, H. (1981) Isolation and amino acid sequence of a morphogenic peptide from *Hydra. Proc. Natl Acad. Sci. USA 78*, 7000-4

Schaller, H.C., Hoffmeister, S. and Bodenmuller, H. (1984) Hormonal control of regeneration in *Hydra*. In: Hoffman, J. and Porchet, M. (eds) *Biosynthesis, metabolism and mode of action of invertebrate hormones*, pp. 5-9. Springer-Verlag, Berlin

Scheller, R.H., Jackson, J.F., McAllister, L.B., Schwartz, J.H., Kandel, E.R. and Axel, R. (1982) A family of genes that codes for ELH, a neuropeptide eliciting a stereotyped pattern of behaviour in *Aplysia. Cell 28*, 707-19

Scheller, R.H., Jackson, J.F., McAllister, L.B., Rothman, B.S., Mayeri, E. and Axel, R. (1983) A single gene encodes multiple neuropeptides mediating a stereotyped behaviour. *Cell 35*, 7-22

Scheller, R., Kaldany, R.R., Kreiner, T., Mahon, A.C., Nambu, J.R., Schaefer, M. and Taussig, R. (1984) Neuropeptides: mediators of behaviour in *Aplysia. Science, Wash. 225*, 1300-8

Schlesinger, D.H., Babirak, S.P. and Blankenship, J.E. (1981) Primary structure of an egg-releasing peptide from the atrial gland of *Aplysia califfornca*. In: Schlesinger, D.H. (ed.) *Symposium on neurohypophyseal peptide hormones and other biologically active peptides*, pp. 137-50. Elsevier North Holland Biomedical Press, New York

Schooneveld, H., Romberg-Privee, H.M. and Veenstra, J.A. (1985) Adipokinetic hormone-immunoreactive peptide in the endocrine and central nervous system of several insect species. A comparative immunocytochemical approach. *Gen. Comp. Endocrinol. 57*, 184-94

Schooneveld, H., Tesser, G.I., Veenstra, J.A. and Romberg-Privee, H. (1983) Adipokinetic hormone and AKH-like peptide demonstrated in the corpora cardiaca and nervous system of *Locusta migratoria* by immunocytochemistry. *Cell Tissue Res. 230*, 67-76

Schot, L.P.C., Boer, H.H., Swaals, D.F. and Van Noorden, S. (1981) Immunocytochemical demonstration of peptidergic neurons in the central nervous system of the pond snail, *Lymnaea stagnalis*, with antisera raised to biologically active peptides of vertebrates. *Cell Tiss. Res. 216*, 273-91

Schramm, M. and Selinger, Z. (1984) Message transmission: receptor controlled adenylate cyclase system *Science, Wash. 225*, 1350-6

Schueler, P.A., Elde, R.P., Herman, W.S. and Mahoney, W.C. (1986) Identification and initial characterization of adipokinetic hormone-like immunoreactive peptides of rat origin. *J. Neurochem. 47*, 133-8

Schwartz, J.C. (1983) Metabolism of enkephalins and the inactivating neuropeptidase concept. *Trends Neurosci. 6*, 45-8

Schwarz, T.L., Lee, G.M.H., Siwicki, K.K., Standaert, D.G. and Kravitz, E.A. (1984) Proctolin in the lobster: the distribution, release and characterisation of a likely neurohormone. *J. Neurosci. 4*, 1300-11

Sedlmeier, D. and Keller, R. (1981) The mode of action of the crustacean neurosecretory hyperglycemic hormone. I. Involvement of cyclic nucleotides. *Gen. Comp. Endocrinol. 45*, 82-90

Seecoff, R.L. and Dewhurst, S. (1974) Insulin is a *Drosophila* hormone and acts to enhance the differentiation of embryonic *Drosophila* cells. *Cell Diff. 3*, 63-70

Shymala, J.R., Nambu, J.R. and Scheller, R.H. (1986) Expression of the egg-laying hormone gene family in the head ganglia of *Aplysia. Brain Res. 371*, 49-57

223

Siegert, K.J. and Mordue, W. (1986) Quantification of adipokinetic hormones I and II in the corpora cardiaca of *Schistocerca gregaria* and *Locusta migratoria*, *Comp. Biochem. Physiol. 84A*, 279-84

Siegert, K., Morgan, P. and Mordue, W. (1985) Primary structures of locust adipokinetic hormones II. *Biol. Chem. Hoppe-Seyler*, 336, 723-7

Siegert, K. and Ziegler, R. (1983) A hormone from the corpora cardiaca controls fat body glycogen phosphorylase during starvation in tobacco hornworm larvae. *Nature (Lond.) 307*, 526-7

Singh, G.J.P. and Orchard, I. (1982) Is insecticide-induced release of insect neurohormones a secondary effect of hyperactivity of the central nervous system? *Pest. Biochem. Physiol. 17.* 232-42

Siwicki, K.K. and Kravitz, E.A. (1984) Proctolin colocalizes with several different transmitters in lobster neurons. *Soc. Neurosci. Abstr. 10*, 152

Smith, W.A., Gilbert, L.I. and Bollenbacher, W.E. (1984) The role of cyclic AMP in ecdysone synthesis. *Molec. Cell. Endocrinol*, 37, 285-94

Smith, W.A., Gilbert, L.I. and Bollenbacher, W.E. (1985) Calcium-cyclic AMP interactions in prothoracicotropic hormone stimulation of ecdysone synthesis. *Molec. Cell. Endocrinol. 39*, 71-8

Smyth, D.G., Zakarian, S., Deakin, J.F.W. and Massey, D.E. (1981) β-Endorphin related peptides in the pituitary gland: isolation, identification and distribution. In: *Peptides of the pars intermedia*, (Ciba Foundation Symposium 81), pp, 79-96. Pitman Medical, London

Snyder, S.H. (1980) Brain peptides as neurotransmitters. *Science, Wash. 209*, 976-83

Stangier, J., Hilbrich, C., Beyreuther, K. and Keller, R. (1986) Isolation and characterisation of a crustacean cardioactive peptide (CCAP) from pericardial organs of the shore crab, *Carcinus maenas. Bull. Soc. Zool. France III*, 28

Starratt, A.N. and Brown, B.E. (1975) Structure of the pentapeptide proctolin, a proposed neurotransmitter in insects. *Life Sci. 17*, 1253-6

Starratt, A.N. and Brown, B.E. (1979) Analogs of the insect myotropic peptide proctolin: synthesis and structure–activity studies. *Biochem. Biophys. Res. Commun. 90*, 1125-30

Starratt, A.N. and Steele, R.W. (1984) *In vivo* inactivation of the insect neuropeptide proctolin in *Periplaneta americana. Insect Biochem. 14*, 97-102

Steele, R.W. and Starratt, A.N. (1985) *In vitro* inactivation of the insect neuropeptide proctolin in haemolymph from *Periplaneta americana. Insect Biochem. 15*, 511-19

Stefano, G.B., Kream, R.M. and Zukin, R.S. (1980) Demonstration of stereospecific opiate binding in the nervous tissue of the marine mollusc, *Mytilus edulis. Brain Res. 181*, 440-5

Stefano, G.B. and Leung, M.K. (1984) Presence of met-enkephalin-Arg[6]-Phe[7] in molluscan neural tissues. *Brain Res. 298*, 362-5

Stefano, G.B. and Scharrer, B. (1981) High affinity binding of an enkephalin analog in the cerebral ganglion of the insect *Leucophaea maderae* (Blattaria) *Brain Res. 225*, 107-14

Stefano, G.B., Scharrer, B. and Assanah, P. (1982) Demonstration, characterisation and localisation of opioid binding sites in the midgut of the insect *Leucophaea maderae* (Blattaria). *Brain Res. 253*, 205-12

Stern, A.S., Lewis, R.V., Kimura, S., Rossier, J., Gerber, L.D., Brink, L., Stein, S. and Udenfriend, S. (1979) Isolation of the opioid heptapeptide Met-enkephalin (Arg[6]-Phe[7]) from bovine adrenal medullary granules and striatum. *Proc. Natl Acad. Sci. USA 76*, 6680-3

Stone, J.V., Mordue, W., Batley, K.E. and Morris, H.R. (1976) Structure of locust

224

adipokinetic hormone, a neurohormone that regulates lipid utilization during flight. *Nature (Lond.) 263*, 207-11

Stone, J.V., Mordue, W., Broomfield, C.E. and Hardy, P.M. (1978) Structure–activity relationships for the lipid mobilizing action of adipokinetic hormone action of adipokinetic hormone. Synthesis and activity of a series of hormone analogues. *Eur. J. Biochem. 89*, 195-202

Sullivan, R.E. and Newcomb, R.W. (1982) Structure function analysis of an arthropod peptide hormone: proctolin and synthetic analogues compared on the cockroach hindgut receptor. *Peptides 3*, 337-44

Taghert, P.H. and Truman, J.W. (1982) Identification of the bursicon-containing neurons in abdominal ganglia of the tobacco hornworm *Manduca sexta. J. exp. Biol. 98*, 385-402

Taghert, P.H., Tublitz, N.J., Truman, J.W. and Goodman, C.S. (1984) Monoclonal antibodies that recognise cardioactive peptides in the moth, *Manduca sexta. Soc. Neurosci. Abstr. 10*, 152

Takeda, S., Vieillemaringe, J., Geffard, M. and Remy, C. (1986) Immunohisto-logical evidence of dopamine cells in the cephalic nervous system of the silkworm *Bombyx mori.* Coexistence of dopamine and α-endorphin-like substance in neurosecretory cells of the suboesophageal ganglion. *Cell Tiss. Res 243*, 125-8

Tang, J., Yang, H.Y.T. and Costa, E. (1984) Inhibition of spontaneous and opiate-modified nociception by an endogenous neuropeptide with Phe-Met-Arg-Phe-NH_2-like immunoreactivity. *Proc. Natl Acad. Sci. USA 81*, 5002-5

Taussig, R., Kaldany, R.R. and Scheller, R.H. (1984) A cDNA close encoding neuropeptides isolated from *Aplysia* neuron Lll. *Proc. Natl Acad. Sci. USA 84*, 4988-92

Thompson, K.L. Decker, S.J. and Rosner, M.R. (1985) Identification of a novel receptor in *Drosophila* for both epidermal growth factor and insulin. *Proc. Natl Acad. Sci. USA 82*, 8443-7

Thorpe, A. and Duve, H. (1984) Immunochemical applications in the study of insect neuropeptides with special emphasis on the peptides of vertebrate type. In: Borkovec, A.B. and Kelly, T.J. (eds) *Insect neurochemistry and neurophysiology*, pp. 197-222. Plenum, New York

Truman, J.W. (1980) Cellular aspects of eclosion hormone action on the CNS of insects. In: Sattelle, D.B., Hall, L.M. and Hildebrand, J.G. (eds) *Receptors for neurotransmitters, hormones and pheromones in insects,* pp. 223-32. Elsevier, North Holland Biomedical Press, Amsterdam

Truman, J.W., Mumby, S.M. and Welch, S.K. (1979) Involvement of cyclic GMP in the release of stereotyped behaviour patterns in moths by peptide hormone. *J. Exp. Biol. 84*, 201-12

Tublitz, N.J. and Truman, J.W. (1985) Identification of neurones containing cardioaccelerating peptides (CAPs) in the ventral nerve cord of the tobacco hawkmoth, *Manduca sexta. J. Exp. Biol. 116*, 395-410

Tublitz, N.J., Taghert, P.H. and Evans, P.D. (1985) A monoclonal antibody acts as a functional blocker of cardioacceleratory peptide activity in the tobacco hawkmoth, *Manduca sexta. Soc. Neurosci. Abstr. 11*, 326

Valentino, K.L., Winter, J. and Reichard, L.F. (1985) Applications of monoclonal antibodies to neuroscience research. *Ann. Rev. Neurosci. 8*, 199-232

Vreugdenhil, E., Geraerts, W.P.M., Jackson, J.F. and Joose, J. (1985) The molecular basis of the neuroendocrine control of egg-laying behaviour in *Lymnaea. Peptides (Fayetteville, NY) 6* (Suppl. 3), 465-70

Walsh, K.A., Ericsson, L.H., Parmelee, D.C. and Titani, K. (1981) Advances in protein sequencing. *Ann. Rev. Biochem. 50*, 261-84

Watson, W.H., Angustine, G.J. and Benson, J.A. (1983) Proctolin and an endogenous proctolin-like peptide enhance the contractivity of the *Limulus* heart, *J. exp. Biol. 103*, 55-73

White, J.D., Stewart, K.D., Krause, J.E. and McKelvy, J.F. (1985) Biochemistry of peptide-secreting neurons. *Physiol. Rev. 65*, 553-606

Williams, C.M. (1967) The present status of the brain hormone. In: Beament, J.W.L. and Treherne, J.E. (eds) *Insects and physiology*, pp. 133-9, Oliver & Boyd, Edinburgh

Witten, J., Schaffer, M.A., O'Shea, M., Cook, J.C., Hemling, M.E. and Rinehart, K.L. (1984) Structure of two cockroach neuropeptides assigned by fast atom bombardment mass spectrometry. *Biochem. Biophys. Res. Commun. 124*, 350-358

Witten, J.L. and O'Shea, M. (1985) Peptidergic innervation of insect skeletal muscle: immunochemical observations. *J. Comp. Neurol. 242*, 93-101

Yang, H.Y.T., Fratta, W., Majane, E.A. and Costa, E. (1985) Isolation, sequencing, synthesis, and pharmacological characterisation of two brain neuropeptides that modulate the action of morphine. *Proc. Natl Acad. Sci. USA 82*, 7757-61

Ziegler, R., Eckart, K., Schwarz, H. and Keller, R. (1985) Amino acid sequence of *Manduca sexta* adipokinetic hormone elucidated by combined fast atom bombardment (FAB)/tandem mass spectrometry. *Biochem. Biophys. Res. Commun. 133*, 337-42

Zukin, R.S. and Zukin, S.R. (1984) The case for multiple opiate receptors. *Trends Neurosci. 7*, 160-4

6

Neuronal Cultures as Experimental Systems

David J. Beadle

INTRODUCTION

Whereas monolayer, neuronal cell culture techniques have made, and continue to make, a major contribution to our understanding of the vertebrate nervous system (Fischbach and Nelson, 1977), little progress has been made using invertebrate cell cultures, which remain in a relatively primitive state (Beadle and Hicks, 1985). However, during the last few years a small number of invertebrate culture systems have been developed in an attempt to solve specific neurobiological problems for which cell culture techniques offer the best hope of success. For example, an *in vitro* preparation of snail neurones has been developed to study the mechanisms underlying neuronal growth (Wong *et al.*, 1981), and cultures of *Aplysia* and leech neurones have been used to investigate the formation and specificity of synaptic connections (Ready and Nicholls, 1979; Camardo *et al.*, 1983). Similarly, insect neuronal cultures have been developed to facilitate pharmacological studies of neurotransmitters and their receptors (Usherwood *et al.*, 1980; Lees *et al.*, 1983) and to permit a genetic analysis of neuronal function (Wu *et al.*, 1983a). All of these culture preparations have been developed specifically to circumvent many of the difficulties that arise during neurobiological investigations of the invertebrate nervous system because of its structural organisation. This chapter outlines the progress that has been made in the use of invertebrate culture preparations for solving specific neurobiological problems, and attempts to demonstrate that invertebrate neurones *in vitro* resemble their *in situ* counterparts sufficiently to warrant their use as experimental models of the intact nervous system.

INVERTEBRATE NEURONAL CULTURE PREPARATIONS

Recently a number of invertebrate neuronal culture preparations have been developed from a range of phyla. The majority of these are pure, monolayer neuronal cultures with little or no contamination from other cell types, and many of them are composed of identified neurones whose *in vivo* properties have been well characterised. Such *in vitro* systems provide obvious advantages as experimental models for a wide range of neurobiological investigations. The major invertebrate preparations are described briefly below although this list is by no means exhaustive.

Periplaneta americana

The dissociated neuronal culture technique for neurones from the cockroach, *P. americana*, was originally devised by Chen and Levi-Montalcini (1970) using whole nerve cords from cockroach embryos, and has recently been modified to improve the degree of differentiation of the *in vitro* neurones (Beadle *et al.*, 1982). The cultures are produced by the mechanical disruption of the nerve cord or brain resulting in the destruction of glial elements leaving a nerve cell culture uncontaminated by other cell types. The technique, which has recently been described in some detail (Dewhurst and Beadle, 1985), involves culturing the dissociated cells in uncoated plastic petri dishes using a modification of the hanging column method (Shields *et al.*, 1975). The neurones are grown in the '5 + 4' medium of Levi-Montalcini *et al.* (1973) consisting of five parts Schneider's revised *Drosophila* medium and four parts Eagles' basal medium containing penicillin and streptomycin. In this medium the cells attach to the substratum and produce neurites at which time they are transferred to a second medium consisting of equal parts Leibovitz's L-15 and Yunker's modified Grace's medium in which they appear to achieve complete structural differentiation (Beadle *et al.*, 1982). Similar techniques have been used to culture neurones from embryos and nymphs of the locust *Schistocerca gregaria* (Giles *et al.*, 1978), pupae of the moth *Spodoptera littoralis* (Hicks *et al.*, 1981), and larvae of the fruit fly, *Drosophila melanogaster* (Wu *et al.*, 1983a).

Immediately following dissociation, insect neurones are spherical in shape and range in size from 2–3µm for *Drosophila* cells to 50 µm or more for nymphal locust or pupal moth cells. The spherical somata rapidly attach to the floor of the culture vessel and produce thin neurites within 24 h (Figures 6.1A and B). Within a few days the neurites make connections with neighbouring cells and a complex array of neuronal processes forms. As the neurones differentiate, the neurites coalesce to form fibre bundles and these produce extensive branching that eventually leads to a dense network of processes connecting clumps of cell bodies (Figures 6.1C and D). A typical culture of cockroach neurones contains up to 50 000 cells and 10 µg of protein, and can be maintained for at least 6 weeks. The neurones

Figure 6.1: Light micrographs of living neurones from dissociated brains of 23-day *P. americana* embryos. (A) After 12 h *in vitro*, neuronal somata (arrows), glial cell nuclei (g1) and debris are present on the plastic substrate. (B) After 5 days *in vitro* approximately 95 per cent of neuronal somata are connected by thin unbranched fibres (arrows) to closely adjacent neighbours. (C) After 8 days *in vitro* exposure to secondary growth medium has resulted in a marked increase in somatic diameter, the appearance of thick fibre bundles (arrows) and an increased amount of branching. (D) After 26 days *in vitro* the neurones are connected by fibre bundles (arrows) that have branched profusely to form a dense neuronal network. Scale bar: 100 μm.

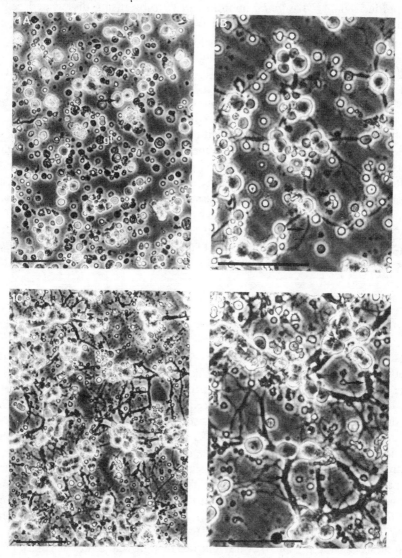

are about 20 μm in diameter and the fibre bundles may reach 15 μm in diameter, containing twenty or more individual processes (Beadle and Lees, 1986). Locust neurones may grow up to 80 μm in diameter and produce neurites 500 μm in length, but they differ from their *in vivo* counterparts in being mainly multipolar rather than producing a single, major process (Giles *et al.*, 1978). In contrast, although cultured *Drosophila* neurones can be maintained for at least 40 days, the cells remain quite small, less than 10 μm in diameter, making physiological experiments difficult to perform (Wu *et al.*, 1983a). Recently an attempt has been made to produce giant *Drosophila* neurones *in vitro* by fusing dissociated cells with polyethylene glycol (Suzuki and Wu, 1984).

Helisoma trivolvis

Techniques for culturing snail neurones have been developed by Wong *et al.* (1981). The cultures are prepared by the mechanical disruption of desheathed ganglia after treatment with proteolytic enzymes. The neurones are grown in a serum-free medium consisting of half-strength Leibovitz's L-15 medium and *Helisoma* saline on polylysine-coated petri dishes. The cultured neurones maintain their spherical shape and are viable for at least two weeks but do not produce neurites unless conditioning factors are present (Figure 6.2). In the presence of these factors extensive neurite sprouting occurs (Wong *et al.*, 1981). This technique has also been extended to other molluscan species (Wong *et al.*, 1983) and has recently been modified to allow the culture of identified snail neurones (Haydon *et al.*, 1985).

Hirudo medicinalis

Ready and Nicholls (1979) and Fuchs *et al.* (1981) have described methods for culturing single or small groups of identified leech neurones. The connective tissue capsule around the ganglion is opened and glia are washed away revealing the neuronal cell bodies. Individual neurones, such as the Retzius cells that cause mucus secretion, and pressure (P) sensory neurones, can be recognised by their size and position, and other cells can be identified by recording from them intracellularly before isolation. Individual neurones are isolated by slipping a fine nylon loop over the soma of the cell, tying tightly around its process and pulling it away from the ganglion. The isolated neurones are cultured in Leibovitz's L-15 medium containing 2 per cent fetal calf serum in plastic petri dishes coated with collagen and polylysine (Figure 6.3). The isolated cells can survive for several weeks and produce several neurites that grow at a rate of 10–20 μm

230

Figure 6.2: *Helisoma* neurones growing *in vitro*. (A) A neurone that has been growing in control medium for 4 days. Limited growth was seen in 1 per cent of these cells. (B) A neurone grown in 12 h conditioned medium (CM) for 4 days. Note the well defined veiling and lack of distinct neurites. (C) A neurone grown in 72 h CM for 4 days. Note the presence of well defined neurites and the extent of outgrowth. Scale bar = 50 μm.

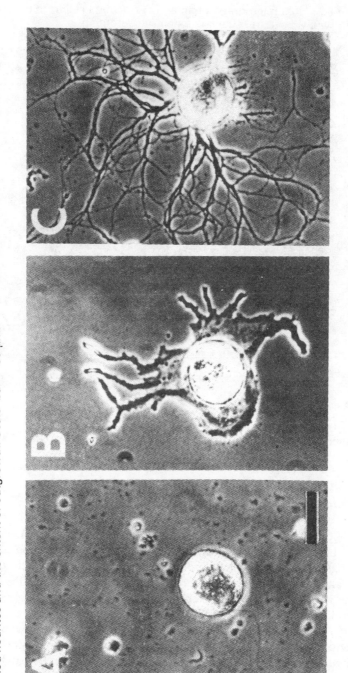

Source: reproduced with permission from Wong *et al.* (1984).

Figure 6.3: Light micrograph of an anterior pagoda cell from the leech after 1 week in culture, showing extensive sprouting

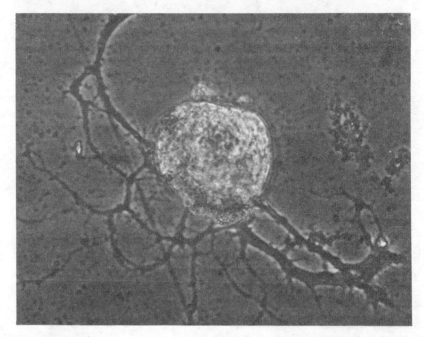

Source: reproduced with permission from M. Pelligrino and M. Simonneau (unpublished).

per day. When two isolated cells are cultured within 50–200 μm of one another, many of the neurites span the area between them and form selective connections (Fuchs *et al.*, 1981).

Aplysia californica

Methods for culturing identified neurones from ganglia of *A. californica* have been described by Dagan and Levitan (1981) and Schacher and Proshansky (1983). Ganglia are treated with proteases prior to desheathing, and individual neurones are isolated from the desheathed ganglia either by manipulation with etched tungsten needles or fine-tipped microelectrodes, or by the nylon filament loop method of Ready and Nicholls (1979). The neurones are cultured in a medium made from equal parts of Leibovitz's L-15 medium and a salt solution so that the final salt concentration is the same as for *Aplysia* saline. In some cases fetal calf serum or sterile *Aplysia* haemolymph is added to the medium (Schacher and Proshansky, 1983). The cultures are maintained either on chick plasma clots or in plastic dishes coated with polylysine. Isolated neuronal cell bodies, ranging in size from 50 to 700 μm, remain viable for up to 6 weeks. In general only the smaller neurones produce neurites, and most cells are multipolar. The neuritic processes produce extensive networks of fibres in the culture vessel and are able to make selective connections (Figure 6.4).

FACTORS AFFECTING NEURITE OUTGROWTH

As a variety of chemical factors have been identified that influence the growth and differentiation of vertebrate neurones (Levi-Montalcini, 1976; Patterson, 1978; Adler et al., 1981; Nishi and Berg, 1981; Schwartz et al., 1982), it is not surprising that similar factors appear to be necessary for the growth of invertebrate neurones. This was first demonstrated by Chen and Levi-Montalcini (1970) who showed that explanted embryonic foregut tissue is necessary for the in vitro fibrillar growth from embryonic neurones of the cockroach. Similarly it has been shown that neurones from Helisoma remain viable in culture for several days but do not produce neurites unless they are co-cultured with intact central ganglionic rings (Wong et al., 1981); that the initiation of neurite outgrowth in Aplysia neurones is enhanced by the presence of haemolymph in the medium (Schacher and Proshansky, 1983); and that the addition of leech blood to the culture medium markedly improves the chances of the formation of chemical synapses between cultured leech neurones (Fuchs et al., 1982). All of these examples suggest that some invertebrate tissues either release or contain chemical factors that are essential for the growth of invertebrate neurones in vitro. In cultures that are grown in the absence of other tissues, such as some insect neuronal preparations, it is likely that the presence of fetal calf serum in the medium provides substitutes for such factors.

In cultures of Helisoma neurones it has been demonstrated that intact brains release a nerve growth-promoting factor into the medium that initiates neurite outgrowth and elongation in the isolated neurones (Figure 6.2). This factor does not appear to influence the survival or electrophysiological properties of the neurones since spherical neurones without processes support normal overshooting action potentials in the absence of this factor (Wong et al., 1981). The neurite growth-promoting factor is released only from nervous tissue and it is adsorbed on to the polylysine-coated surface of the culture dishes where it influences the growth of neurones in those dishes. The ability of conditioned culture dishes to support neurite outgrowth is abolished by treatment with chymotrypsin, or heating for 5 min at 100°C, and is reduced by the inhibition of protein synthesis with anisomycin. Conditioned dishes are unaffected by DNase or RNase. This suggests that nervous tissue of Helisoma produces and releases a macromolecular factor that is proteinaceous in nature and that

233

Figure 6.4: Dark-field photomicrographs of *Aplysia* neurones after 4 days in culture. (A) L10 with two LUQ cells. The LUQ cell on the left (LUQ1) was plated with an axon whereas the LUQ cell on the right (LUQ2) was plated initially without an axon. Electrophysiological examination of the synapses formed showed the dual component P.S.P. on LUQ1 and the single slow component P.S.P. on LUQ2. (B) L10 with two RUQ cells. All cells were initially plated without axons. No chemical connections formed. Note that the processes of L10 overlap with those of the RUQ cells. (C) L10 with LUQ and RUQ cells. All cells were plated initially with axons. L10 made synapses only with LUQ cells even though RUQ cell processes overlap with L10 processes

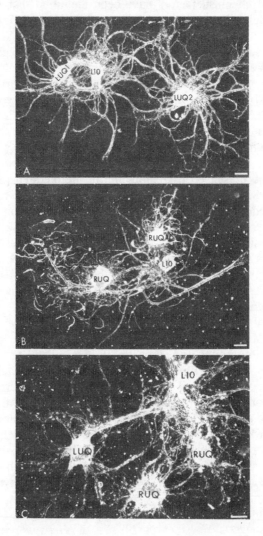

Source: reproduced with permission from Camardo *et al.* (1983).

234

promotes neurite outgrowth in isolated *Helisoma* neurones (Wong *et al.*, 1984). Furthermore, this factor will also promote neurite outgrowth from neurones of closely related species such as *Biomphalaria* but not from more distantly related species such as *Aplysia* (Wong *et al.*, 1983). This molluscan growth-promoting factor also appears to be quite distinct from those that are present in the haemolymph of *Aplysia* and that appear to promote outgrowths in isolated *Aplysia* neurones (Schacher and Proshansky, 1983). Passing haemolymph over a series of polylysine-treated dishes does not significantly affect its growth-promoting influence, suggesting that it is a soluble factor and not a surface-attached one. It may be that molluscan neurones are influenced by a number of chemical factors that affect their survival, growth and metabolism, and it is likely that similar factors will be identified in other invertebrates. Neuronal cultures provide ideal experimental systems for identifying and investigating the nature and mode of action of such factors since the investigator has complete control of the environment in which the neurones are growing.

ELECTRICAL PROPERTIES OF CULTURED INVERTEBRATE NEURONES

Many of the passive electrical membrane properties and voltage-dependent ionic currents that have been characterised in *in vivo* neurones are retained by neurones that are growing *in vitro*. The resting potential and input resistance of cultured neurones is generally similar to that of their *in situ* counterparts (Dagan and Levitan, 1981; Fuchs *et al.*, 1981; Wong *et al.*, 1981; Giles and Usherwood, 1985; Lees *et al.*, 1985) although there are exceptions. The resting potential of cultured leech neurones falls immediately following isolation but returns to normal within 1 h. The resting potentials recorded from identified neurones, −50 to −70 mV, are generally higher than those reported *in situ*, with the highest being −80 to −100 mV for a cultured L motor neurone compared with −40 mV *in situ* (Fuchs *et al.*, 1981). In contrast, the resting potentials of nymphal locust neurones in culture are lower than those recorded from neurones *in vivo* or from freshly dissociated neurones (Usherwood *et al.*, 1980). The input resistance of locust neurones *in vitro*, 10–40 MΩ, is similar to that obtained from freshly dissociated neurones whereas for leech and snail neurones it is generally higher than that recorded *in vivo*, with values as high as 240 MΩ being recorded in 3-day-old leech cultures (Fuchs *et al.*, 1981).

Many of the neurones in invertebrate culture preparations are capable of producing all-or-none action potentials that are indistinguishable from those of their *in vivo* counterparts. About 30 per cent of locust neurones produce spontaneous action potentials *in vitro*, and many others that do not spike spontaneously can be induced to do so by the injection of appropriate current pulses (Usherwood *et al.*, 1980). A number of *Aplysia* neurones exhibit spontaneous regular firing in culture, although of eight R15 neurones from which recordings have been made none showed the highly characteristic regular bursting pattern normally observed *in situ* (Dagan and Levitan, 1981). Retzius cells and mechanosensory cells in leech cultures produce action potentials that are identical to their counterparts in the ganglion, and trains of impulses in cultured sensory cells are followed by a prolonged after-hyperpolarisation resembling that seen in normal ganglia. In contrast, L motor neurones in culture produce action potentials about six times larger than those of L cells *in situ* (Fuchs *et al.*, 1981). The excitable characteristics displayed by embryonic cockroach neurones growing *in vitro* undergo a sequence of developmental changes (Lees *et al.*, 1985) that are remarkably similar to those described for insect neurones during normal embryogenesis (Goodman and Spitzer, 1980, 1981). During the first 12 days in culture the cells cannot be stimulated to produce action potentials, the characteristic membrane response to the injection of depolarising current pulses being delayed rectification (Figure 6.5A). However, after this period increasing numbers of cells can be stimulated to produce spikes (Figure 6.5B) and after about three weeks in culture about 20 per cent of them exhibit spontaneous activity (Figure 6.5C) some of which may have a synaptic origin. All of the spikes recorded in these neurones are abolished by 1 µM tetrodotoxin (Figure 6.5D) and are refractory to concentrations of cobalt ions known to block calcium channels, suggesting that they are sodium spikes. In contrast, in *Drosophila* neurones *in vitro* no action potentials can be evoked during the injection of depolarising pulses unless strontium or barium ions are added to the bathing medium, suggesting that these are calcium spikes (Wu *et al.*, 1983a).

When cultured cockroach neurones are held under whole-cell voltage-clamp conditions, voltage-dependent outward potassium currents are revealed. By using the drugs 4-aminopyridine and tetraethylammonium ions it has been shown that these cells possess a fast transient or I_A current and a delayed rectifying or I_K current (Beadle and Lees, 1986). Similarly cultured R15 neurones from *Aplysia* have been shown to possess an inwardly rectifying or I_R potassium current whose conductance is increased by the neurotransmitter, serotonin (Lotshaw and Levitan, 1985). When cockroach neurones are studied with the patch-clamp technique, the presence of individual outward currents is revealed and the frequency and amplitude of these increases with increasing membrane depolarisation (Figure 6.6). These outward currents have a mean channel opening time of about 1.5 ms and a single channel conductance of 13 to 16 pS and they almost certainly represent the voltage dependent delayed rectifying current demonstrated in these cells. Cultured *Drosophila* neurones also possess a delayed rectifying current with unit conductance of about 7 pS and a mean

Figure 6.5: Examples of membrane properties of cockroach neurones growing *in vitro*. (A) Delayed rectification in response to current injection in neurones less than 12 days *in vitro*. (Resting membrane potential: −49 mV; pulse width; 120 ms.). (B) A train of overshooting spikes evoked by intracellular injection of current in a neurone after 19 days *in vitro*. (Resting membrane potential: −60 mV; pulse width 120 ms.) (C) Spontaneous train of spikes from a neurone after 23 days in culture. (Resting membrane potential: −68 mV; sweep duration: 770 ms.) (D) The effect of sodium blockade on evoked spikes in a single neurone after 14 days *in vitro*. Upper trace: control response. Lower trace: after 30 s exposure to 1 μM tetrodotoxin. (Resting membrane potential: −64 mV; pulse width 80 ms.)

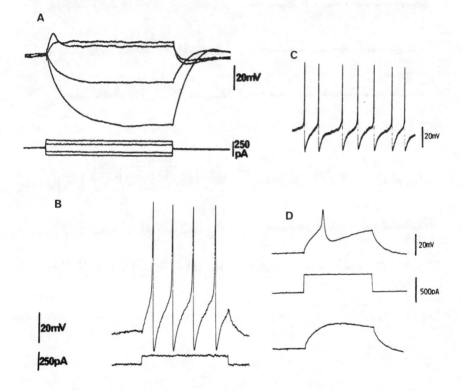

Source: modified from Lees *et al.* (1985).

channel opening time of 10–15 ms (Sun and Wu, 1984). In addition these neurones possess a calcium-dependent potassium current and an inward rectifer (Ganetsky and Wu, 1985). Since mutant nerve cells, such as the eag[1] mutant with a reduced delayed rectifying current and the Hk[1] mutant with abnormal pacemaker activities, can be readily cultured, a genetic analysis of these membrane currents, using cell culture techniques, should be possible (Sun and Wu, 1985).

237

Figure 6.6: Patch-clamp recordings of ionic currents in cockroach neurones bathed in normal physiological saline. (A) Cockroach neurones after 5 days *in vitro*. Upward deflections of the trace represent outward currents. The membrane potential was held at +30, +50, +70 and +90 mV from the resting membrane potential. (B) Cockroach neurones after 20 days *in vitro*. The membrane potential was held at +80 mV from the resting potential.

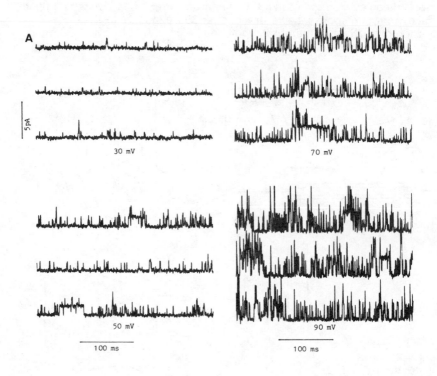

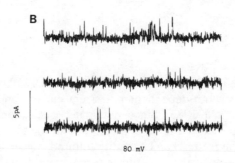

Source: from Beadle and Lees (1986).

238

SYNAPSE FORMATION *IN VITRO*

The organisation of the invertebrate nervous system, in which the formation of synaptic contacts is confined to deep within the neuropile, does not facilitate the study of synapses. The formation of synapses under culture conditions permits many aspects of synaptogenesis and synaptic function to be studied more conveniently, as has been successfully demonstrated with vertebrate preparations (Furshpan *et al.*, 1976; Fischbach and Nelson, 1977; Patterson, 1978). Electrical coupling between neurones has been demonstrated in both leech and *Aplysia* cultures (Dagan and Levitan, 1981; Fuchs *et al.*, 1981). When pairs of leech cells are plated in direct apposition, both Retzius and L cells become coupled, and as in the animal these connections are non-rectifying. Also, sensory cells and motor neurones become electrically coupled and the spread of current shows clear rectification resembling the properties of these connections within the ganglion. Although close apposition of cell pairs facilitates coupling, cells at a distance are also able to form electrical connections by way of processes (Fuchs *et al.*, 1981). The connections formed between isolated cells *in vitro* did not appear to represent random coupling since Retzius cells would form connections with other Retzius cells but not with P sensory cells whereas the latter did not become electrically coupled with each other but did form electrical synapses with L cells (Fuchs *et al.*, 1981). This suggests that some degree of inherent cell–cell recognition occurs *in vitro*. In general, the connections formed *in vitro* mirror those found in the animal, but occasionally apparently new connections are formed such as that between Retzius cells and L cells that occurs in culture but has not been discerned in the ganglion (Fuchs *et al.*, 1981).

Leech and *Aplysia* neurones will also form chemical synapses when grown in culture (Fuchs *et al.*, 1982; Camardo *et al.*, 1983). In *Aplysia* cultures L10 neurones will make specific synapses with appropriate target neurones but will not make them with inappropriate ones even though the non-target neurones possess receptors to the neurotransmitter released by this cell. L10 is a cholinergic interneurone that exerts a dual inhibitory effect on cells L2 to L6 left upper quadrant cells (LUQ cells). When L10 neurones are cultured with LUQ cells, chemical synapses are established in about 80 per cent of the cultures after 3 to 5 days. In all cases the connection is inhibitory and the physiological characteristics of the dual inhibitory connection are retained *in vitro* as long as the LUQ cells are plated with their initial axons intact (Figure 6.7). In contrast, if L10 neurones are co-cultured with RUQ cells, no evidence of chemical synapses is found even though RUQ cells possess cholinergic receptors. Finally, if L10 cells are cultured with both LUQ and RUQ cells, they are able to form specific chemical synapses only with the LUQ cells and do not interact with the RUQ cells (Figure 6.4). These results indicate that L10 cells in culture can

239

Figure 6.7: Synaptic activity between *Aplysia* neurones in culture. L10 forms chemical synapses with LUQ cells plated with axons (A to D) or without axons (E). (A) A single spike in L10 elicits an i.p.s.p. on to the LUQ cell after 3 days in culture. The latency of the response is approximately 20 ms and the duration is 800 ms. (B) The i.p.s.p. is reversed when the LUQ cell is hyperpolarised to −50 mV below the resting potential (RP = −50 mV), and −25 mV below the reversal potential. (C) L10 forms a dual component p.s.p. on the LUQ cell after 5 days in culture. A single spike elicits the fast p.s.p. and repetitive firing elicits a fast p.s.p. and, with a latency of 2 to 3 s, a slow i.p.s.p. (D) The fast component of the p.s.p. is sensitive to intracellular chloride ion concentration. In (D$_1$), repetitive stimulation of L10 elicits a fast component accompanied by a slow component in the LUQ cell impaled with a potassium citrate electrode. In D$_2$ the same LUQ cell is impaled with a potassium chloride electrode. The fast i.p.s.p. is reversed at the resting potential but the slow p.s.p. is unaffected. (E) L10 forms only the slow component p.s.p. on a LUQ cell plated without its axon after 5 days in culture. A single spike in L10 elicits no response from the LUQ cell. With repetitive firing, a slow response can be seen at the resting potential (left trace). This response is nulled when the LUQ cell is hyperpolarised to −20 mV (centre trace)

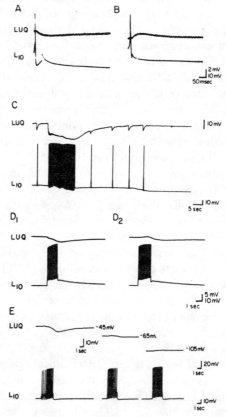

Source: reproduced with permission from Camardo *et al.* (1983).

240

select the appropriate target neurone when presented with both correct and incorrect target cells (Camardo *et al.*, 1983). Furthermore, L10 will form appropriate chemical connections with different target cells in culture such that the postsynaptic potential is appropriate for each follower cell. Hence, action potentials in cultured L10 cells produce inhibitory postsynaptic potentials (i.p.s.p.s) in LUQ cells, excitatory postsynaptic potentials (e.p.s.p.s) in R15 cells, and an e.p.s.p. followed by an i.p.s.p. in cell L7 (Schacher, 1983).

Chemically mediated synaptic transmission has also been demonstrated between Retzius cells and P sensory cells in cultures of leech neurones (Fuchs *et al.*, 1982). When these cells are placed immediately adjacent to one another in the culture dish, transmission is observed between about 60 per cent of the pairs. Impulses in the Retzius cell evoke a hyperpolarisation of 1–2 mV in the P sensory neurone when potassium acetate electrodes are used, but these responses are depolarising with a potassium chloride electrode suggesting that the potentials in the P sensory cell arise in part from an increase in chloride conductance. Several lines of evidence indicate the chemical nature of transmission from Retzius to sensory cell. To begin with, the synaptic potentials in the P sensory cells arise after a delay of about 5 ms and take about 50 ms to reach their peak. Secondly the transmission is influenced by the concentrations of calcium and magnesium in the bathing medium, ions known to affect transmitter release (Katz and Miledi, 1965). Increasing calcium augments the synaptic potential whereas increasing magnesium reversibly abolishes all synaptic potentials. Finally, transmission has been shown to be unidirectional, there being no evidence of transmitter release from P sensory cells to Retzius cells (Fuchs *et al.*, 1982). The transmitter released at these synaptic terminals is almost certainly 5-hydroxytryptamine (5HT) since Retzius cells in culture have been shown to maintain their ability to synthesise, store and release 5HT, and 5HT has been shown to elicit small, slow hyperpolarisations from P sensory neurones *in vitro* (Henderson, 1981, 1983). Furthermore, structures resembling synapses have been identified in cultures of Retzius and P sensory neurones with accumulations of dense-core vesicles clustered close to the presynaptic membrane in the Retzius cells (Henderson *et al.*, 1983).

Although ultrastructural profiles with the characteristics of synapses have been identified in insect neuronal cultures (Hicks *et al.*, 1981; Beadle *et al.*, 1982), there is as yet no physiological evidence of synaptic transmission in these cultures. Direct attempts to demonstrate synaptic activity in locust neuronal cultures have been unsuccessful (Usherwood *et al.*, 1980; Giles and Usherwood, 1985). However, when cockroach neurones are co-cultured with myoblasts and myocytes from embryonic cockroaches, the muscle cells differentiate to form myotubes and the neurones make connections with them. When these connections are examined by electron microscopy, they are found to have ultrastructural characteristics typical of

Figure 6.8: (A) Electron micrographs of nerve–muscle co-cultures of *P. americana*. A typical presynaptic terminal with presynaptic vesicles and an area of synaptic contact (arrow) (left micrograph) and a type of presynaptic terminal found only in co-cultures containing cerebral ganglia with vesicles that appear to be neurosecretory (right micrograph). M: Muscle cell, v: synaptic vesicles. (B) Spontaneous e.p.s.p.s recorded from a myotube grown in nerve–muscle co-cultures. The amplitude of the postsynaptic potentials is dependent on the cell membrane potential. Calibration: 20 mV, 100 ms

A

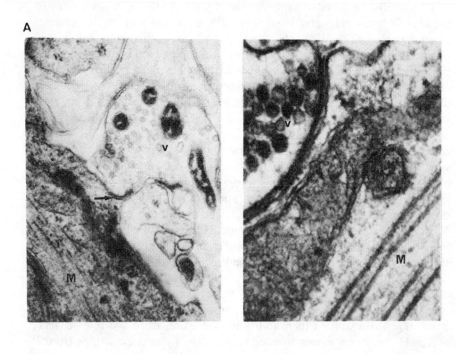

B

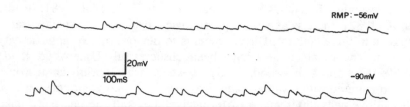

in vivo insect neuromuscular junctions (Figure 6.8A). Furthermore, when electrophysiological recordings are made from myotubes in these co-cultures, postsynaptic excitatory potentials are revealed (Figure 6.8B). Such potentials are not seen in myotubes cultured in the absence of neurones (Bermudez *et al.*, 1986). It would appear that although synapses are not formed in insect neuronal cultures, synaptic connections do occur if the neurones are cultured with developing muscle. *In vitro* preparations of invertebrate neurones and muscles can be seen to provide excellent models for a range of studies of synaptic structure and physiology.

TRANSMITTER UPTAKE STUDIES

Acetylcholine

Cultures from both *P. americana* and *D. melanogaster* nerve cords appear to contain large populations of differentiated cholinergic neurones since these cells can accumulate [³H]choline from the culture medium, and synthesise and accumulate acetylcholine (Wu *et al.*, 1983a; Beadle *et al.*, 1984; Bermudez *et al.*, 1985), properties that have been used to character-ise the cholinergic system in vertebrate nervous tissue (Simon *et al.*, 1976; Burgess *et al.*, 1978; Pomerai and Carr, 1982). Cockroach neurones *in vitro* possess both a high- and a low-affinity choline uptake system with the former having a K_m of 0.57 ± 0.28 µM and a V_{max} of 2.99 ± 0.69 pmol/10min/culture. The high-affinity system is very dependent on sodium ions as their replace-ment with lithium reduces uptake by 87 per cent. Increasing concentrations of potassium ions also reduce choline uptake, probably by an effect on the membrane potentials. The accumulation of [³H]choline appears to be unaffected by the metabolic inhibitors sodium azide and 2,4-dinitrophenol, although iodoacetamide, a glycolysis inhibitor, reduces uptake by about 50 per cent. Hemicholinium-3 is a potent inhibitor of the high-affinity system, with as little as 1 µM reducing uptake by 80 per cent and 100 µM by more than 95 per cent. Cultured *Drosophila* neurones can synthesise acetylcho-line with an efficiency comparable to the intact larval CNS (Wu *et al.*, 1983a,b), and in cockroach neuronal cultures 34 per cent of the total accumulated radioactivity is present as [³H]acetylcholine after the neurones have been exposed to 0.5 µM [³H]choline (Bermudez *et al.*, 1985).

Barker *et al.* (1982) have identified a factor that is produced by the central nervous system of the snail, *Helisoma*, that stimulates choline uptake and metabolism in cultured neurones. When buccal ganglia from *Helisoma* are maintained *in vitro* in a defined medium, they can accumu-late [³H]choline from that medium and synthesise [³H]acetylcholine and other metabolites. When buccal ganglia are grown in a brain-conditioned medium, there is an increased accumulation of these metabolites with average values of 1.7-fold, 3.7-fold and 2.6-fold for acetylcholine, phosphorylcholine and lipid, respectively. The total uptake of [^{3}H]choline is increased by 1.7-fold on average. This choline metabolism stimulating factor is quite distinct from the neurite growth-promoting factor that is also released from *Helisoma* central nervous tissue, since it does not bind to polylysine-treated substrates but remains in the supernatant and also appears to have a different basis of production and release (Barker *et al.*, 1982). Trophic factors that affect the metabolism of choline have also been identified in vertebrate neuronal cultures (Varon *et al.*, 1979; Nishi and Berg, 1981).

Serotonin

Individual, isolated Retzius cells from the leech maintained in L-15 medium can accumulate [^{3}H]tryptophan or [^{3}H]hydroxytryptophan from the culture medium and synthesise [^{3}H]5HT. The synthesis of 5HT from tryptophan demonstrates the pressure of tryptophan hydroxylase and 5HTP decarboxylase in the cultured Retzius cells. Individual, cultured Retzius cells also accumulate approximately 100 times more [^{3}H]5HT from the culture medium than cultured non-serotonergic cells, and the accumulation is markedly reduced by 10–20 μM chlorimipramine, a blocker of 5HT uptake (Gerschenfeld *et al.*, 1978). This suggests that cultured Retzius cells have a specific uptake mechanism for 5HT (Henderson, 1983). When these cells are incubated with [^{3}H]tyrosine, [^{3}H]glutamate or [^{3}H]choline, they do not synthesise detectable amounts of octopamine, dopamine, noradrenaline, gamma-aminobutyric acid or acetylcholine.

5HT synthesised and accumulated by cultured Retzius cells can also be released in a manner that suggests that the *in vitro* neurones possess the cellular machinery for transmitter release. Retzius cells depolarised by potassium ions or direct stimulation release [^{3}H]5HT into the culture medium, and this release is calcium-dependent and can be blocked by magnesium. Since freshly isolated Retzius cells do not exhibit this release mechanism, it would appear that it develops during the period that the neurone is in culture (Henderson, 1983).

Gamma-aminobutyric acid (GABA)

Autoradiographic experiments show that about 20 per cent of cultured cockroach neurones can accumulate [^{3}H]GABA from their culture medium (Figure 6.9) and that this accumulation is highly dependent on the presence of sodium ions (Beadle and Lees, 1986). Although no bio-

Figure 6.9: [³H]GABA autoradiography. Light-field micrographs of neuronal cultures from the cockroach after 14 days *in vitro*. Cells were incubated for 15 min in 0.5 μM [³H]GABA in a physiological saline, thoroughly washed and fixed prior to treatment for autoradiography using standard techniques. Label is confined to the neurites and somata of a highly specific group of cells (approximately 20 per cent of those present). The absolute Na⁺ dependence of this labelling and its occurrence after prolonged washing suggests a high-affinity uptake of the probe and not binding to membrane receptors for GABA.

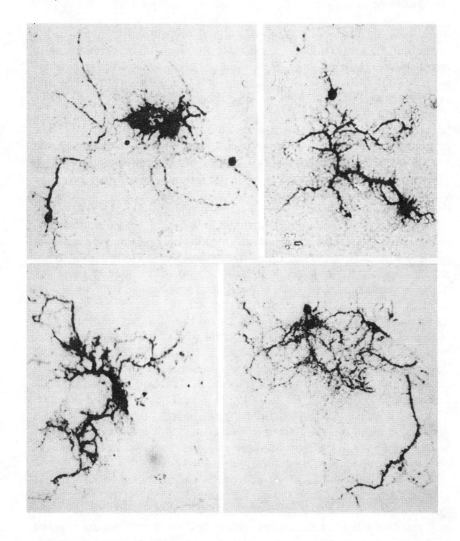

Source: from Beadle and Lees (1986).

chemical data are yet available on this uptake system, it seems likely that GABAergic neurones are present in these cultures since GABA uptake is characteristic of such neurones (Iversen, 1984).

NEUROTRANSMITTER PHARMACOLOGY

Acetylcholine

Many cells in invertebrate neuronal cultural preparations respond to the application of acetylcholine. These responses tend to be depolarising in insect cultures (Usherwood *et al.*, 1980; Lees *et al.*, 1983; Giles and Usherwood, 1985) but may be depolarising or hyperpolarising in leech and *Aplysia* neurones (Camardo *et al.*, 1983; Pelligrino and Simonneau, 1984). The majority of nymphal locust neurones growing *in vitro* are depolarised by acetylcholine but only at concentrations as high as 100 μM. These neurones are therefore 100 times less sensitive than freshly dissociated somata taken from the adult locust central nervous system, suggesting a possible decline in the population density of extrajunctional receptors during the culture period (Giles and Usherwood, 1985). In contrast, 90 per cent of cultured embryonic cockroach neurones respond to 5 μM ACh when applied by pressure ejection with depolarisations that are often sufficiently large to evoke all-or-none action potentials or trains of spikes in impaled somata (Lees *et al.*, 1983; Beadle and Lees, 1986). Qualitatively similar responses are obtained to pressure ejection of nicotine at 0.1–1 μM. These responses, in cockroach neurones, are antagonised by 10^{-7}M mecamylamine and 10^{-6}M *d*-tubocurarine and by high concentrations of muscarinic antagonists, indicating the essentially nicotinic nature of these responses. The snake toxin, α-bungarotoxin (α-BTX), also abolishes these responses in the majority of cockroach neurones at a concentration of 25 μM, although a few cells appear to have nicotinic receptors that are insensitive to BTX since they still respond to ACh and nicotine after 3 h incubation in the toxin (Lees *et al.*, 1983). The apparent reversal potential for these responses calculated from the relationship between the evoked potential changes and the ACh-mediated change in membrane resistance is in the range −15 to −20 mV, and the major ion involved in the conductance increase appears to be sodium (Lees, 1984).

Identified leech and *Aplysia* neurones *in vitro* respond to ACh in a manner similar to that seen *in situ*. For example, both left and right upper quadrant cells from *Aplysia* respond to the ionophoresis of ACh on to the cell soma or regenerated neurites with a dual inhibitory response, a fast i.p.s.p. followed by a slow one (Camardo *et al.*, 1983), whereas Retzius and P sensory cells from the leech are depolarised by ACh (Pelligrino and Simonneau, 1984). In anterior pagoda cells (Ap cells), which are hyper-

246

polarised by ACh, the response appears to be mediated by an increase in chloride conductance, reversing at −65 mV, and shows no sign of desensitisation. The response is blocked by 10^{-3}M d-tubocurarine and 10^{-6}M α-BTX (Pelligrino and Simonneau, 1984).

The currents underlying these responses in insect neurones have recently been investigated using the patch-clamp technique. Addition of 5 μM carbamylcholine (CCh) at resting potential to the saline contained in the patch pipette evokes small, inward currents of about 2 pA amplitude and 1–2 ms duration in cultured cockroach neurones. No such currents are seen using an electrode without CCh. The amplitude and opening time of these currents increases with increasing hyperpolarisation of the cell membrane. With CCh in the patch pipette, two conductance states of the ACh channel are observed with single channel conductances of 18 pS and 48 pS (Figure 6.10A) (Beadle $et\ al.$, 1987). When these neurones are held under voltage clamp conditions and 5 μM CCh is applied to the neuronal membrane, an inward current is evoked that is accompanied by an increase in current noise. When this noise is analysed by fluctuation analysis, the resulting spectra are best fitted with a single Lorentzian component with a corner frequency of 30 Hz. This suggests a single class of ACh channels with a mean channel opening time of approximately 5 ms (Beadle $et\ al.$, 1987) (Figure 6.10B). The obvious discrepancy in these studies between the single channel analysis and the noise analysis may be due to channel activation on neuritic processes away from the soma. Single channel currents have also been recorded from $Drosophila$ neurones growing in culture in response to 100 nM ACh. The mean channel opening time was from 2 to 3.5 ms and the single channel conductance was approximately 9 to 25 pS (Wu $et\ al.$, 1983).

The acetylcholine receptors that have been identified on cultured cockroach neurones have been further characterised using ligand binding techniques. The majority of cockroach neurones $in\ vitro$ possess a specific binding component for $[^{125}\text{I}]\alpha$-BTX with many of the properties expected of a nicotinic receptor (Beadle $et\ al.$, 1983, 1984). At nM concentrations $[^{125}\text{I}]\alpha$-BTX binds specifically to the neurones and under equilibrium conditions the specific component is saturable, non-specific binding accounting for approximately 10 per cent of the total binding. Scatchard analysis of the binding data reveals an apparent dissociation constant of 3.51×10^{-9}M and a maximal number of binding sites of 42 fmol/culture. The pharmacological specificity of the binding component is essentially nicotinic, with the most potent competitors for the BTX binding sites being d-tubocurarine and nicotine with IC_{50} of 3×10^{-7}M and 4×10^{-6}M, respectively, and one of the least potent being muscarine. Autoradiography reveals that the $[^{125}\text{I}]\alpha$-BTX binding sites are distributed over both the cell bodies and the axonal processes in these cells (Lees $et\ al.$, 1983). Using the ionophoretic application of ACh to map receptor sites, Usherwood $et\ al.$

247

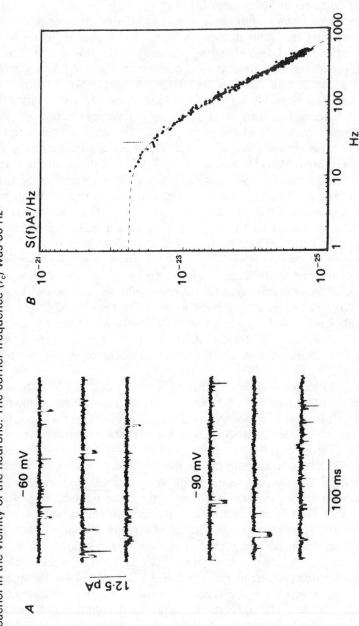

Figure 6.10: (A) Patch-clamp recordings of single-channel activity induced by 5 μM carbachol applied on to the membrane of a culture cockroach neurone. (B) Spectrum of the current fluctuations corresponding to pressure application of 50 μM carbachol in the vicinity of the neurone. The corner frequence (f_c) was 30 Hz

Source: from Beadle *et al.* (in press).

(1980) found a random distribution of receptors over the somata of cultured locust neurones, whereas Pelligrino and Simonneau (1984) found that a region of increased sensitivity to ACh occurs at the base of a growing sprout with a gradient of sensitivity decreasing towards the tip of the neurite in leech Ap cells. Characteristically, the base is three to five times more sensitive to ACh than the soma or the growth cone.

Gamma-amino butyric acid (GABA)

The majority of cells in insect neuronal cultures appear to possess receptors for the putative inhibitory neurotransmitter, GABA. All seven neurones tested in nymphal locust cultures responded to the bath application of GABA at concentrations of 10^{-3}M or above. The response of locust neurones was usually a hyperpolarisation of $10-20$ mV accompanied by a decrease in input resistance, but one cell gave a biphasic response (Giles and Usherwood, 1985). Similarly, when applied on to cockroach neuronal somata at 10^{-4}M by pressure pulses, GABA inhibits spontaneous activity and produces conductance increases in approximately 60 per cent of cells tested (Figure 6.11A). The responses are voltage dependent, being hyper-polarising in the majority of cells with a reversal potential in the range -65 to -75 mV. However, depolarising responses are seen in cells with very large resting potentials or can be induced by injecting chloride ions through the recording electrode into the impaled cell. When cells are held under voltage-clamp conditions, the currents underlying these responses can be seen. They are almost linear with respect to holding potential, reversing at about -15 mV, which is close to the equilibrium potential for chloride ions when comparing the chloride concentration in the bathing saline (221 mM) with that in the intracellular patch solution (114 mM). In all cells tested, the GABA responses are completely and reversibly blocked by 10^{-5}M picro-toxin (Figure 6.11B), but they are insensitive to 10^{-5}M bicuculline (Figure 6.11C). The current evoked by the application of 50 μM GABA on to the neuronal membrane is accompanied by an increase in current noise, and this noise has been analysed with fluctuation analysis methods. The spectra can be reasonably well fitted with a single Lorentzian component with a corner frequency of 14.5 Hz at a holding potential of -50 mV, indicating a single class of ion channels with an opening time of about 11 ms (Beadle and Lees, 1986).

5-Hydroxytryptamine (5HT)

Pressure (P) sensory neurones of the leech *in vitro* respond to pressure pulses of 5HT (Henderson, 1983; Pelligrino and Simonneau, 1984). Small

249

Figure 6.11: Responses of cockroach neurones *in vitro* to pressure application of GABA. (A) Spontaneous action potentials recorded from a neurone maintained for 22 days *in vitro*. 10^{-3}M GABA applied as 100-ms or 600-ms pulses (arrows) inhibits spiking and hyperpolarises the cell by up to 11 mV (RMP: -45 mV) (B) At 10^{-5} M pressure ejected picrotoxin (horizontal bar) blocked the conductance and polarity changes associated with 1-s pulses of 10^{-4} M GABA (arrows). The response was partially recovered after cessation of toxin application (RMP: -50 mV). In some cells 10^{-5} M picrotoxin evoked a small depolarisation possibly by reducing passive chloride permeability (lower trace). (C) Spontaneous activity in a 22-day neurone. Responses to 1-s pulses of 10^{-4} M GABA (arrows) were unaffected by concurrent application of 10^{-5}M bicuculline (horizontal bar) (RMP: -48 mV), upper trace. Bicuculline-insensitive depolarising responses to 1-s GABA pulses in a 14-day neurone (RMP: -75 mV), lower trace

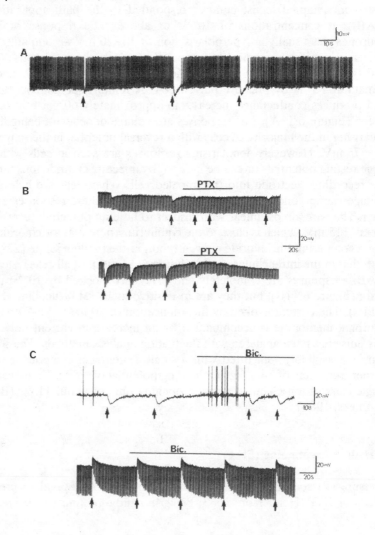

pulses produce a slow hyperpolarisation that desensitises rapidly, and longer pulses or trains of pulses produce a faster, depolarising response that does not desensitise. The hyperpolarisation is reversed by the intracellular injection of chloride, but the depolarisation appears to be independent of chloride concentration (Henderson, 1983). The sensitivity of the neuronal membrane of the P sensory cell mirrors that of the Ap cell to ACh, with the largest responses occurring at the base of sprouting neurites (Pelligrino and Simonneau, 1984). Identified *Aplysia* neurones growing in culture also respond to 5HT. R15, isolated in primary culture, responds to 5HT by increasing the conductance of an inwardly rectifying potassium current, I_R. Forskolin, an activator of adenylate cyclase, also increases the conductance of I_R in these cells and other neurones. The abdominal ganglion cell R2 and the buccal ganglion cell B1 also respond in a qualitatively similar manner to 5HT and forskolin (Lotshaw and Levitan, 1985).

DISCUSSION

Invertebrate animals have been popular specimens for neurobiological studies because of their relatively simple nervous systems, their stereotyped behaviour patterns and the occurrence of large, uniquely identifiable neurones in their ganglia. For many years these advantages have obscured the need to develop *in vitro* systems of invertebrate nervous tissue, and consequently preparations such as synaptosomes and neuronal cultures, which have yielded such a wealth of information on the vertebrate nervous system, have not been available until very recently to invertebrate neurobiologists. Yet as the study of the nervous system becomes increasingly concerned with its cellular and molecular aspects, the advantages of these *in vitro* preparations may become crucial to the solution of specific neurobiological problems that are difficult if not impossible to solve with the intact nervous system. Neuronal cultures present a particularly attractive preparation for the experimental neurobiologist because of the degree of manipulation to which they can be subjected and the accessibility of their membranes and synaptic connections. However, attempts to relate the behaviour of cultured neurones to that of the intact nervous system must be treated with caution since the dissociation process by which cultures are produced is a catastrophic event that may have severe effects on the cells, and the validity of using them must depend on the extent to which they resemble *in vivo* neurones (Beadle and Hicks, 1985).

The invertebrate neuronal culture preparations that have been the subject of this chapter appear to retain many of their normal characteristics during their period of growth *in vitro*. Retzius cells and mechanosensory cells from the leech retain many of their normal membrane properties

251

when grown in culture, and are able to form specific electrical synapses similar to those reported *in vivo* (Fuchs *et al.*, 1981). *Aplysia* L10 neurones growing in culture develop chemical synaptic connections with other cells exhibiting both the specificity and pharmacology of those described in the intact nervous system (Camardo *et al.*, 1983; Schacher, 1983). Cockroach neurones *in vitro* possess transmitter uptake systems and receptor molecules identical to those found *in situ* (Lees *et al.*, 1983; Bermudez *et al.*, 1985) and their excitable characteristics undergo a sequence of developmental changes similar to that observed during embryogenesis (Lees *et al.*, 1985). As a consequence these preparations can be confidently used as reliable models of the intact nervous system.

All of these neuronal culture preparations have been developed to help solve particular neurobiological problems, and they have provided valuable information on such subjects as neuronal growth control, the specificity of synapse information and the pharmacology of neurotransmitter receptors. Unrestricted access to the cell membrane of these neurones and the ability to manipulate their environment make these cultures particularly suitable for modern neurobiological studies. High-resolution 'patch-clamp' recording of transmembrane currents through single channels and the use of monoclonal antibodies to label specific cell surface molecules are facilitated by the availability of such preparations. Neuronal cultures also offer the best opportunity for genetic manipulation and a genetic analysis of neurobiological events.

ACKNOWLEDGEMENTS

The author wishes to acknowledge financial support from the Agricultural and Food Research Council, the Science and Engineering Research Council and the Wellcome Foundation.

Acknowledgements are also due to my colleagues Dr G. Lees, Miss I. Bermudez and Miss A. Mendez in England and Dr C.A. Beadle, Dr Y. Pichon and Dr T. Shimahara in France.

REFERENCES

Adler, R.M., Manthorpe, M., Skaper, S. and Varon, S. (1981) Polyornithine-attached neurite-promoting factors. Culture sources and responsive neurons. *Brain Res.* 206, 129-44

Barker, D.L., Wong, R.G. and Kater, S.B (1982) Separate factors produced by the CNS of the snail *Helisoma* stimulate neurite outgrowth and choline metabolism in cultured neurons. *J. Neurosci. Res. 8*, 419-32

Beadle, C.A., Beadle, D.J., Pichon, Y. and Shimahara, T. (1987) Patch-clamp and noise analysis studies of cholinergic properties of cultured cockroach neurones. *J. Physiol. 371*, 145P

Beadle, D.J., Botham, R.P. and Lees, G. (1983) Acetylcholine receptors on cultured insect neurones. *J. Physiol. 340*, 59-60P

Beadle, D.J. and Hicks, D. (1985) Insect nerve cell culture. In: Kerkut, G. and Gilbert, L. (eds) *Comprehensive insect biochemistry, physiology and pharmacology, 5*, pp. 181-212. Pergamon Press, Oxford

Beadle, D.J., Hicks, D. and Middleton, C. (1982) Fine structure of *Periplaneta americana* neurones in long-term culture. *J. Neurocytol. 11*, 611-26

Beadle, D.J. and Lees, G. (1986) Insect neuronal cultures: a new tool in insect neuropharmacology. In: Ford, M.G. *et al.* (eds) *Neuropharmacology and neurobiology*, pp. 423-44. Ellis Horwood, Chichester

Beadle, D.J., Lees, G. and Botham, R.P. (1984) Cholinergic neurones in neuronal cultures from *Peiplaneta americana*. In: Borkoven, A.B. and Kelly, T.J. (eds) *Insect neurochemistry and neurophysiology*, pp. 317-20. Plenum Press, New York

Bermudez, I., Lees, G., Middleton, C., Botham, R. and Beadle, D.J. (1985) Choline uptake by cultured neurones from the central nervous system of embryonic cockroaches. *Insect Biochem. 15*, 427-34

Bermudez, I., Lees, G., Botham, R.P. and Beadle, D.J. (1986) Myogenesis and neuromuscular junction formation in cultures of *Periplaneta americana* neurones and myoblasts. *Dev. Biol. 116*, 467-76

Burgess, E.J., Atterwill, C.K. and Prince, C.K. (1978) Choline acetyl transferase and the high affinity uptake of choline in corpus striatum of reserpinised rats. *J. Neurochem. 31*, 1027-33

Camardo, J., Proshansky, E. and Schacher, S. (1983) Identified *Aplysia* neurons form specific chemical synapses in culture. *J. Neurosci. 3*, 2614-20

Chen, J.S. and Levi-Montalcini, R. (1970) Axonal growth from insect neurons in glia-free cultures. *Proc. Natl Acad. Sci. USA 66*, 32-9

Dagan, D. and Levitan, I.B. (1981) Isolated identified *Aplysia* neurons in cell culture. *J. Neurosci. 1*, 736-40

Dewhurst, S. and Beadle, D.J. (1985) Cell and tissue culture from the insect nervous system. In: Breer, H. and Miller, T.A. (eds) *Neurochemical techniques in insect research*, pp. 207-22. Springer Verlag, Berlin

Fischbach, G.D. and Nelson, P.G. (1977) Cell culture in neurobiology. In: Brookhart, J.M. and Mountcastle, V.B. (eds) *Handbook of physiology, 1*, pp. 719-774. American Physiological Society, Bethesda, MD

Fuchs, P.A., Nicholls, J.G. and Ready, D.F. (1981) Membrane properties and selective connections of identified leech neurones in culture. *J. Physiol. 316*, 203-24

Fuchs, P.A., Henderson, L.P. and Nicholls, J.G. (1982) Chemical transmission between individual Retzius and sensory cells of the leech in culture. *J. Physiol. 323*, 195-210

Furschpan, E.J., MacLeish, P.R., O'Lague, P.H. and Potter, D.D. (1976) Chemical transmission between rat sympathetic neurons and cardiac myocytes developing in microcultures. *Proc. Natl Acad. Sci. USA, 73*, 4225-9

Ganetsky, B. and Wu, C.F. (1985) Genes and membrane excitability in *Drosophila*. *Trends Neurosci. 8*, 322-6

Gerschenfeld, H.M., Hamon, M. and Paupardin-Tritsch, D. (1978) Release of endogenous serotonin from identified serotonin neurones and the physiological role of serotonin reuptake. *J. Physiol. 274*, 265-78

Giles, D.P., Joy, R.T. and Usherwood, P.N.R. (1978) Growth of isolated locust neurones in culture. *J. Physiol. 276*, 74P

Giles, D.P. and Usherwood, P.N.R. (1985) Locust nymphal neurones in culture: a new technique for studying the physiology and pharmacology of insect central neurones. *Comp. Biochem. Physiol. 80C*, 53-9

253

Goodman, C.S. and Spitzer, N.C. (1980) Embryonic development of neurotransmitter receptors in grasshoppers. In: Sattelle, D.B. *et al.* (eds) *Receptors for neurotransmitters, hormones and pheromones in insects*, pp. 208-14. Elsevier/North Holland Biomedical Press, Amsterdam

Goodman, C.S. and Spitzer, N.C. (1981) The development of electrical properties of identified neurones in grasshopper embryos. *J. Physiol. 313*, 369-84

Haydon, P.G., Cohan, C.S., McCobb, D.P., Miller, H.R. and Kater, S.B. (1985) Neuron-specific growth cone properties as seen in identified neurons of *Helisoma. J. Neurosci. Res. 13*, 135-47

Henderson, L.P. (1981) Serotonergic transmission between isolated leech neurones in culture. *Neurosci. Abstr. 7*, 597

Henderson, L.P. (1983) The role of 5-hydroxytryptamine as a transmitter between identified leech neurones in culture. *J. Physiol. 339*, 309-24

Henderson, L.P., Kuffler, D.P., Nicholls, J. and Zhang, R.-J. (1983) Structural and functional analysis of synaptic transmission between identified leech neurones in culture. *J. Physiol. 340*, 347-58

Hicks, D., Beadle, D.J. and Giles, D.P. (1981) Ultrastructure of dissociated neurones from pupae of *Spodoptera littoralis* growing in culture. *Int. J. Insect Morphol. Embryol. 10*, 399-407

Iversen, L.L. (1984) Amino acids and peptides: fast and slow chemical signals in the nervous system? *Proc. Roy. Soc. Lond. B. 221*, 245-60

Katz, B. and Miledi, R. (1965) The effect of calcium on acetylcholine release from motor nerve terminals. *Proc. Roy. Soc. Lond. B. 161*, 493-503

Lees, G. (1984) A pharmacological study of cultured neurones from *Periplaneta americana*. Ph.D. Thesis, Council for National Academic Awards, London

Lees, G., Beadle, D.J. and Botham, R.P. (1983) Cholinergic receptors on cultured neurones from the central nervous system of embryonic cockroaches. *Brain Res. 288*, 49-59

Lees, G., Beadle, D.J., Botham, R.P. and Kelly, J.S. (1985) Excitable properties of insect neurones in culture: a developmental study. *J. Insect Physiol. 31*, 135-43

Levi-Montalcini, R. (1976) The nerve growth factor: its role in growth, differentiation and function of the sympathetic adrenergic neuron. *Progr. Brain Res. 45*, 235-57

Levi-Montalcini, R., Chen, J.S., Seshan, K.R. and Aloe, L. (1973) An *in vitro* approach to the insect nervous system. In: Young, D. (ed.) *Developmental neurobiology of arthropods*, pp. 5-36. Cambridge University Press, London

Lotshaw, D.P. and Levitan, I.B. (1985) Effects of serotonin on cultured identified *Aplysia* neurons. *J. Gen. Physiol. 86*, 18a

Nishi, R. and Berg, D.K. (1981) Two components from eye tissue that differentially stimulate the growth and development of ciliary ganglion neurons in cell culture. *J. Neurosci. 1*, 505-13

Patterson, P.H. (1978) Environmental determination of autonomic neurotransmitter function. *Ann. Rev. Neurosci. 1*, 1-17

Pelligrino, M. and Simonneau, M. (1984) Distribution of receptors for acetylcholine and 5-hydroxytryptamine on identified leech neurones growing in culture. *J. Physiol. 352*, 669-84

Pomerai, D.I. and Carr, A. (1982) GABA and choline accumulation by cultures of chick embryo neuroretinal cells. *Expl. Eye Res. 34*, 553-63

Ready, D.F. and Nicholls, J.F. (1979) Identified neurons isolated from leech CNS make selective connections in culture. *Nature* (Lond.) *182*, 67-9

Schacher, S. (1983) *Aplysia* neurons from synaptic connections in dissociated cell culture. *Neurosci. Res. Progr. Bull. 20*, 875-7

Schacher, S. and Proschansky, E. (1983) Neurite regeneration by *Aplysia* neurons in dissociated cell culture: modulation by *Aplysia* hemolymph and the presence of the initial axonal segment. *J. Neurosci. 3*, 2403-13

Schwartz, M., Mizrachi, Y. and Kimhi, Y. (1982) Regenerating goldfish retinal explants: induction and maintenance of neurites by conditioned medium from cells originated in the nervous system. *Devel. Brain Res. 3*, 21-8

Shields, G., Dubendorfer, A. and Sang, J.H. (1975) Differentiation *in vitro* of larval cell types from early embryonic cells of *Drosophila melanogaster. J. Embryol. Exp. Morphol., 33*, 159-75

Simon, J.R., Atweh, S. and Kuhar, M.J. (1976) Sodium-dependent high affinity choline uptake: a regulatory step in the synthesis of acetylcholine. *J. Neurochem. 26*, 909-22

Sun, Y-A. and Wu, C-F. (1984) Voltage-dependent single-channel currents in dissociated CNS neurons of *Drosophila. Neurosci. Abstr. 10*, 689

Sun, Y-A. and Wu, C-F. (1985) Genetic alterations of single channel potassium currents in dissociated central nervous system neurons of *Drosophila. J. Gen. Physiol. 86*, 16a

Suzuki, N. and Wu, C-F. (1984) Fusion of dissociated *Drosophila* neurons in culture. *Neurosci. Res. 1*, 437-42

Usherwood, P.N.R., Giles, D. and Sutar, C. (1980) Studies of the pharmacology of insect neurones *in vitro*. In: *Insect neurobiology and pesticide action*, pp. 115-28. Society of Chemical Industry, London

Varon, S., Manthorpe, M. and Adler, R. (1979) Cholinergic neuronotrophic factors: 1. Survival, neurite outgrowth and choline acetyltransferase activity in monolayer cultures from chick embryo ciliary ganglia. *Brain Res. 173*, 29-45

Wong, R.G., Hadley, R.D., Kater, S.B. and Hanser, C. (1981) Neurite outgrowth in molluscan organ and cell cultures: the role of conditioning factors. *J. Neurosci. 1*, 1008-21

Wong, R.G., Martel, E.C. and Kater, S.B. (1983) Conditioning factor(s) produced by several molluscan species promote neurite outgrowth in culture. *J. exp. Biol. 105*, 389-93

Wong, R.G., Barker, D.L., Kater, S.B. and Bodnor, D.A. (1984) Nerve-growth promoting factor produced in culture media conditioned by specific CNS tissues of the snail *Helisoma. Brain Res. 292*, 81-91

Wu, C-F., Berneking, J.M. and Barker, D.L. (1983a) Acetylcholine synthesis and accumulation in the CNS of *Drosophila* larvae: analysis of *Shibire^{ts}*, a mutant with a temperature-sensitive block in synaptic transmission. *J. Neurochem. 40*, 1386-96

Wu, C-F., Suzuki, N. and Poo, M-M. (1983b) Dissociated neurons from normal and mutant *Drosophila* larval central nervous system in cell culture. *J. Neurosci. 3*, 1888-99

255

7

Neurotoxins*

Eliahu Zlotkin

INTRODUCTION

Various animals employ toxic substances in order to capture their prey or to deter their enemies (Zlotkin, 1973). They can be subdivided into (a) venomous animals — which possess the proper instrumentation for stinging–piercing and the time- and site-directed introduction of their mixtures of toxic substances (defined as venoms) into the circulation or tissues of their prey or opponent; (b) poisonous animals which are devoid of the apparatus for stinging and employ toxic substances for defensive purposes (Blum, 1981).

In nature, whenever chemical means are employed for prey immobilisation, they are always associated with venomous stinging apparatuses. The active components in venoms are predominantly proteins and polypeptides which, through their covalent structures and the resulting spatial arrangements, possess unique pharmacological specificities such as site-directed, selective neurotoxicity. A rapid paralysis of an animal is achieved through an effect on its nervous system. Thus animal venoms (such as those of snakes, scorpions, sea anemones, etc.) serve as an important source of polypeptide neurotoxins.

Neurotoxin is defined as a compound which affects the function of excitable tissues due to a specific recognition and binding affinity for critical sites located in these tissues. Thus, neurotoxins possess high binding affinities (at or below the nanomolar concentration) for low-capacity sites, enabling them to exert their paralytic effects in extremely low doses. Neurotoxins that satisfy the above definition are also included among the so-called defensive substances. These substances are substantially low molecular weight organic molecules such as simple hydrocarbons, cyclic compounds, heterocyclic alkaloidal compounds, steroidal alkaloids, etc. which serve as blockers and modifiers of ion-conducting structures in excit-

---

*Dedicated to Dr Shlomo Zlotkin on his 85th birthday.

256

able membranes and serve as important tools for the study of their chemistry and function (Albuquerque and Daly, 1977; Catterall, 1980; Hucho and Ovchinnikov, 1983; Witkop and Brossi, 1984).

The intensive and successful employment of neurotoxins as pharmacological tools in the study of vertebrate mammalian neurochemistry and neuropharmacology is exemplified and reviewed in the next section. As shown (see Discussion) the above has stimulated a similar approach to arthropod neuropharmacology expressed in the employment of identical toxins for the characterisation of insect axonal membranes, cholinergic receptors, ion channels, etc.

With this background the present review is substantially dedicated to certain unique aspects exclusively related to arthropod neuropharmacology, namely the insect and crustacean selective neurotoxins. As such this chapter serves as an expanded, modified and updated version of a previous article (Zlotkin, 1985). It is hoped that this approach will encourage the development of special pharmacological tools for the study of unique processes in arthropod neurophysiology and offer new strategies for research into and design of selective pesticides.

NEUROTOXINS IN VERTEBRATE NEUROCHEMISTRY

This section is mainly dedicated to neurotoxins derived from animal venoms. Bacterial toxins are thus excluded.

Ion channel toxins

Ion channels appear to be composed of a selectivity filter, determining which ion may pass and which is excluded, and a gate which determines, in response to a signal, when an ion may pass. The signal moving the gate may be an electrical one — a change in membrane potential (such as the axonal action potential Na^+ channel) — or a chemical one (such as a transmitter-activated postsynaptic Na^+ channel or the Ca^{2+} activated K^+ channels). This section is dedicated to neurotoxins that interact with channels possessing the two kinds of gating mechanism.

Sodium channel toxins

The recent impressive progress that has been made in the elucidation of the chemical basis of the function of voltage-sensitive sodium channels involved in action potential generation in nerve, heart and skeletal muscle (Catterall, 1984, 1985) should be attributed to the skilful employment of various neurotoxins.

The heterocyclic guanidines tetrodotoxin (TTX) and saxitoxin (STX)

257

derived from certain aquatic animals represent the selective blockers of the ion selectivity filter (Hille, 1971, 1975; Catterall, 1980). The guanidinium moieties and the unusually acidic hydroxyls (Figure 7.1), common features of the two molecules, have been shown to have similar stereochemical orientation (Figure 7.1; Hille, 1975) and are supposed to represent points of attachment to the receptor sites of Na^+ channels (Catterall, 1980; Hu and Kao, 1986) of both toxins. These toxins have played an important role in the characterisation of Na^+ channels concerning their ion selectivity, voltage dependence and localisation in the membrane (Narahashi et al., 1964, 1966, 1967; Nakamura et al., 1965; Hille, 1968), and their density and distribution in the axonal membrane (Ritchie et al., 1976; Catterall and Morrow, 1978; Strichartz et al., 1979) and in monitoring isolation, chemical characterisation and purification of the sodium channels derived from rat brain, electroplax of the electric eel and rat skeletal muscle (Henderson and Wang, 1972; Benzer and Raftery, 1973; Hartshorne et al., 1982; Miller et al., 1983; Tanaka et al., 1983; Hartshorne and Catterall, 1984).

Figure 7.1: Structure of some Na^+ channel toxins: heterocyclic guanidines and some lipid-soluble toxins

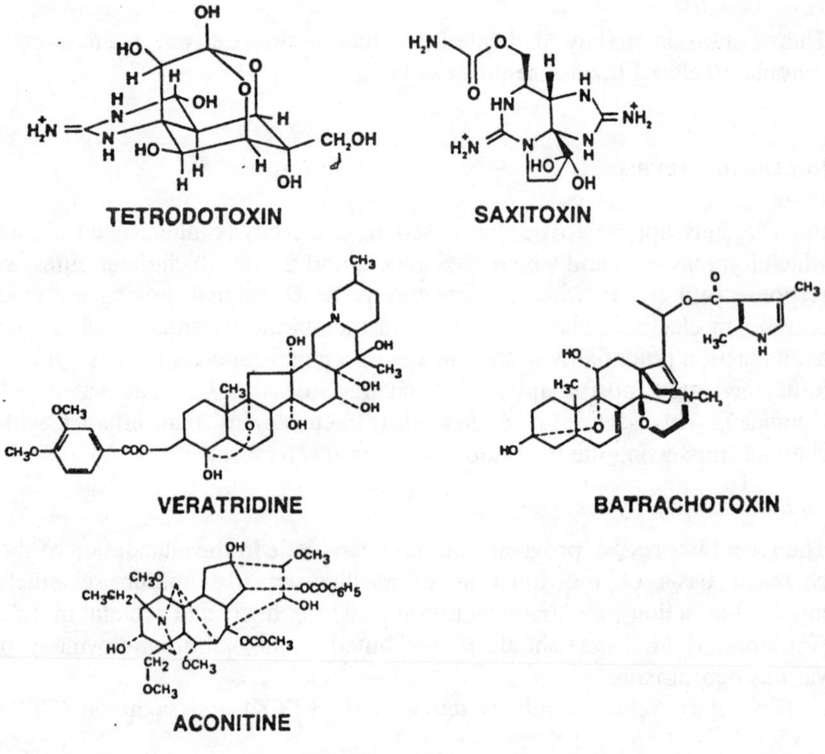

TETRODOTOXIN

SAXITOXIN

VERATRIDINE

BATRACHOTOXIN

ACONITINE

258

The second category of sodium-channel-affecting compounds are the lipid-soluble alkaloids that modify activation and inactivation (Catterall, 1980, 1984) such as veratridine, batrachotoxin and aconitine (Figure 7.1), originating from plants in the family Melanthaceae, from frogs of the genus *Phyllobates* and from plants of the genus *Aconitum*, respectively. Their chemistry and pharmacology have been reviewed previously (Ohta *et al.*, 1973; Herzog *et al.*, 1974; Schmidt and Schmitt, 1974; Catterall, 1975a,b; Albuquerque and Daly, 1976). These substances were shown to depolarise excitable membranes by blocking sodium inactivation and shifting its activation to more negative potentials. The combination of these two effects causes a fraction of the ion channels to be persistently activated at the resting membrane potential. The increase in sodium permeability induced by these toxins is inhibited by tetrodotoxin (Ulbricht, 1969; Catterall, 1975a) thus indicating that their effects are mediated through the same channels that generate the action potential.

The North African scorpion venom α-toxins affecting vertebrates (M_r 7 kDa) (Rochat *et al.*, 1970; Zlotkin *et al.*, 1978) and the sea anemone neurotoxins (M_r 3–5 kDa) (Schweitz *et al.*, 1981) comprise the third group of Na^+ channel toxins. In spite of the lack of any detectable sequence homologies between the above two groups of toxins, both affect sodium conductance in a similar manner expressed in: (1) inhibition of sodium inactivation (Romey *et al.*, 1975; Low *et al.*, 1979); (2) synergistic interaction with the lipid-soluble toxins in the persistent sodium activation (Catterall, 1975b, 1977a,b; Jacques *et al.*, 1978); (3) voltage-dependent binding (Catterall and Beress, 1978; Ray *et al.*, 1978) and (4) occurrence of a mutual competitivity in the binding of the scorpion and sea anemone toxins (Couraud *et al.*, 1978; Ray *et al.*, 1978; Catterall, 1979), suggesting common receptors. It has been concluded (Catterall, 1980) that the above toxins bind to voltage-sensing or gating structures of sodium channels. It is noteworthy that the scorpion venom α-toxins affecting vertebrates (Couraud and Jover, 1984) were also employed as tools to detect and localise sodium channels in neuronal cells and tissues (Berwald-Netter *et al.*, 1981; Catterall, 1981) and chemically characterise rat brain major channel components by photoaffinity labelling (Beneski and Catterall, 1980; Darbon *et al.*, 1983; Sharkey *et al.*, 1984).

The fourth neurotoxin receptor site on sodium channels is occupied by the so-called β-scorpion toxin affecting vertebrates, which is derived from the American scorpions of the genera *Centruroides* and *Tityus* (Couraud and Jover, 1984). In their basic chemical properties the α-toxins closely resemble the β-toxins. However, the latter differ in their pharmacology by: (a) affecting sodium activation (Cahalan, 1975; Couraud *et al.*, 1982); (b) possessing binding sites which are distinct from those of the α-toxins; (c) possessing a voltage-independent binding; and (d) in contrast to the α-scorpion toxins, in not demonstrating any cooperativity with the lipid-soluble toxins such as veratridine (Jover *et al.*, 1980a; Jaimovich *et al.*, 1982; Darbon *et al.*, 1983).

It was recently shown (Bidard *et al.*, 1984) that ciguatoxin, an oxygenated polyether compound derived from dinoflagellates and accumulated in the bodies of reef fish (Baguis *et al.*, 1980), acts selectively on voltage-dependent Na^+ channels in nerve and muscle cells as well as synaptic terminals. This compound increases sodium permeability which is blocked by TTX and enhanced by various toxins such as veratridine pyrethroids and sea anemone and scorpion toxins. A combination of electrophysiological NA^+ flux and binding assays has revealed that ciguatoxin belongs to a new class of sodium channel toxins (Bidard *et al.*, 1984).

Potassium channel toxins

Three polypeptide neurotoxins derived from arthropod venoms were shown to selectively affect potassium conductance in excitable membranes.

The first corresponds to fraction II-11 isolated from the venom of the Mexican scorpion *Centruroides noxius* (M_r 7000 — Possani *et al.*, 1981) which was shown to specifically suppress the K^+ permeability in a dose-dependent manner when assayed in a voltage-clamped giant axon (Carbone *et al.*, 1982).

The second is apamin (basic, 18 amino acids, 2 disulphide bridges) which has a well known sequence (Habermann, 1971; Van Rietschoten *et al.*, 1975) and which was recently shown to serve as a specific blocker of the calcium-dependent potassium permeability in excitable cells (Hugues *et al.*, 1982 a,b). Neuroblastoma cells (Hugues *et al.*, 1982a) and rat-brain primary cultures were shown to possess high-affinity apamin binding sites with a K_D of 12–22 and 60–120 pM and maximal capacity of 12 and 3–8 pmol/mg of membrane protein respectively. Apamin's receptor in rat brain is a protein component with an M_r of about 28000 (Hugues *et al.*, 1982b), and its endogenous equivalent was isolated and pharmacologically identified in pig brain (Fosset *et al.*, 1984).

The third polypeptide is charybdotoxin (M_r 7000) isolated from the venom of the scorpion *Leiurus quinquestriatus* which was shown to reversibly block the apamin-nonsensitive Ca^{2+}-dependent K^+ channels in mammalian skeletal muscles (Miller *et al.*, 1985; Abia *et al.*, 1986).

Calcium channel toxins

Maitotoxin (MTX), isolated from toxic dinoflagellates and fishes inhabiting tropical seas, is considered to be a non-peptidic substances of a large molecular weight (Yasumoto *et al.*, 1979). This toxin was shown to induce a series of effects in various preparations (Takahashi *et al.*, 1982, 1983; Ohizumi and Yasumoto, 1983; Ohizumi *et al.*, 1983; Miyamoto *et al.*, 1984; Kobayashi *et al.*, 1985) which were enhanced by increasing Ca^{2+} concentration, abolished in Ca^{2+}-free medium and inhibited by calcium ion

260

blockers. It was finally concluded that maitotoxin serves as a specific activator of voltage-sensitive Ca^{2+} channels in excitable membranes. The pharmacology of maitotoxin was essentially mimicked by a polypeptide (M_r = 26000) toxin (tessulatoxin) isolated from the venomous apparatus of a marine snail *Conus tessulatus* (Kobayashi *et al.*, 1983).

Modification of calcium ion conductance was shown to be a primary effect of certain so-called presynaptic toxins which affect transmitter release. This aspect is presented in the next section.

Presynaptic toxins

Modifiers of Ca^{2+} conductance

Several toxic peptides were shown to affect neurotransmitter release through an action on calcium channels.

WCgTX, a peptide isolated from the venom of a marine snail *Conus geographicus*, was shown to (a) irreversibly block stimulus-evoked release of transmitter in the frog neuromuscular junction by suppressing Ca^{2+} entry into the nerve terminal; and (b) suppress irreversibly the Ca^{2+} component of the action potential in the dorsal root ganglion neurones from an embryonic chick (Kerr and Yoshikami, 1984; Ohizumi *et al.*, 1986).

In contrast, activation of presynaptic calcium channels was claimed to be the primary cause of the stimulation of neurotransmitter release induced by the partially purified polypeptide, β-leptinotarsin-h (M_r 57000) derived from the haemolymph of the beetle *Leptinotarsa haldemani* (McClure *et al.*, 1980; Crosland *et al.*, 1984).

The release of ACh from nerve terminals by the venom of a polychaete annelid *Glycera convoluta* (Manaranche *et al.*, 1980) was attributed to a glycoprotein (M_r 300000) shown to bind to the presynaptic plasma membrane and trigger Ca^{2+} entry into the nerve terminal either through activation of presynaptic Ca^{2+} channels or through a possible channel-forming action (Kagan *et al.*, 1982; Morel *et al.*, 1983).

Toxins with phospholipase A_2 activity

The four most characterised toxic phospholipases A_2 derived from elapid snake venoms which inhibit the evoked release of neurotransmitter are β-bungarotoxin, notoxin, taipoxin and crotoxin, which, in spite of their differences in structure and complexity, demonstrate comparable enzymic and pharmacological activities (see review by Howard and Gundersen, 1980). At a frog neuromuscular junction β-bungarotoxin, the most extensively studied, induces an initial inhibition, followed by a substantial increase and final irreversible blockage of transmitter release (Abe *et al.*, 1977; Llados *et al.*, 1980).

The dependence of the neurotoxicity on the phospholipase activity of

261

the above toxins has been demonstrated first by replacing Ca^{2+} (which is needed for neurotransmission and activation of enzymic activity) by Sr^{2+} (which can replace Ca^{2+} only for neurotransmission), and secondly by the employment of a specific alkylating inhibitor of phospholipase A_2, p-bromophenacyl bromide (Wernike *et al.*, 1975; Abe *et al.*, 1976, 1977; Strong *et al.*, 1976; Marlas and Bon, 1982). The molecular basis of this relation between the neurotoxic and enzymic activities is not fully understood but it is evident from the data that have accumulated and have been analysed (Howard and Gundersen, 1980) that the above toxins affect neurotransmission by modifying the ionic conductance at the presynaptic membrane. It may thus be suggested that the site specificity of these toxins follows from their unique binding affinity to certain structural features of the presynaptic membrane, and their action is probably due to hydrolysis of membrane phospholipids at critical sites. Such hydrolysis may structurally affect ion channels or membrane-bound ionic pumps, or alternatively form some toxic reaction products.

Latrotoxin

The stimulatory presynaptic effect of the venom of *Latrodectus* (black widow) spiders (Clark *et al.*, 1970; Longenecker *et al.*, 1970) is attributed to a single protein component of 110–130 kDa (devoid of sugar residues and enzymic activity) named α-latrotoxin (α-LTX) (Frontali *et al.*, 1976). The isolated toxin was described as oligomeric (monomer M_r 75 000), with the dimer and tetramer having biological activity. Acidic conditions cause the oligomer to dissociate. Amino acid composition and tryptic peptides of the toxin were also reported (Salikhov *et al.*, 1982a,b).

At all vertebrate synapses so far studied, irrespective of the nature of the transmitter involved, α-LTX elicited its typical massive stimulation of neurotransmitter release (Grasso and Senni, 1979; Paggi and Toschi, 1977; Tzeng *et al.*, 1978; Baba and Cooper, 1980; Meldolesi, 1982). α-LTX was shown to: (1) bind to neuronal tissues with a high affinity (K_D 0.2–4.0 nM) and a low capacity of 2200 and 300 binding sites per square micrometre in synaptosomes, PC12 cells (neurosecretory cell line derived from rat pheochromocytoma) and neuromuscular junction, respectively (Meldolesi *et al.*, 1983); (2) induce a depolarisation of the neuronal membrane attributable to Na^+ entry but in a TTX non-sensitive manner (Nicholls *et al.*, 1982); (3) induce a large increase in Ca^{2+} influx which was only partially inhibited by the well known calcium blocker, verapamil. These effects may account for the toxin's transmitter-releasing action. There are presently two explanations for the above effect of α-LTX on the ionic conductance at the presynaptic membrane. First, the toxin acts as a cation ionophore, as supported by experiments with artificial lipid bilayer membranes (Finkelstein *et al.*, 1976; Krasilnikov *et al.*, 1982). Secondly, following a more recent study with PC12 cells (Vincentini and Meldolesi,

1984), the toxin appears as an inducer of phosphoinositide breakdown, which is known as an event coupled to the activation of receptors of various hormones and transmitters and may operate through Ca^{2+} influx or redistribution. It was therefore suggested that the α-LTX binding site is a receptor coupled across the membrane to the phosphoinositide hydrolysing system (Vincentini and Meldolesi, 1984).

Postsynaptic toxins

Blockers of nicotinic cholinergic receptors

The main lethal and paralytic action of elapid (cobra and krait) and hydro-phiid (sea snakes) venoms is substantially due to a series of basic poly-peptides which function as potent curaremimetic blockers of nicotinic cholinergic receptors, thus irreversibly paralysing the skeletal musculature of vertebrates. This is the most studied group of neurotoxins, and for detailed information the reader is directed to the following reviews and books: Lee (1972, 1979a,b); Tu (1974, 1977); Rosenberg (1978); Eaker and Wandstrom (1980); and Hucho and Ovchinnkov (1983). Certain essential aspects will be presented briefly here.

The amino acid sequences of about a hundred of these toxins have been determined. They can be subdivided into 'short' (60–62 amino acids, four disulphide bridges) and 'long' toxins (71–74 amino acids, five disulphide bridges). Both groups share closely resembling allocation of the disulphide bridges, conserved residues (Tyr 25, Trp 29, Asp 31, Arg 37, Gly 38, Gln 44, Pro 50 and Asn 69) and an obvious structural homology in their spatial arrangement (Saenger et al., 1983). A series of site-directed chemical modifications of the α-cobratoxin have indicated that the interaction with the cholinergic receptor includes a multipoint attachment in which the single tyrosine and the area around the extra disulphide bond do not participate (Chibber et al., 1983). From a neuropharmacological point of view, the strong affinity ($K_D \approx 10^{-12}$ M) and slow rates of dissociation of receptor–toxin complexes have made the neurotoxins extremely useful as labelling reagents for isolation, purification and assay of nicotinic receptors (Changeux et al., 1970). α-Bungarotoxin (74 a.a.; 5 S–S: M_r 8 kDa) isolated from the venom of Bungarus multicinctus (Mebs et al., 1972) has played a major role. The ACh receptor from Torpedo is a membrane protein, each molecule comprising five polypeptide chains: two chains (M_r = 40000) and three homologous chains of M_r 50000 (β), 60000 (γ) and 65000 (δ) (Raftery et al., 1980) of which the α-subunits seem to carry the specific recognition site of the toxin (Oblas et al., 1983).

With this background, the occurrence of a new class of cholinergic blockers is curious. Recently the structures of three toxins from the fish-eating snail Conus geographicus (Gray et al., 1981) and one toxin from

263

Conus magus (McIntosh *et al.*, 1982) were reported. These are short homologous peptides (13–15 amino acids), containing two disulphide bonds. They paralyse vertebrates by blocking the acetylcholine receptor at the neuromuscular junction (Gray *et al.*, 1981). The pharmacological resemblance between these conotoxins and the snake venom α-toxins, which are about five times as big, deserves attention and may contribute to an understanding of the molecular features related to the interaction with the cholinergic receptor.

Glutaminergic blockers

It has been suggested that *L*-glutamate serves as an excitatory transmitter in crustacean and insect neuromuscular junctions (Usherwood *et al.*, 1968; Gerschenfeld, 1973) as well as in synapses of the vertebrate central nervous system (Curtis and Johnston, 1974; Krnjevic, 1974). The final and further identification of this transmitter in nervous systems, as well as the study of the chemical nature of the respective postsynaptic receptors, demands the availability of specific high-affinity glutaminergic blockers. It appears that this necessity has recently been satisfied through a series of findings indicating the occurrence of glutaminergic blockers in various spider venoms.

A fraction (JSTX) isolated (by gel filtration chromatography) from the venom of the spider *Nephila clavata* when assayed on a crustacean neuromuscular junction was shown to: (1) selectively suppress, in a dose-dependent manner and with exponential kinetics, the excitatory postsynaptic potential without affecting the inhibitory postsynaptic potentials, (2) block the ionophoretically applied glutamate postsynaptic potential; but failed to affect the aspartate-induced depolarisation — finally suggesting the specific blockage of glutamate receptors in crustacean neuromuscular junctions (Kawai *et al.*, 1982a; Abe *et al.*, 1983). The above JSTX fraction was also shown to specifically block glutamate receptors in the mammalian brain (Kawai *et al.*, 1982b) and was employed as a tool to identify glutaminergic transmission in a squid giant synapse (Kawai *et al.*, 1983b) and in acoustico-lateralis receptors of a marine catfish (Nagai *et al.*, 1984). The occurrence of glutaminergic blockers was demonstrated in the venom of fourteen additional araneid spiders (Kawai *et al.*, 1983a; Michaelis *et al.*, 1984; Usmanov *et al.*, 1985; Quicke, 1985). In a recent study, the specificity of glutaminergic blockage by spider venoms was further emphasised by their inhibition of the high-affinity sodium-independent *L*-glutamate binding in rat brain synaptic membranes and their derived glutamate binding glycoprotein. These data may indicate that the spider venoms interact with glutamate recognition sites related to the physiological glutamate receptors (Michaelis *et al.*, 1984).

The chemical nature of the active components is still unclear and there exist some discrepancies concerning their molecular weight. The active compounds derived from *Nephila clavata, Argiope trifasciata* and *Araneus gemma* were shown to be heat resistant and of an extremely low molecular weight in the range 500–1000 (Kawai *et al.*, 1983a,b; Quicke, 1985). On the other hand, Michaelis *et al.* (1984) claim that in the venom of *Araneus gemma* the active components are of an estimated minimal molecular weight of 7 400 and several multimers.

NEUROTOXINS SPECIFICALLY AFFECTING INSECTS

The search for insect-selective toxins directs our attention to venomous animals which prey on insects and employ their venom against insects. These venomous animals are themselves terrestrial arthropods and represent all the major divisions of this large phylum.

Insect-selective components in *Bracon* venom

Biological function of venom

In the terebrant groups of solitary parasitic wasps the venom is primarily employed for securing food for their offspring. For details concerning biology and prey–predator interactions of terebrant wasps the reader is directed to Beard (1963, 1978) and Steiner (1986). The venom system of Braconidae has received the greatest attention. Parasitism by Braconidae may be either internal (in free-living hosts) or external (in hosts that live in confined quarters). The external parasitoids may be subdivided into those that induce a temporary paralysis and those that induce a permanent one. The latter were studied in detail through the venom of *Bracon hebetor, B. brevicornis* and *B. gelechiae* (Beard, 1952, 1978). *Bracon* wasps require for the development of their offspring lepidopterous larvae such as *Ephestia, Plodia* or *Galleria.* The female wasp paralyses its host by injecting venom through its stinging apparatus and feeds upon the haemolymph released from the wound made in the host integument. The male wasp does not feed on the host's haemolymph. Oviposition is independent of the stinging and feeding behaviour, the eggs being deposited on or very near the paralysed larva. After leaving the egg, the wasp larvae cling to the host and they too feed upon the haemolymph by puncturing the integument (Beard, 1952).

Selective toxicity to insects

Braconid and other terebrant parasitic wasps appear to be incapable of stinging vertebrate animals. Their selective toxicity to insects is indicated by the following data:

265

(1) Injection of venom-gland extracts of *Bracon* (= *Microbracon* = *Habrobracon*) *hebetor* to insects belonging to 48 species representing eight orders has revealed a clear preference for Lepidoptera and even a distinction among various insects belonging to the same order (Drenth, 1974b; Piek and Spanjer, 1986).

(2) The extremely potent toxicity of the authentic venom to susceptible insects is expressed as a dilution of 2×10^8 still effective (Beard, 1952) or 26.8×10^{-10} ml inducing 100 per cent paralysis (Tamashiro, 1971).

(3) The venom induces a flaccid paralysis of long duration (weeks) within 15 min after the sting (Piek *et al.*, 1982b), which is a consequence of (a) a presynaptic blockage of the excitatory glutaminergic transmission (Walther and Rathmayer, 1974; Piek *et al.*, 1982b) attributed to the prevention of the vesicle exocytosis (Walther and Reinecke, 1983); (b) the normal persistence of the inhibitory (GABAergic) neuromuscular transmission.

(4) Neurophysiological studies have revealed that the venom of *B. hebetor* affected neither the cholinergic transmission in vertebrate neuromuscular preparations (Deitmer, 1973; Piek *et al.*, 1982b) nor the glutaminergic transmission (Gerschenfeld, 1973) of spider, crab and crayfish (Rathmayer and Walther, 1976).

Chemistry of the toxic components

The protein nature of the *Bracon* venom active principles was established by assays of heat inactivation, non-dialysability, precipitability by ammonium sulphate (Beard, 1952, 1960) and inactivation by proteolytic enzymes (Tamashiro, 1971). The difficulty in purifying the active component(s) is due to their extreme lability in various conditions (Drenth, 1974a; Visser *et al.*, 1983; Piek and Spanjer, 1986).

An extract of whole animals was submitted to a process of purification specified in the legend to Figure 7.2. The third and last step of column chromatography on QAE-Sephadex (Figure 7.2) resulted in two active fractions (A and B) with a total recovery of about 40 per cent of the toxic activity of the starting material. The two fractions were shown to be impure by disc electrophoresis. Analytical gel chromatography has shown that peaks of paralysing activities of fractions A and B corresponded to molecular weights of 42 000 and 57 000, respectively (Spanjer *et al.*, 1977). Both fractions affected the frequency of the miniature excitatory postsynaptic potentials (JM e.p.s.p.s) without modifying their amplitude and the resting potential of muscle membrane, indicating a presynaptic effect strongly resembling that of the crude venom (Piek *et al.*, 1982b).

A recent effort at purification (Visser *et al.*, 1983), using the same starting material and certain modifications in the procedure, has led to the isolation of A- and B-MTX the purity of which was assessed by disc

266

Figure 7.2: The isolation of two paralysing fractions from a homogenate of the wasp *Microbracon hebetor*. Column of 15 × 2 cm filled with QAE-Sephadex and equilibrated and eluted with 0.3 M ammonium carbonate, pH 9.2. Material for separation comprised 86 mg of substance lyophilised in the presence of sucrose obtained from a homogenate of 10 g of female wasps by two preceding steps. Batchwise separation on DEAE-Sephadex A50 and gel filtration on a column of Sephadex G-100. In the present run a flow rate of 10 ml h^{-1} was employed. The arrow indicates the start of a linear gradient of molarity up to 1 M of ammonium carbonate; ———, absorbance at 280 nm; •-----• toxic activity assayed by paralysis of *Galleria mellonella* larvae.

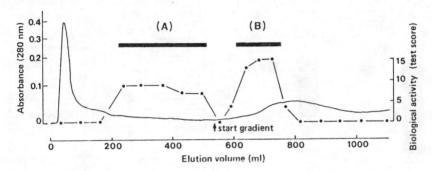

Source: from Spanjer *et al.* (1977).

electrophoresis (Figure 7.3). The relative amino acid compositions of the two toxins are presented in Table 7.1.

To summarise, A-MTX and B-MTX are labile proteins with molecular weights of 43 700 and 56 700 and pI of 6.85 and 6.62, and with an increase of 18 and 28 times in the specific toxicity and recovery of 0.6 and 1.7 per cent of the biological activity, respectively. The toxins are inactivated by proteolytic enzymes, by a tryptophan-selective reagent and by a reducing reagent. The relative amino acid compositions of both toxins show great similarity (Table 7.1). The biological effects of the two toxins are similar to those of the crude extract (Visser *et al.*, 1983).

Components in *Philanthus* venom toxic to insects

Function and action of the venom

As in terebrants (discussed earlier) the aculeate solitary wasps employ their venom primarily for supplying food to their offspring. In contrast to the terebrants which encounter prey in the range of two to three orders of magnitude larger than themselves, the relatively large aculeate forms hunt prey that are roughly the same size as themselves.

267

Figure 7.3: Analytical polyacrylamide gel electrophoresis of *Microbracon hebetor* toxins. Photograph: staining for proteins. Histograms: localisation of biological activity. (a) Crude preparation of A-MTX after step 4; (b) purified A-MTX after step 6; (c) crude preparations of B-MTX after step 4: (d) purified B-MTX after step 6. Amounts of 182, 25, 308 and 23 μg of proteins, respectively, were applied to the gels. Staining with Coomassie Brilliant Blue R-250. Histograms show distribution of biological activities eluted from gel slices after electrophoresis of purified toxin (step 6) preparations; 24 μg of protein (30500 paralysing units) of A-MTX and 22 μg of protein (35500 paralysing units) of B-MTX were applied to the gels. The peak in the activities coincides with the protein band in parallel gels. The anode is on the lower end of the gels.

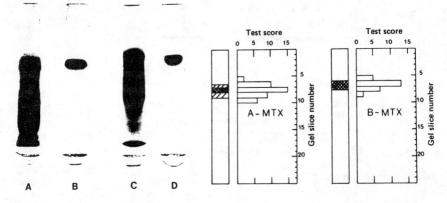

Source: from Visser *et al.* (1983).

The Sphecidae are the most intensively studied among the aculeate solitary wasps (Rathmayer, 1962a, b, 1978; Steiner, 1986). In most cases the prey, composed of insects and spiders, is more or less deeply paralysed by the injection of the venom. Once the wasp has grasped the prey with its legs, it immediately tries to insert the sting. Wasps which prey on Hymenoptera or adult holometabolous insects typically apply only one sting, through either the thin intersegmental membranes on the base of the legs or the cervical integument between head and prothorax. Sphecids preying on Orthoptera or larval insects, particularly caterpillars, typically sting several times. The insertion of the sting is guided by tactile stimuli (Steiner, 1962) sensed by mechanoreceptors located on its sheath (Rathmayer, 1962a).

Among the Sphecidae the venom of *Philanthus triangulum* has been studied extensively. In spite of the fact that *P. triangulum* preys exclusively upon honeybees, it can in practice paralyse, by forced stinging, all other insects so far tested. These include 37 genera from 15 families from seven orders, as well as spiders (Rathmayer, 1962b, 1978). When further compared with *Bracon*, the *Philanthus* venom is much more diverse in its pharmacological action. As in the case of *Bracon*, the venom of *Philanthus*

Table 7.1: Relative amino acid compositions of A-MTX and B-MTX

	Molar ratios[a]	
	A-MTX	B-MTX
1/2 Cys[b]	0.42±0.07	0.37±0.02
Asp	1.69±0.02	1.62±0.04
Thr	0.66±0.06	0.72±0.02
Ser	0.68±0.09	0.68±0.01
Glu	1.48±0.07	1.50±0.03
Pro	0.76±0.04	0.66±0.02
Gly	1.14±0.05	1.17±0.05
Ala	1.00	1.00
Val	0.72±0.04	0.79±0.01
Met[c]	0.24±0.03	0.32±0.01
Ile	0.55±0.03	0.61±0.01
Leu	0.90±0.03	0.99±0.04
Tyr	0.36±0.04	0.40±0.04
Phe	0.54±0.03	0.54±0.03
Lys	0.86±0.03	1.03±0.03
His	0.32±0.01	0.40±0.01
Arg	0.46±0.01	0.46±0.02

Data represent the averages ± SEM of duplicate determinations of two preparations of both toxins
[a] Relative to Ala.
[b] As cysteic acid after oxidation.
[c] As methionine sulphone after oxidation.
Tryptophan was present but was not determined.
Source: from Visser *et al.* (1983).

induces a flaccid paralysis which is substantially due to a presynaptic block-age of neuromuscular transmission (May and Piek, 1979) and does not affect the visceral musculature (Rathmayer, 1966; Piek and Simon-Thomas, 1969); unlike the former, the latter affects both excitatory and inhibitory transmission (Piek *et al.*, 1971). In addition, the venom of *Philanthus* is able to elucidate postsynaptic (Piek *et al.*, 1980a, 1982a) as well as central (Piek *et al.*, 1980b, 1982a) neurotoxic effects (see next section) which were not detected with *Bracon* venom (Piek *et al.*, 1982b). To summarise, it appears that sphecid venoms possess a relative specificity to insects but may affect other arthropods as well.

Toxic components of Philanthus *venom*

As described in the preceding section, *Philanthus* venom exhibits a diverse pharmacology which may suggest chemical complexity.

In the study of Spanjer *et al.* (1982), lyophilised methanol extracts of isolated abdomens of *Philanthus* were exposed to a sequence of fraction-ation steps as specified in the legend to Figure 7.4. In spite of the absence of any systematic quantitative data concerning the recovery of substance and biological activity in the above study, the following information could have been obtained:

Figure 7.4: Isolation of pharmacologically active substances from abdomens of *Philanthus triangulum* wasps. (a) Lyophilised methanol extract of 2500 abdomens (~ 8 g of dry substance) was charged on a column (45 × 3 cm) filled with Sephadex G-100, equilibrated and eluted with distilled water. The section of the elution pattern indicated by the bar contains the bee-paralysing activity. This section was pooled and lyophilised. (b) The toxic fraction obtained in (a) was charged on a column (51.6 × 4 cm) filled with Sephadex G-25, equilibrated and eluted with ammonium acetate 0.2 M, pH 4.75 buffer. The different sections of the elution pattern (A–G) possessed the following activities: B, paralysis of honeybees (part of the activity); A+B+C, this combination contained the total paralytic activity; C, this possessed the axion depolarising and trans-synaptic blockage at the cockroach's sixth abdominal ganglion. (c) The recombined fractions A+B+C obtained in (b) were charged on a column (40 × 2.5 cm) filled with the ion exchanger SP Sephadex equilibrated and eluted with 0.4 M ammonium acetate, pH 4.75 buffer. A linear gradient of molarity up to 0.7 M was applied (vertical arrow). The different sections (fractions) of the elution pattern were pooled as indicated by the bars (α–ε) and possessed the following activity: $\gamma + \delta$, part of the bee-paralysing activity; $\beta + \gamma + \delta$, enhancement of paralytic activity; α, depolarisation and blockage in the sixth abdominal ganglion (Piek *et al.*, 1982). Fractions β and δ were further purified by thin-layer chromatography, resulting in single ninhydrin-positive spots. The final products are now commonly named β-philanthotoxin (β-PTX) and δ-philanthotoxin (δ-PTX).

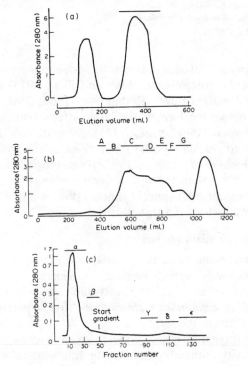

Source: from Spanjer *et al.* (1982).

(1) The active factors are extremely low molecular weight substances.

(2) The paralytic activity of the starting material is a consequence of both a combination of effects of separate components and a cooperative interaction among some of them.

(3) Components β, γ and δ blocked neuromuscular transmission. The δ-component, however, was shown to block ionic channels postsynaptically (Clark et al., 1980; Piek, 1982).

(4) The α-component is the one that is responsible for the central effects (Spanjer et al., 1982; Piek and Spanjer, 1986) but it was recently identified as acetylcholine which probably originates from various body parts (Piek et al., 1983).

(5) It is not clear from the above studies whether all the isolated components are integral constituents of the venom since no comparative studies with the milked venom or venom reservoirs have so far been performed.

The final purification of the toxins β-PTX and δ-PTX from the β- and δ-preparations (Spanjer et al., 1982) was performed by the aid of high-performance liquid chromatography (HPLC) on Lichrosorb RP18 columns in 0.005 mol per litre hydrochloric acid, 2% methanol solution (Piek and Spanjer, 1986). Using field desorption mass spectrometry, the molecular weights for β- and δ-PTXs of 243 and 435 respectively were determined. Using the method of exact mass determination by electron impact mass spectrometry, a molecular formula of $C_{13}H_{29}N_3O$ and $C_{23}H_{41}N_5O_3$ was calculated for β- and δ-toxins, respectively. Nuclear magnetic resonance (NMR) spectra show some similarity in structure of these toxins and they both reveal a polyamine character (Piek et al., 1985; Piek and Spanjer, 1986). To summarise, the venom of P. triangulum contains two agonists, O-acetylcholine and glutamate, and three toxins which are antagonists, called β-, γ- and δ-PTX. The three toxins cause a postsynaptic block of glutamate-induced depolarisation with an activity ratio of 1:30:100 for the β-, γ- and δ-toxins, respectively (Spanjer et al., 1982).

In spite of the uncertainty concerning the biological origin of these toxins (venom reservoir or body tissues), from a purely pharmacological point of view they are interesting. The neuropharmacological complexity of these toxins was recently clarified through a series of electrophysiological studies, and the findings may be summarised as follows:

(1) The venom P. triangulum paralyses insects mainly by affecting synaptic transmission in both the insect CNS and neuromuscular system.

(2) δ-PTX exerts a diversity of actions on synaptic transmission:

(a) Blockage of open cation (Na^+) channels is activated by junctional or extra-junctional glutamate receptors in locust muscle fibres in a semi-irreversible manner (Clark et al., 1982) and the locust procto-

271

deum (Dunbar and Piek, 1983).

(b) Blockage of ion channel activated by the nicotinic acetylcholine receptor as demonstrated first by the block of transmission from cercal nerve to the giant interneurones in the sixth abdominal ganglion of *Periplaneta americana* and the ionophoretically evoked *O*-acetylcholine potentials of the giant neurones (Piek *et al.*, 1984), and secondly by the reduction of the twitch tension and tetanic contraction in the rat phrenic nerve–diaphragm preparation which was attributed to the blockage of the acetylcholine-receptor-activated ion channels. The latter was expressed in the reduction of their temperature and voltage sensitivity and in the shortening of the decay time of the end-plate currents (Van Wilgenburg *et al.*, 1984).

(c) Inhibition of glutamate uptake in locust muscle nerve terminals and glial cells and hippocampus slices of the rat without effect on GABA uptake (van Marle *et al.*, 1985, 1986).

(3) β-Philanthotoxin has no paralysing activity when injected into the honeybee worker. However, it potentiates the paralysing activity of γ- and δ-philanthotoxins. In the isolated locust muscle, β-philanthotoxin inhibits the recovery from desensitisation by glutamate (Piek *et al.*, 1985).

(4) It was suggested (Piek *et al.*, 1985) that the efficacy of the venom of the *Philanthus* wasp arises from a synergistic action of the agonists and antagonists. The channel-blocking activity of δ-toxin is facilitated by their opening either through the occurrence of *O*-acetylcholine (in the CNS) or the inhibition of the re-uptake of glutamate (in the neuromuscular junction). Moreover, glutamate uptake inhibition may cause desensitisation, which would be facilitated by the inhibition of the recovery from desensitisation caused by the β-toxin (Piek *et al.*, 1985).

Factors in black widow spider venom selectively toxic to insects

Latrodectus *spiders and their venom*

All the spiders (about 25 000 known species) are venomous with the sole exception of the small family of Uloboridae, which has secondarily lost its venom (Glatz, 1969). Being devoid of powerful pedipalps, possessing a soft easily pierced body and weak chelicerae, many groups of spiders avoid direct close contact with their prey. In these spiders the initial stage of the mechanical immobilisation, performed by other venomous arthropods with the aid of powerful legs or pedipalps, is accomplished with the aid of the web and the complementary wrapping of the trapped prey by viscous silk. The mechanically immobilised prey is now bitten by the chelicerae and finally paralysed before being eaten. This procedure was observed in the black widow spider (*Latrodectus mactans* and *L. m. tredecimguttatus*

272

(Theridiidae: Bearg, 1959; Bettini and Maroli, 1978) and other web spiders.

Because of their medical importance the venom of the black widow spiders was, and still is, the most intensively studied. *Latrodectus* venoms were found to be lethal and paralytic to a wide range of vertebrate as well as invertebrate animals (Table 7.2; Maroli *et al.*, 1973). The crude venom was also shown to affect strongly different neuronal and neuromuscular preparations derived from Crustacea (Grasso and Paggi, 1967; Kawai *et al.*, 1972) as well as Mollusca (Gruener, 1973). As discussed earlier, the neurotoxicity of the *Latrodectus* venoms to vertebrates is attributed to α-latrotoxin which induces a depletion of every neurotransmitter so far studied. The same fundamental phenomenon, neurotransmitter depletion, was also shown to occur with cholinergic (D'Ajello *et al.*, 1971) and glutaminergic (Cull-Candy *et al.*, 1973; Griffiths and Smyth, 1973; Ornberg *et al.*, 1976) systems of insects. These data might suggest that the vertebrates and the arthropods are affected by the same factor(s) in the crude venom. This aspect is treated in the next section.

Insect toxic factors in the venom of Latrodectus

The existence of distinct toxic factors in the venom of *Lactrodectus mactans tredecimguttatus* specifically affecting insects and mammals was demonstrated long ago by Frontali and Grasso (1964). Desalted (by Sephadex G-25) venom gland extracts were fractionated by two separate methods: (a) preparative vertical column electrophoresis; (b) column chromatography by Sephadex G-100 followed by an anion exchanger (DEAE Sephadex A50). Both methods of separation have enabled the distinction to be made between protein fractions (in the molecular weight range of 60–80 kDa) each possessing one of the following activities: fast reversible paralytic effects on flies (Musca *domestica*); slow progressive

Table 7.2: Toxicity of *Latrodectus mactans tredecimguttatus* venom to amphibians, birds, insects and mammals

Animals	LD$_{50}$ values	
	(mg[a] per animal)	(mg[a] per kg)
Frog	2.18 ±0.49	145 ±32
Canary	0.085±0.03	4.7 ±1.7
Blackbird	0.42 ±0.19	5.9 ±2.7
Pigeon	0.11 ±0.025	0.36±0.07
Chick	0.19 ±0.04	2.1 ±0.57
Cockroach	0.015	2.7
Housefly	0.000013	0.6
Guinea pig	0.028	0.075
Mouse	0.013	0.90

[a]mg of protein (Lowry's method).
Source: from Maroli *et al.* (1973).

and irreversible paralysis of flies; toxicity (paralytic and lethal) to guinea pigs.

The fractions with fast and slow paralytic effects on flies, obtained by column electrophoresis (LV_1 and LV_2 respectively), were, on the basis of body weight, 30 and 66 times respectively less toxic towards guinea pigs. On the other hand, the fraction LV_3 was 750 times less toxic towards houseflies when compared with guinea pigs. Frontali and Grasso (1964) concluded that this considerable difference in toxicity might be related to some structural or functional differences between mammals and insects.

A more recent study was dedicated to the purification, from the venom of *Latrodectus*, of the factor responsible for enhancement of transmitter release and depletion in a vertebrate system (Frontali *et al.*, 1976). This factor was obtained by gel filtration (Sephadex G-200) and ion exchange (DEAE-Sephadex A-50) chromatography in succession, and was shown to consist of at least four protein components that are similar in their molecular weights (about 130000) and isoelectric points (5.2–5.5) and are immunologically indistinguishable. These components do not contain sugar residues and are devoid of lipolytic and proteolytic activity. Similar results were obtained in a parallel study employing similar methods (Grasso, 1976). In the course of the above study (Frontali *et al.*, 1976), factors specifically affecting insects (lethality to houseflies, effects on the cockroach heart) were shown to differ from those affecting vertebrates (mice lethality, frog neuromuscular preparation), thus confirming the earlier findings of Frontali and Grasso (1964). Furthermore, there exists a clear distinction between the factors that affect insects and those that act on crustaceans, as demonstrated by their excitatory effect on neurones of the crayfish stretch receptor. The data are presented in Table 7.3. It has been suggested (Frontali *et al.*, 1976) that the following relationships exist between the present and previous work (Frontali and Grasso, 1964): B_5 is included in LV_3; LV_1 and LV_2 are included in C, and E_2 is included in LV_4. According to the data presented in Table 7.3, it may be concluded that, in spite of the fact that the crustacean active fractions E and E_2 were not assayed in insects and the insect toxic fraction C_3 was not assayed in the crustacean preparation, a distinction between insect and crustacean was demonstrated at the level of the Sephadex G-200 fraction C which was not active in the crustacean preparation. A clear distinction between factors in the venom of the black widow spider affecting vertebrates and those active in arthropods was reached at the level of the Sephadex G-200 separation; fraction B was not active in arthropods. Neither the insect (C_3, Table 7.3) nor the crustacean (E_2, Table 7.3) toxic fractions were pure, as proven by gel electrophoresis (Frontali *et al.*, 1976). However, these fractions (Fritz *et al.*, 1980) correspond to molecular weights of about 120 and 65 kDa for fractions C and E, respectively. It was also shown that the vertebrate toxic component B_5 (α-latrotoxin) (Frontali *et al.*, 1976) was inactive in a lobster

Table 7.3: Qualitative effects of various fractions obtained from the chromatographic separation of the venom of *Latrodectus mactans tredecimguttatus* on different biological systems (Frontali *et al.*, 1976)

	Frog neuromuscular junction	Mouse toxicity	Housefly toxicity	Cockroach heart beat	Crayfish stretch receptor
Whole extract	+	+	+	+	+
A	−	−	−	−	−
B	+	+	−	−	−
C	+	+	+	+	−
D	−	−	−	−	−
E	−	−	ND	ND	+
B_5	+	+	−	−	−
C_3	−	−	+	+	ND
C_5	+	+	+	−	ND
E_1	−	−	ND	ND	−
E_2	−	−	ND	ND	+

+ = Active; − = inactive; ND = not determined.

Remarks: Fractions A, B, C, D and E were obtained by a Sephadex G-200 fractionation of venom gland extract. Fraction B_5 is derived from the above SG-200 fraction B and corresponds to the α-latrotoxin toxic to vertebrates. Fraction C_3 and C_5 are derived from the SAG-200 fraction C and were obtained by anion exchange chromatography (DEAE-Sephadex A50). Fractions E_1 and E_2 are derived from the SG-200 fraction E and were obtained by the same method as fractions C_3 and C_5.

neuromuscular preparation (Fritz *et al.*, 1980) which was responsive to fraction E.

Ornberg *et al.* (1976) have separated the venom gland extracts of the American black widow spider *Latrodectus mactans* on a discontinuous polyacrylamide gel electrophoresis. Eleven fractions were extracted from gel sections and assayed for toxicity by injection into cockroach nymphs and by application to cockroach neuromuscular preparation. One of the fractions, a slowly migrating protein with a molecular weight of 125 kDa, produced a slow permanent paralysis of the test animals and accounted for the major actions of crude venom by inducing a large transient rise in the miniature end-plate potential frequency and eventual synaptic blockage. With the above information (Frontali and Grasso, 1964; Frontali *et al.*, 1976; Fritz *et al.*, 1980) it may be assumed that this factor corresponds to the insect toxin component of the venom. It was suggested (Fritz *et al.*, 1980) that this component seems to be a constituent of fraction C (Frontali *et al.*, 1976) as this fraction has the corresponding molecular weight and is the only one to affect houseflies and cockroaches (Table 7.3). If we further extend our conclusion, it appears that three different components in the *Latrodectus* venom are capable of inducing transmitter release, but each demonstrates specificity to a distinct group of animals, even distinguishing between the closely related insects and crustaceans.

Insect-selective neurotoxins derived from scorpion venoms

In the present section the following abbreviations will be employed: *AaIT*, *AaCT* and *AaMT2* are the insect, crustacean and mammalian II toxins, respectively, derived from the venom of the scorpion *Androctonus australis*; *AmIT*, the insect toxin from the venom of *Androctonus mauretanicus*; *BjIT1* and *2*, the insect toxins 1 and 2 from the venom of *Buthotus judaicus*; *LqIT1* and *2*, the insect toxins 1 and 2 from the venom of *Leiurus quinquestriatus quinquestriatus*; *SmIT1* and *2*, the insect toxins 1 and 2 from the venom of *Scorpio maurus palmatus*.

Scorpions and their venom

According to numerous morphological (Vachon, 1952) as well as some biochemical (Goyffon and Kovoor, 1978), characteristics, scorpions are divided into two groups: buthoids and chactoids. The buthoids, which include the dangerous, medically important scorpions, comprise 40 per cent of the known species and consist of a single family, the Buthidae. The chactoids consist of five families and are, in general, medically insignificant. The vast majority of the chemical and pharmacological studies of scorpion venoms were performed with buthoid venoms (Goyffon and Kovoor, 1978; Zlotkin *et al.*, 1978), including their employment as pharmacological tools for the study of excitability in biological systems (Catterall, 1980) and as sources for the excitatory and depressant insect toxins mentioned subsequently.

In scorpions the venom apparatus is located in the last bulbous segment (telson) of the body. Scorpions are predominantly nocturnal hunters which move slowly and do not see or hear their prey, which is comprised substantially of insects. Instead scorpions possess receptors on their legs which are extraordinarily sensitive to subtle disturbances of the soil (Vachon, 1953; Bearg, 1961; Brownell, 1984). This mechanism, however, enables them to detect the target prey at distances of no more than 8–10 cm away (Brownell, 1984). The initial contact with the prey is performed with the large pedipalps. The employment of the sting is dependent on the prey's vigour and dimensions (Rosin and Shulov, 1963; Stahnke, 1966). A strong active prey is stung by either one or several successive quick and well-directed thrusts, and is held by the stretched pedipalps as far as possible from the scorpion's body until it is immobilised (Rosin and Shulov, 1963; Stahnke, 1966; Bucherl, 1971). The immobilised prey is soon torn and crushed into fine particles which are ingested in the form of a fluid suspension. It is thus obvious that a device for paralysing the prey at the earliest moment of contact is essential. This object is achieved by the use of venoms which possess highly specific chemical adaptations.

There is an apparent resemblance between the effects of scorpion venom on arthropods and those on mammals. This is expressed in the excitatory symptoms of envenomation and the action of scorpion venom on neuromuscular preparations of an insect, a crustacean (Parnas and Russell, 1967; Parnas *et al.*, 1970) and a mammal (Katz and Edwards, 1972; Brazil *et al.*, 1973; Zlotkin *et al.*, 1978). In all the above preparations scorpion venom caused muscular stimulatory effects due to presynaptic excitation at the level of the exposed nerve endings resulting in the release of the corresponding transmitter. This similarity in action could have led to the assumption that the effect of scorpion venoms on mammals and arthropods is due to the same chemical factors in the crude venom.

By the aid of assay of contraction paralysis of blowfly larvae (Figure 7.5A) it was found that there was no correlation between larva paralysis and mice lethal potencies of 16 different Buthidae venoms (Zlotkin *et al.*, 1971a), suggesting that the two toxic activities may result from different factors. This possibility was supported by the finding that the AaMT1 and 2 toxins, which are strongly lethal to mammals and isolated according to the criterion of mice lethality (Rochat *et al.*, 1967; Miranda *et al.*, 1970), were completely inactive to blowfly larvae (Zlotkin *et al.*, 1971b). The final proof of the diversity between factors affecting mammals and those paralysing insects in scorpion venoms was obtained by starch gel electrophoretical separation of the venom of *A. australis* (Zlotkin *et al.*, 1971b) and six other Buthidae venoms (Zlotkin *et al.*, 1972b). It has been found that three of the above venoms contain more than one fraction that cause contraction paralysis of larvae. These fractions were lethal to fly larvae and readily inactivated by trypsin, thus demonstrating their protein nature (Zlotkin *et al.*, 1972b).

The above early findings led to the detection and isolation of the scorpion venom insect toxins (see below). The existence of an additional group of toxins, the crustacean toxins, was initially demonstrated by the fact that, in contrast to the crude venom of *A. australis*, its derived insect toxin (AaIT) and mammalian toxin were unable to affect an isopod (terrestrial crustacean) or a scorpion, suggesting the presence of discrete neurotoxins (Zlotkin *et al.*, 1972a) specifically affecting these arthropods.

Excitatory insect toxins: chemistry

The above mentioned contraction–paralysis assay of fly larvae (Zlotkin *et al.*, 1971a; Figure 7.5A) enabled the detection and isolation of the first group of the so-called excitatory insect toxins — AaIT, AmIT, BjIT1 and LqIT1 (Table 7.4; Zlotkin *et al.*, 1971c, 1979; Lester *et al.*, 1982; Zlotkin *et al.*, 1985). Their purification was based on a sequence of steps composed of water extraction, dialysis (AaIT, AaMT), recycling (AaIT, AaMT, LqIT1) or conventional (BjIT1) Sephadex G-50 gel filtration column chromatography and diverse manipulations of ion exchange column chromatography employed in equilibrium or molarity gradient elutions. Their chemical purity was assessed by column chromatography, disc

Figure 7.5: Responses of *Sarcophaga faculata* blowfly larvae to the injection of various scorpion venoms and their derived insect toxins. (A) Typical contraction paralysis as induced by various buthid scorpion venoms and the excitatory AaIT, AmIT, BjIT1 and LqIT1 toxins (see Table 7.4). (B) Typical flaccid extended paralysis as induced by the crude venom of *S. m. palmatus* and its derived toxins (see p. 276) as well as the depressant insect toxins BjIT2 and LqIT2 (Table 7.4).

electrophoresis, analytical isoelectrofocusing and amino acid analyses. These are single-chained polypeptides composed of about 70 amino acids, with an M_r of 7.5–8.0 kDa, and a pI ranging from 6.5 to 8.4.

The AaIT was the first insect toxin isolated, and its chemistry is so far the most intensively studied, resulting in its full covalent structure (Darbon *et al.*, 1982). The amino acid sequence of AaIT was established by means of extensive phenylisothiocyanate degradation in a liquid sequencer programmed either with a 'quadrol programme' for native or modified protein or a 'dimethyl benzylamine/parvalbumin programme' for peptides. Because of the low solubility of the S-alkylated toxin in quadrol buffer, the sequence was finally determined through peptide degradation. The primary

Table 7.4: The amino acid composition of the insect selective scorpion venom toxins compared with some mammalian toxins derived from the same venoms

Amino acid	Insect toxins								Mammalian toxins		
	AaIT[a]	AmIT[b]	BjIT1[c]	BjIT2[c]	LqIT1[d]	LqIT2[d]	SmIT1[e]	SmIT2[e]	AaMT2[f]	LqMT5[f]	SmMT[g]
Threonine	11	11	8	7	9	8	2	3	8	10	2
Serine	4	3	3	4	2	5	0	4	3	1	2
Glutamic acid	6	5	3	4	5	5	3	2	2	3	2
Proline	3	3	4	4	4	7	3	1	4	4	2
Glycine	4	4	7	10	4	13	2	2	7	7	3
Alanine	3	3	4	4	4	1	1	2	3	3	2
Half-cystine	8	8	6	6	8	8	4	6	8	8	8
Valine	3	4	3	4	6	1	2	2	4	2	1
Methionine	0	1	0	0	0	0	0	0	0	0	1
Isoleucine	2–3	3	4	2–3	2	1	2	3	1	2	0
Leucine	5–6	4	2	2	6	4	2	1	2	2	2
Tyrosine	1	6	5	3	7	4	1	2	7	6	1
Phenylalanine	7	1	1	2	1	1	0	0	1	2	0
Lysine	1	7	8	5	9	7	4	3	5	8	5
Histidine	1	1	1	1	2	0	0	0	2	0	0
Arginine	1	1	1	2	1	3	2	3	1	3	1
Tryptophan	1	1	3	8	0	4	0	0	1	2	0
Total	66–68	67	67	68–69	71	72	30	36	64	65	32

Abbreviations: IT and MT: insect and mammalian toxins, respectively.

Aa, Am, Bj, Lq, and Sm indicate the following names of scorpions from the venoms of which the above toxins are derived: *Androctonus australis, A. mauretanicus, Buthotus judaicus, Leiurus quinquestriatus* and *Scorpio maurus palmatus*, respectively.

References: [a] Zlotkin et al. (1971c); [b] Zlotkin et al. (1979); [c] Lester et al. (1982); [d] Zlotkin et al. (1985); [e] Lazarovici et al. (1982); [f] Miranda et al. (1970); [g] Lazarovici et al. (1982).

structure of the AaIT and the main steps of the enzymic and sequential degradation are presented in Figure 7.6. The tryptophan residue, which was initially spectrophotometrically determined in the native toxin, was not found in the sequence determination. The molar extinction coefficient (at 277 nm) of the carboxymethylated derivative is in perfect accordance with the six tyrosine residues present. The unusual spectral properties of the native toxin can be attributed either to a special conformation of the native toxin or to the presence of a non-covalently bound chromophore in the native toxin which is lost upon denaturation.

The position of the four disulphide bridges was determined by two methods. The first of these involved amino acid analysis of proteolytic peptides before and after performic oxidation. The second employed partial labelling of the half-cystine residues with [^{14}C]iodoacetic acid and differentiation of the paired amino acids by determination on tryptic peptides of the specific radioactivities of the S-[^{14}C]carboxymethylated phenylthiohydantoin cystines. Of the four disulphide bridges present in both the AaMT2 (Kopeyan *et al.*, 1974) and AaIT (Figure 7.7) three are in homologous positions. The fourth differs in that one of the two half-cystines involved, Cys^{12} in AaMT2, shifts to Cys^{38} in the insect toxin. On the basis of several considerations (Darbon *et al.*, 1982) concerning the primary sequence of scorpion venom toxins, it is likely that the position of the disulphide bridges is the same for all scorpion neurotoxins active on mammals. This conclusion is strongly supported by a recent study demonstrating that the locations of disulphide bridges of mammalian toxin II derived from the venom of the scorpion *Buthus occitanus tunetanus* are identical with those of AaMT2 (Gregoire and Rochat, 1983).

Action. When assayed on a locust hindleg *extensor tibiae* muscle and its motor nerve (Walther *et al.*, 1976; Rathmayer *et al.*, 1978), it was shown that the muscular stimulation induced by AaIT is due to an excitatory pre-synaptic action on the motor axons. The axonal effect of AaIT was expressed in the induction of repetitive firing (Figure 7.8) without affecting the time course of the induced action potentials. It was also shown that the various peripheral branches of the motor nerve serve as the primary target of the insect toxin (Walther *et al.*, 1976). This aspect was exemplified most recently by autoradiographical studies in which the *extensor tibiae* muscle was treated with [^{125}I]AaIT (L. Fishman unpublished).

The causal origin of the AaIT toxins' induced repetitive firing was clarified through a study on an isolated giant axon from the central nervous system of the cockroach in both current and voltage-clamp experiments using a double oil-gap single-fibre technique (Pelhate and Zlotkin, 1981, 1982). It was concluded that the repetitive activity induced by AaIT results from the voltage-dependent modulation of sodium inactivation coupled with an increase in sodium permeability. The voltage-clamp studies have

280

Figure 7.6: The primary structure of the AaIT. The amino acid positions were determined by automatic degradation of the carboxymethylated protein. Peptides 'T' and 'C' were obtained after tryptic and chymotrypic digestion, respectively

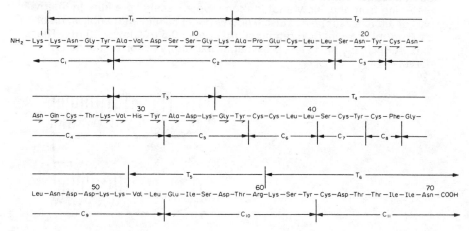

Source: from Darbon *et al.* (1982).

Figure 7.7: Comparison of the location of the disulphide bridges of the AaIT (IT) and the AaMT2 (MT) toxins

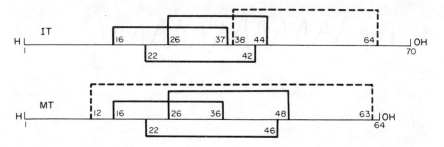

Source: from Darbon *et al.* (1982).

also indicated that, in contrast to the partial suppressory effect on the potassium current induced by the *A. australis* mammal and crustacean toxins, AaIT had no effect on this current (Pelhate and Zlotkin, 1982). The excitatory toxins, BjIT1 (Lester *et al.*, 1982) and LqIT1 (Zlotkin *et al.*, 1985), when assayed on the cockroach isolated axonal preparation, have yielded data closely resembling those on AaIT. Their induced repetitive firing is attributable to the increase in the activatable sodium conductance and the slowing of its turning off.

281

Figure 7.8: Neuromuscular effect of the AaIT on the locust hindleg *extensor tibiae* nerve muscle preparation. (A) Start of two trains of slow e.s.p.s. synchronously recorded from two distant locations on the muscle 50 min after application of 0.15 g/ml of insect toxin. Calibration: 10 mV, 100 ms, (B) Spontaneous neuromuscular activity caused by insect toxin. Synchronous recording from the 'slow' motor axon (upper trace) and a fibre in the main part of the *extensor tibiae* muscle (lower trace) 65 min after application of c.0.5 µg/ml of toxin. Calibration: 50 ms; upper 0.5 mV, lower 10 mV. These data (A,B) clearly indicate a presynaptic excitatory action on the motor nerve.

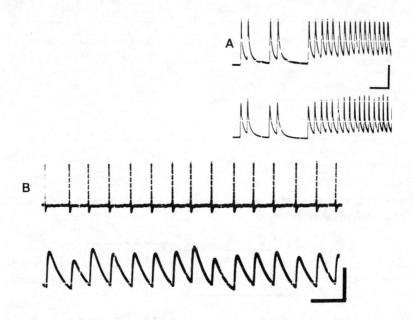

Source: from Walther *et al.* (1976).

Depressant insect toxins

The existence of depressant flaccidity-inducing insect toxic components in Buthidae scorpion venoms was suggested by the progressive increase in the yield of the total contractive activity during the purification process of the excitatory (contractive) toxin (Zlotkin *et al.*, 1971c, 1979, 1985) and the occurrence of mixed-intermediate symptoms between the two opposite forms of response presented in Figure 7.5 induced by the venom of *Buthotus judaicus*.

The systematic fractionation of the *B. judaicus* venom by column chromatography has revealed that the above complex symptomatology follows from the existence of two insect toxins: an excitatory one (BjIT1 — p. 276) and a second toxin (BjIT2 — Lester *et al.*, 1982) which induces a gradual progressive paralysis of fly larvae in an extended flaccid form (Figure 7.5B) and a slow progressive paralysis and eventual death of locusts. BjIT2 is about 36 times more toxic than the crude venom according to the paralytic potency in fly larvae. It is composed of 69 amino acids and has an estimated molecular weight of 7894, a characteristic amino acid composition (Table 7.4) and a pI value of 8.30.

A second depressant flaccidity-inducing toxin (LqIT2) was recently isolated from venom of the scorpion *L. q. quinquestriatus* (Zlotkin and Gordon, 1985). It is a basic polypeptide (M_r 7930; pI 8.3) which accounts for about 0.7–0.8 per cent of the dry weight of the crude venom. Concerning its amino acid composition (Table 7.4), on the one hand it demonstrates certain characteristics of many other different scorpion toxins such as the high content of lysine, the absence of methionine and histidine and the possession of eight half-cystines supposed to constitute four disulphide bridges (Zlotkin *et al.*, 1978; Rochat *et al.*, 1979), and on the other hand it resembles the flaccidity-inducing toxin BjIT2 (Lester *et al.*, 1982; Table 7.4) by its relative abundance of glycine and tryptophan.

The progressively developing flaccidity of the BjIT2 and LqIT2 toxins can be explained through the data obtained with the *in vitro* cockroach axonal preparation (Lester *et al.*, 1982; Zlotkin *et al.*, 1985) expressed in two main effects. The first was a progressive irreversible depolarising action as shown by the continuous decrease in the value of the resting membrane potential. The second effect was the evident progressive suppression of the sodium current expressed in the clear reduction of the sodium conductance. The combination of these two effects explains the progressive block of the evoked action potentials. The blocking effect of the depressant toxins on axonal excitability may account for the induction of flaccidity and progressive paralysis in the test animals.

Cooperative insect toxins

It appears to us that the so far poorly investigated chactoid venoms (see p. 276) may serve as an additional source of interesting pharmacological substances. Preliminary data (Zlotkin *et al.*, 1972c, 1973) have indicated that the venom of the chactoid scorpion *Scorpio maurus palmatus* (Scorpionidae) possesses an extremely low toxicity to mammals and an unusual symptomatology in insects. In contrast to the immediate contraction paralysis of larvae (Figure 7.5A) induced by the Buthidae scorpion venoms (Zlotkin *et al.*, 1971a), this venom induces a slow progressive flaccidity (Figure 7.5B). This has directed our attention to the insect toxic factors in this *Scorpio* venom, resulting in the isolation and chemical–pharmacological characterisation of several insect toxic components.

It has been shown (Lazarovici *et al.*, 1982) that the toxicity of the crude venom to insects is due to three separate groups of substances, the so-called cytotoxins, phospholipases and neurotoxins. the neurotoxic fraction contains two factors, the fast reversibly paralytic (IP) and the slow lethal (IT). The slow lethal factor is composed of two toxins (SmIT1 and 2). These were obtained by an additional step of column chromatography, and their purity was assessed by column chromatography, disc electrophoresis, isoelectrofocusing, analytical ultracentrifugation and amino acid analysis. SmIT1 and 2 are two polypeptides possessing unique amino acid compositions with molecular weights of 3232 and 3964 and pI 8.8 and 9.2, and are thought to contain two and three disulphide bridges, respectively (Table 7.4). Each of the two separated toxins has demonstrated only a slight increase in its specific toxicity and a very low recovery of toxicity when compared with the crude venom. A clear cooperative interaction was demonstrated between the fast paralytic and lethal (IP and IT, respectively) fractions as well as between the two insect toxins, resulting in an evident recovery of the original toxicity of the crude venom to insects.

The synergic interaction between the SmIT1 and 2 has obtained a unique expression when assayed on the isolated cockroach axon under voltage-clamp conditions (Lazarovici et al., 1982). The combination of the SmIT1 and 2 caused a reversible blockage of both the sodium and potassium currents. This may explain the specific symptomatology and the mechanism of paralysis of these toxins. The fact that the blockage of both ions is the consequence of cooperative interaction may suggest certain structural associations and/or similarity between the sodium and potassium channels in insect axons. Such a possibility deserves further experimentation.

Other kinds of toxins derived from scorpion venoms

The amino acid sequence of a peptide (Amm P_2, 35 amino acids) isolated from the venom *A. mauretanicus*, possessing toxicity to insects, has been reported (Rochat et al., 1979). When tested on fly larvae it was shown to be at least three orders of magnitude less toxic than the AmIT (see above and Table 7.4) derived from the same venom (E. Zlotkin and J.P. Rosso, unpublished) and its activity was completely abolished following treatment with an antiserum prepared against the AaIT (H. Rochat, personal communication) which closely resembles the AmIT. It was thus concluded that the toxicity of Amm P_2 peptide was due to its contamination by the potent AmIT present in the same venom. Five peptides (I_1–I_5) toxic to insects isolated from the venom of the Asian scorpion *Buthus epeus* (Grishin et al., 1976; Zhdanova et al., 1977) and its European subspecies (Grishin et al., 1982) have been reported. These peptides are composed of 34–36 amino acids cross-linked by four disulphide bridges. The full amino acid sequence of two of these peptides has been reported (Zhdanova et al., 1977; Grishin et al., 1982; Ovchinnikov and Grishin, 1982). These peptides closely resemble the above Amm P_2 peptide in their primary structure. The toxicity of the purified toxins was reported to exceed 3 μg per cockroach, but no systematic information concerning the recoveries, yield and specific toxicities during the process of purification, as well as comparative assays with vertebrates or other arthropods, was reported. The specificity and the pharmacological significance of the above *B. epeus* peptides remain to be clarified.

A certain diversification from the pattern shown by the Buthinae venoms (Table 7.4) is demonstrated by the venom of the North American scorpion *Centruroides sculpturatus*. A purification process based on column chromatography and monitored by mice lethality has resulted in the isolation of four toxins (I–IV) (McIntosh and Watt, 1972). Three additional components, the so-called variants 1–3, were isolated and shown to possess low toxicity to mice and chickens, as well as crickets (Babin *et al.*, 1974). Amino acid sequences have been reported for toxin I and the three variants (Babin *et al.*, 1974, 1975). The toxicities of the different components of *C. sculpturatus* venom are presented in Table 7.5 (Watt *et al.*, 1978). As shown (Table 7.5), a clear group specificity is demonstrated by toxin III which is 13 times more potent than the crude venom in its toxicity to chicks and is 60 times weaker than the latter in insects. A reversed preference is shown by component B 140-1 which is more toxic than the crude venom to insects but is about seven times weaker than the latter to chicks. Other components demonstrate a rather relative specificity. The three variants (1–3) are evidently much less active than the crude venom in both test animals, with a certain preference for the insects. These insect active components did not demonstrate the typical strong increase in their specific toxicities as shown by the different insect toxins presented in Table 7.4. This may suggest either a partial inactivation during the process of purification or a cooperative interaction with other venom components which may account for the relatively high toxicity of the crude venom.

Table 7.5: Toxicities of eight toxins isolated from *Centruroides sculpturatus* venom

Toxin	Toxicitis (mg/kg)	
	Chicks (LD_{50})	Crickets (Ed_{50})
Toxin IV	0.018±0.0023	1.86± 0.31
Toxin III	0.017±0.0025	69.0 ±11.0
Toxin I[a]	0.53 ±0.10	37.0 ± 6.0
B 140–1	1.60 ±0.30	0.86± 0.25
B 144–4 (6 Cys/2)	0.30 ±0.68	4.3 ± 1.3
Variant 1[a]	> 20	12.0 ± 2.6
Variant 2[a]	> 20	21.5 ± 4.6
Variant 3[a]	> 20	255.0 ±38.0
Whole venom	0.22 + 0.02	1.20± 0.09

[a] Amino acid sequences have been reported.
Toxicities are expressed as LD_{50} or ED_{50} ± 95% confidence range.
SourceL from Watt *et al.* (1978).

Toxin receptor interactions: binding studies

The information presented below mainly corresponds to the insect toxin derived from the venom of the scorpion *Androctonus australis* (AaIT) which was successfully radioiodinated (Gordon *et al.*, 1984), and therefore served as an essential tool for a series of drug–receptor binding studies.

The selectivity to insects of the action of AaIT and related toxins was assessed by three separate experimental systems:

(1) Lethality assays which included ten species of insects (representing four different orders of holo- and hemimetabolous insects), arachnids, (scorpions) and crustaceans (three species) and mammals (laboratory mice). In these assays AaIT was shown to be toxic only to insects (Zlotkin *et al.*, 1972a, and unpublished).

(2) Assays on nerve–muscle preparations of a mammal, a crustacean, an arachnid and an insect — indicating the exclusive action of AaIT only on the insect nerve–muscle preparation (Rathmayer *et al.*, 1978).

(3) Binding assays to preparations of insect nervous tissue, insect non-innervated tissues, crustacean nervous tissues (Teitelbaum *et al.*, 1979; Zlotkin *et al.*, 1979) and rat brain synaptosomes (Gordon *et al.*, 1984). It was shown that a specific binding of the radioiodinated insect toxin occurred only in the insect nervous tissue.

To summarise, the selectivity of the AaIT and related toxins to insects is based on their exclusive affinity to the insect nervous tissue. It was decided, therefore, to study in quantitative terms the interaction of the AaIT with the insect neuronal membranes through radioligand binding assays.

In order to study the interaction of AaIT with insect neuronal membranes, the preparation of synaptic plasma vesicles derived from osmotically shocked locust synaptosomes (Breer and Jeserich, 1980) was devised (Gordon *et al.*, 1982; Zlotkin and Gordon, 1985). The pharmacological functionality of this preparation was demonstrated by its ability to catalyse active neurotransmitter uptake (Gordon *et al.*, 1982; Breer, 1983a) and to maintain a membrane potential in a modifiable manner (Gordon *et al.*, 1984).

The binding of $[^{125}I]$AaIT to the locust synaptosomal plasma membrane vesicles under kinetic and equilibrium conditions is presented in Figure 7.9. It has been shown that $[^{125}I]$labelled AaIT binds specifically and reversibly to a single class of non-interacting binding sites of high affinity ($K_D = 1.2$--3 nM) and low capacity (1.2–2.0 pmol/mg protein) (Figure 7.9c). The values of the rate of association (Figure 7.9a) and dissociation (Figure 7.9b) constants (k_1, k_{-1}) are, respectively, 1.36×10^6 M^{-1} s^{-1} and 1.89×10^{-3} s^{-1}, and are in a good accordance with the equilibrium dissociation constant (Gordon *et al.*, 1984). The use of various ionophores and changes in external potassium concentration, shown to modify the membrane

Figure 7.9: The binding of the [^{125}I]AaIT to locust synaptosomal membrane vesicles under kinetic and equilibrium conditions. (a) Time course of [^{125}I]AaIT association to the insect synaptosomal membrane vesicles. A: To determine the association kinetics, binding medium containing 1.5 μM [^{125}I]AaIT with (○) or without (●) the unlabelled toxin (1 μM) was used. B: The kinetics of [^{125}I]AaIT binding was linearised according to a pseudo first order equation as specified by Gordon *et al.* (1984) and Zlotkin and Gordon (1985). The association constant was calculated as $k = 1.36 \times 10^6$ M^{-1}s^{-1}. (b) Time course of dissociation of [^{125}I]AaIT from the insect synaptosomal membrane vesicles. A: This was determined by addition of a large excess of unlabelled toxin (1 μM) followed by sampling of the reaction mixture at the indicated time intervals (●). Non-specific binding (○) was similarly determined except that the unlabelled toxin (1 M) was present in the incubation medium. B: Dissociation was linearised according to a first order equation as specified by Gordon *et al.* (1984), and the slope of the first order plot is equivalent to $k = 1.89 \times 10^{-3}$ s^{-1}. The $t_{1/2}$ for loss of specifically bound radioligand is calculated as in $2/k_{-1}$ and corresponds to 6 min. (c) An equilibrium saturation assay of [^{125}I]AaIT binding to locust synaptosomal membrane vesicles. A: Conditions were as specified by Gordon *et al.* (1984). (▲) Total binding; (●) non-specific binding; (○) specific binding. B: Scatchard anlaysis of data obtained in (A) resulting in the equilibrium dissociation constant (K^*_D) = 1.19 nM (which is in good accordance with the value obtained by the kinetic assays (k_1/k_{-1}) and the maximum number of binding sites (B_{max}) = 1.37 pmol/mg membranal protein

(a)

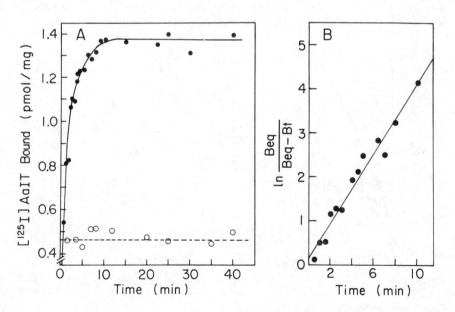

287

Figure 7.9 continued

(b)

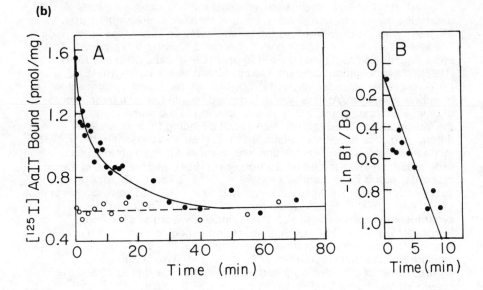

(c)

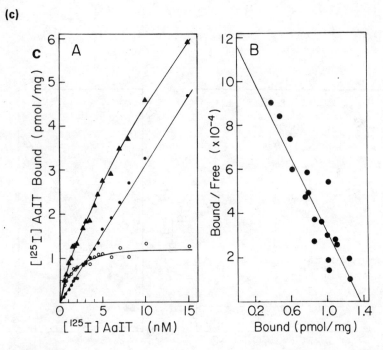

Source: from Gordon *et al.* (1984).

potential of the present neuronal preparation, did not affect the binding of [^{125}I]AaIT, thus indicating its membrane voltage independence. Vera-tridine, tetrodotoxin, sea anemone toxin and the scorpion toxins specific for vertebrates did not affect the binding of [^{125}I]AaIT. Furthermore, the scorpion toxins were devoid of specific binding to the present insect neuronal preparation (Gordon et al., 1984).

Series of equilibrium–saturation binding assays with the [^{125}I]AaIT to synaptosomal plasma membrane vesicles derived from locust brains, locust ventral nerve cords, cricket CNS, fly larvae CNS and fly heads indicate the existence of an evident homogeneity in the AaIT binding constants among various insect neuronal membranes. AaIT binds to single classes of non-interacting binding sites of high affinities (K_D = 1.2–3 nM) and low capacities (0.5–2.0 pmol/mg of membrane protein)(Gordon et al., 1985).

With this background it may be concluded that the possession of an AaIT binding site appears to be a unique property of the insect neuronal membrane. With regard to the pharmacological significance of this binding site, it is noteworthy that an indication of its possible relation to the insect sodium channels has come from the above voltage-clamp data which suggest a specific effect of AaIT and related toxins on sodium conductance in an isolated insect axonal preparation. Additional evidence supporting this notion was given by recent studies on the binding of saxitoxin (STX) to insect neuronal membranes, indicating an obvious identity between the binding capacity of [^{3}H]STX and [^{125}I]AaIT to the locust and fly neuronal membranes (Gordon et al., 1985).

On the basis of the information concerning the mode of action of the excitatory and depressant insect toxins derived from buthid scorpion venoms so far presented in this chapter, it may be concluded that the two groups of toxins, which elicit contrasting symptoms at the level of the whole animal, both affect sodium conductance and differ mainly in their opposite effects on the activatable sodium conductance in the insect neuronal membrane. It was therefore decided to determine whether the two groups of toxins shared the same binding site. This aspect was treated by a series of binding assays on the competitive displaceability of [^{125}I]AaIT by the various excitatory and depressant insect toxins from its binding sites in the insect neuronal membrane (Gordon et al., 1984; Zlotkin et al., 1985). The data are presented in Table 7.6. As shown, the two groups of toxins share an additional common property: they are able to competitively displace the [^{125}I]AaIT with K_D values closely resembling those of the labelled and unlabelled AaIT. This suggests that the two differ-ent kinds of insect toxin either share a common binding site or possess separate but mutually related interacting binding sites probably associated with the insect neuronal sodium channel. Further experimental clarification is required to distinguish between these possibilities (Gordon et al., 1984; Zlotkin et al., 1985).

289

Table 7.6: The binding affinities of excitatory and depressant insect toxins to the insect synaptosomal membrane vesicles[a]

Toxins assayed		$K_{0.5}$ (nM)	K_D (nM)
Excitatory toxins	AaIT	2.75	1.30
		2.20	1.00
	BjIT1	4.60	2.20
	LqIT1	0.83	0.37
Depressant toxins	BjIT2	2.75	1.30
	LqIT2	4.30	1.90

[a] Data obtained from assays on competitive displaceability of the [125I] AaIT from its binding sites in the locust synaptosomal membrane vesicles (Gordon *et al.*, 1984; Zlotkin *et al.*, 1985).
K_D values were calculated according to the following equation:

$$K_D = \frac{K_{0.5}}{1 + ([L^*]/K^*_D)}$$

(Weiland and Molinoff, 1981) where $K_{0.5}$ is the apparent concentration of the competing unlabelled compound which inhibits specific binding by 50% at equilibrium, $[L^*]$ is the concentration of free labelled ligand, and K^*_D is the equilibrium dissociation constant of the labelled ligand, which is 1.2 nM for the [125I]AaIT (Figure 7.9c).

To summarise, the insect-selective neurotoxins derived from scorpion venoms illustrate certain unique features in the function of insect neuronal membranes. The pharmacological and applicational aspects of this pheno-menon are discussed later.

NEUROTOXINS SPECIFICALLY AFFECTING CRUSTACEANS

The crustacean-toxins derived from scorpion venoms

Crustacean toxin from the venom of a buthid scorpion: detection and purification

The first hint that scorpion venom might contain a toxic factor specifically active to crustaceans was given by the fact that the insect toxin (AaIT) and the mammal toxins I and II derived from the venom of *Androctonus australis* were inactive when injected into an isopod (a terrestrial crusta-cean, *Armadillidium vulgare*, of the Malacostraca) in strong contrast to the crude venom which induced a quick paralysis (Zlotkin *et al.*, 1972a). The existence of factors specifically affecting crustaceans was demonstrated in the venoms of *A. australis* (Zlotkin *et al.*, 1972a) and *A. mauretanicus* (Zlotkin *et al.*, 1979). The first step in the column chromatography which served for the separation of the mammal toxins (Miranda *et al.*, 1970) and the AaIT insect toxin (Zlotkin *et al.*, 1971c) has enabled the isolation of factors specifically affecting crustaceans. Recycling Sephadex G-50 chromatography resulted in the full recovery of the toxicity of crustaceans and the discovery of its occurrence in three separate fractions (Zlotkin *et al.*, 1972a; Figure 7.10).

Fraction E which possessed the highest specific toxicity to crustaceans also mimicked the crude venom in the induction of lethality to crayfish and the excitatory blockage of its stretch receptor. The insect and mammal toxins (AaIT and AaMT2, respectively) derived from the same venom were inactive to the crayfish (Pansa *et al.*, 1973). Fraction R_2 (Figure 7.10) was submitted to two steps of separation by column chromatography, finally resulting in the purified so-called crustacean toxin (AaCT) (Zlotkin *et al.*, 1975). The final product is a polypeptide composed of 70 amino acids (M_r = 8190) devoid of methionine, histidine and phenylalanine, rich in arginine and thought to contain five disulphide bridges. The final product is 250 times more toxic to isopods than the crude venom, and it represents about 18 per cent of its toxicity. The amino acid composition of AaCT is presented later (Table 7.8).

Figure 7.10: Elution pattern of the venom of *A. australis* on recycling gel filtration on Sephadex G-50. Four columns of 3.2 × 100 cm in series equilibrated and eluded by 0.1 M ammonium acetate pH 8.5–8.6 buffer. Flow rate 60 ml/h. The mixture submitted to fractionation is the water extract of 2 g of crude venom. Fractions A, B, C, D and E are non-toxic to mice and to insects. Fraction LT contains the AaIT insect toxin. Fraction R_1 contains the mammal toxins I and III, and R_2 mammal toxin II. Vertical arrows and numbers correspond to the beginning of the consecutive cycles. Fractions of the elution curves indicated by the full line are collected. Only the material marked by the dotted line is recycled. The total toxicity to isopods was recovered and it was distributed among fractions R_1, E and R_2.

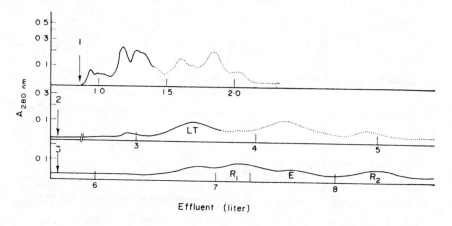

Source: from Zlotkin *et al.* (1972a)

Action and specificity. The action of AaCT was studied on a nerve–muscle preparation of a crayfish. AaCT induces a sequence of phenomena expressed in the increase of excitatory junction potentials (e.j.p.s; Figure 7.11A), evoked repetitive activity (Figure 7.11A), spontaneous repetitive activity (Figure 7.11B), and finally a partial block of excitability of the axon. As shown in Figure 7.11B, this repetitive muscular activity is entirely due to repetitive firing of the motor nerve. To summarise, the excitatory paralysis induced by the AaCT is due to the general stimulation of the skeletal musculature which follows from a presynaptic effect on the level of the motor axons (Rathmayer *et al.*, 1977).

The axonal effects of AaCT on the lateral and median giant axons in the abdominal nerve cord of the crayfish were studied. They resulted in great prolongations of the action potential as well as oscillations in their plateau

Figure 7.11: The effect of AaCT on the nerve–muscle preparation of the opener muscle of isolated walking leg of the crayfish *Astacus leptodactylus*. (A): Effect of 0.026 µg/ml crustacean toxin on e.j.p.s evoked by single stimuli. (a) Control, (b) 25 s, (c) 45 s, and (d) 50 s after toxin application. Calibration bars: 2 mV for (a); 10 mV for (b)–(d); 100 ms. (B): Correlation of e.j.p.s and activity of the motor axons. Simultaneous intracellular recording of repetitive e.j.p.s (upper trace) upon a single stimulus and extracellular recording of action potential (lower trace) from a branch of the motor axon 12 mm proximal to the muscle recording site: 0.2 µg/ml crustacean toxin. Calibrations: 20mV for upper, 400 µV for lower trace; 1 s.

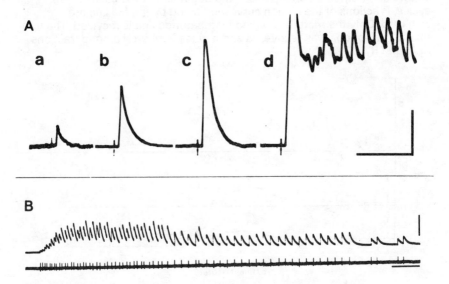

Source: from Rathmayer *et al.* (1977)

phases (Rathmayer *et al.*, 1977). The above axonal effects of the AaCT were mimicked by the *A. australis* mammal toxin I (AaMT1), the AaMT2 and AaIT, however, were inactive. The effect of the AaCT and AaMT1 toxins on the prolongation of the action potentials in the giant axons is in good agreement with results obtained with crude venom from different scorpion species (*Leiurus quinquestriatus, Buthus occitanus*) and with mammal toxin I from *A. australis* on different preparations. These venoms have been found to slow or remove Na^+ inactivation without affecting the kinetics of the Na^+ activation and also to reduce the K^+ outward current (Koppenhofer and Schmidt, 1968; Narahashi *et al.*, 1972; Schmitt and Schmidt, 1972; Romey *et al.*, 1975).

It has to be assumed that the prolongation of the action potentials and the repetitive activity superimposed on their plateaux observed in the giant axons also take place in the peripheral motor axons and in their terminals. The prolongation of the action potential should increase transmitter output drastically, leading to enhanced and prolonged e.j.p.s and to vast summating activity of m.e.j.p.s.

When considering the question of selectivity or degree of animal-group specificity in the action of AaCT, we are directed to a series of studies in which the AaCT, AaMT1, AaMT2 and AaIT toxins were assayed in four different neuromuscular preparations (of a mammal, a crustacean, an arachnid and an insect). Data are presented in Table 7.7 and they indicate that:

(1) The crustacean and the mammal toxins should be considered as possessing only a relative specificity to different organisms. The pharmacological diversity of these substances may represent either multiplicity of active sites or differential affinities to identical receptor sites in different organisms. It is noteworthy that the presence of a component such as the crustacean toxin, which by itself is non-lethal to mammals but is able to perform an excitatory action in mammalian systems, may possess certain pharmacological as well as pharmaceutical advantages.

(2) The discrepancy between the relative potencies of the crustacean and the mammal II toxins for the locust nerve–muscle preparation, as compared with their extremely low ability to paralyse the whole animal by injection (Table 7.7), may suggest that the animal group specificity of the above toxins may perhaps also be attributed to their specific resistance or susceptibility to inactivation processes in the bodies of the respective animals.

(3) There exists a close resemblance between the response of the crayfish and the spider nerve–muscle preparation to the different toxins (Table 7.7). This may indicate certain structural and functional similarities between the neuromuscular systems of these two classes of arthropods, differentiating them from insects.

293

Table 7.7: The relative activity of different scorpion toxins on several nerve–muscle preparations[a] as well as on insect paralysis

Toxic material[b]	Insect[c]	Crustacean[d]	Arachnid[e]	Mammal[f]	Insect[g] paralysis
Crude venom	1	1	1	1	1
Insect toxin	115	No effect	No effect	No effect	25
Crustacean toxin	18	127	67	0.7	0.7
Mammal toxin I	7	27	11	1	5.7
Mammal toxin II	2	4	5	25	0.2

[a] Based on a comparison of the average minimal doses causing an evoked repetitive muscular response. Numbers express the activity in relation to that of the crude venom.
[b] The different toxins were purified from the crude venom of the scorpion *Androctonus australis* Hector (Miranda *et al.*, 1970; Zlotkin *et al.*, 1971c, 1975).
[c] Locust leg *extensor tibiae* preparation (Walther *et al.*, 1976).
[d] Crayfish walking-leg dactylus opener preparation (Rathmayer *et al.*, 1977).
[e] Claw closer muscle preparation in the leg of a mygalomorph spider (Ruhland *et al.*, 1977).
[f] Guinea pig ileum smooth muscle preparation (Tintpulver *et al.*, 1976).
[g] Determined by injection into the body cavity of second and third instar locust larvae.

(4) As mentioned previously AaIT represents substances demonstrating a high degree of selectivity, being able to diversify between related groups of arthropods. This may indicate certain differences in the structure and function of the axonal membranes.

Toxins from venom of a chactoid scorpion affecting Crustacea

The venom of the scorpion *Scorpio maurus palmatus*, which was shown to possess various components affecting insects (including the so-called, cooperative insect toxins (Lazarovici *et al.*, 1982)) contains several polypeptides specifically paralysing and killing various crustaceans, such as isopods and crayfish (Lazarovici *et al.*, 1984). Gel filtration chromatography (Figure 7.12) enabled the isolation of two fractions toxic to Crustacea; the first (M_r 14 kDa Frac. I, Figure 7.12) with a phospholipase activity (it possesses about 40 per cent of the crude venom's lethality to isopods and is also toxic to insects but not to mammals), and the second (M_r 3–4 kDa Frac. II, Figure 7.12) was devoid of enzymic activity and possessed about 3 and 8 per cent of the crude venom lethality and paralytic potency to isopods, respectively. This fraction (II, Figure 7.12) also possesses the cooperative insect toxins (see Lazarovici *et al.*, 1982).

Additional fractionation of fraction II (Figure 7.12) by the aid of cation exchange column chromatography has resulted in the final purification of three low molecular weight polypeptides designated as CT2, CT3a and CT3b which possessed about 0.1, 0.17, 0.06, and 0.3, 0.9, 0.2 per cent of lethal potency and paralytic potency of the crude venom to crustaceans, respectively (Lazarovici *et al.*, 1984). Their purity was assessed by disc electrophoresis, analytical ultracentrifugation, isoelectrofocusing and amino acid analysis (Table 7.8). These basic polypeptides are characterised by very similar and unique amino acid compositions of 31 to 34 amino acids with a molecular mass of about 3.5 kDa and a deficiency in methionine, leucine, phenylalanine, histidine and tryptophan (Table 7.8). The present low molecular weight polypeptides cannot be considered as being selective to crustaceans. Their toxicity is rather relatively specific, and when applied in doses higher by about one order of magnitude they may be toxic to insects as well as to mammals.

The lethal potency of these polypeptides was increased about 10- to 20-fold when applied in the presence of small sublethal doses of the phospholipase Sephadex G-50 fraction I (Figure 7.12). From this fraction a lethal phospholipase which contained 37 per cent of the total venom phospholipase activity and 11 per cent of its toxicity to isopods was purified. This phospholipase (designated as A_1b) consists of 125 amino acids (M_r 14581) and is a hydrophobic, acidic protein composed of two isozymes (pI 4.7 and 4.9). This enzyme demonstrates an A2-type positional specificity (EC 3.1.1.4) with pH and temperature optima of 7.5–8.0 and 40–50°C,

Figure 7.12: Gel-filtration chromatography of *S. m. palmatus* venom. The lyophilised water extract corresponding to 410 mg of Lowry protein was charged on two columns (198 × 1.71 cm) connected in series, each filled with Sephadex G-50 fine (Pharmacia, Sweden) and equilibrated and eluted by 0.1 M ammonium acetate, pH 8.5, buffer. The flow rate was 15 ml/h and fractions of 10 ml were collected. DB indicates the location of the void volume as determined with Blue Dextran. The marked areas correspond to fractions toxic to crustaceans (I and II). Full circles represent absorbance at 280 nm, open circles correspond to the phospholipase activity. The vertical arrows indicate the elution of molecular weight markers: A, trypsin; B, lysozyme; C, phospholipase P_3 from *Naja mossambica* venom; D, insulin. The numbers correspond to the molecular mass in kDa

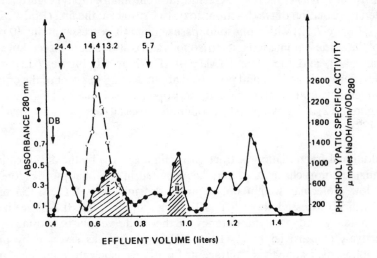

Source: from Lazarovici *et al.* (1984).

respectively, and high calcium requirements. The lethal potency of the above basic polypeptides is evidently increased by the addition of low, sublethal doses of the pure phospholipase. It remains to be clarified whether the augmented lethality caused by the combined action of the basic polypeptides and the phospholipase component is due to a simultaneous action on separate target organs or is a consequence of a direct association resulting in the formation of a new toxic product as in the case of crotoxin complex (Howard and Gundersen, 1980).

Factors in black widow spider venom toxic to Crustacea

The data presented in Table 7.3 and the accompanying text indicate a distinction among factors, derived from *Latrodectus* venom, specifically

Table 7.8: Amino acid composition of the polypeptide toxins derived from the venom of the scorpion *S. m. palmatus* compared with the insect (IT1) and mammal (MT) toxins derived from the same venom and the crustacean toxin (CT) from the venom of the buthid scorpion *A. australis*[a]

Amino acid	S. m. palmatus					A. australis[c]
	CT2	CT3a	CT3b	IT1[b]	MT[b]	CT
Aspartic acid	1.95(2)	2.94(3)	2.21(2)	2	2	6
Threonine[d]	2.57(3)	2.13(2)	2.45(2)	0	2	4
Serine[d]	3.33(3)	3.10(3)	3.40(3)	1	2	5
Glutamic acid	1.18(1)	1.14(1)	1.05(1)	3	2	10
Proline	1.10(1)	1.71(2)	1.22(1)	3	2	2
Glycine	3.70(4)	2.98(3)	2.70(3)	2	3	4
Alanine	2.07(2)	2.08(2)	1.92(2)	1	2	2
Half-cystine[d]	7.16(8)	7.53(8)	8.15(8)	4	8	10
Valine	0.98(1)	1.00(1)	0.82(1)	2	1	4
Methionine	0	0	0	0	1	0
Isoleucine[e]	1.02(1)	1.03(1)	0.91(1)	1	0	2
Leucine[e]	0	0	0	2	0	2
Tyrosine[d]	1.45(1)	1.68(2)	1.48(1)	1	1	4
Phenylalanine	0	0	0	0	0	0
Lysine	3.90(4)	4.70(5)	3.66(4)	4	5	4
Histidine	0	0	0	0	0	0
Arginine	2.00(2)	1.15(1)	1.67(2)	2	1	9
Tryptophan[f]	0	0	0	0	0	2
Total	33	34	31	28	32	70
Molecular weight	3507	3745	3349	3232	3478	8191
$E^{1\%}_{cm}$ 280 nm	5.53	8.75	5.79	5.13	5.57	20.90

[a] The data are an average of four determinations (two for 20- and two for 70-h hydrolyses).
[b] Taken from Lazarovici and Zlotkin (1982).
[c] According to Zlotkin *et al.* (1975).
[d] Extrapolated to zero hydrolysis time.
[e] Determined after 70 h of hydrolysis.
[f] Spectrophotometrically determined.

affecting insects, crustaceans and vertebrates: (a) Sephadex G-200 fraction C is toxic to insects but does not affect the crustacean stretch receptor preparation (Frontali *et al.*, 1976). (b) The latter as well as the lobster neuromuscular preparation were affected by the Sephadex fraction E (Fritz *et al.*, 1980) which induced the typical presynaptic effects expressed in initial potentiation of excitatory and inhibitory postsynaptic potentials as well as a large increase in the frequency of m.e.p.s.p.s which were later abolished (Fritz *et al.*, 1980). (c) The vertebrate toxic component B_5 (α-latrotoxin) was inactive in a lobster neuromuscular preparation.

To summarise, the Sephadex G-200 fraction E and its derived DEAE-Sephadex A_{50} fraction E_2 (see Table 7.3 and accompanying text; Frontali *et al.*, 1976) represent the factor(s) selectively affecting crustaceans in the venom of the black widow spider.

Factors in the venoms of sea anemones toxic to Crustacea

Toxic substances derived from sea anemones

Sea anemones belong, together with hydroids, jellyfish and corals, to the phylum Coelenterata. All members of this phylum possess unique stinging organelles of several types, called nematocysts, from which the paralysing venom is injected into their prey. The latter is composed of small zooplanktonic crustaceans and small fish. Because of the extreme difficulty in obtaining undischarged nematocysts and collecting their authentic venom (Blanquet, 1968; Hessinger and Lenhoff, 1973) the vast majority of studies have used homogenates of whole isolated tentacles or even whole bodies. As a matter of fact it is unclear whether the various toxins so far described represent the nematocyst venom *per se*, or are of a non-nematocyst origin.

With this approach a long series of low molecular weight basic proteins were isolated from a range of sea anemones possessing various activities such as cytolysis, haemolysis (Bernheimer and Avigad, 1981, 1982; Lafranconi *et al.*, 1984), proteinase inhibition (Mebs *et al.*, 1983) and cardio-neurotoxicity (Beress, 1982). The latter are the dominant group, and include the Na^+ channel modulators described earlier which can be subdivided into two categories. The first includes short toxins of 2000–3000 Da (such as toxins III and IV of *Anemonia sulcata* (Beress *et al.*, 1977). The second, the major group, includes the long toxins of 4000–6000 Da (such as toxins I and II of *Anemonia sulcata* (Wunderer *et al.*, 1976; Wunderer and Eulitz, 1978; Figure 7.13).

Specific toxicity to crustaceans

The purification process of eight toxins derived from four different sea anemones was monitored by toxicity assays on mice and crabs (Schweitz *et*

298

al., 1981). These are basic polypeptides of a molecular weight in the range 3–6 kDa.

Data concerning the various biological activities of the different toxins are summarised in Table 7.9. All toxins studied demonstrate an obvious toxicity to crabs which varies by a maximum factor of 20 whereas differences varying by a factor of more than 2000 were observed for LD_{50} values on mice. Four out of the eight toxins possess a relatively very low toxicity to mammals (AS_I, AS_{III}, SG_I and AP_I) and thus a relative specificity to crustaceans. AX_I, on the other hand, is more toxic to mammals than to crustaceans. There exists a striking correlation between their binding affinity to rat brain synaptosomes (K_D), their efficiency (ED_{50}) in increasing the rate of Na^+ uptake in mouse neuroblastoma cells and their toxicity to mice (LD_{50}, Table 7.9). Such correlation was also shown to exist for the recently purified (Schweitz *et al.*, 1985) *Radianthus paumotensis* toxin II (RP_{II}; Figure 7.13) which, in contrast to RP_I, RP_{III} and RP_{IV}, possessed a very low toxicity to mice (4200 µg/kg). Conversely, the results in Table 7.9 show that values of K_D, ED_{50} and LD_{50} of the eight toxins for mice are not correlated with their toxicity to crabs and even vary in the opposite direction. For example, AX_{II} is the toxin that has the highest activity (K_D, ED_{50}, LD_{50}) in mammals (tissue or membranes), and the lowest toxicity to crabs.

The only exception in the above general observation in the series of sea anemone toxins studied by Schweitz *et al.* (1981) is AS_{II}, which is highly active on both mammals and crustaceans and as indicated earlier was shown to exert the same effect on sodium conductance in Crustacea as well as in vertebrate excitable membranes (Bergman *et al.*, 1976; Romey *et al.*, 1976, 1980). This toxin was submitted to a series of chemical modifications in order to study the residues involved in its toxic action on crabs and mice

Table 7.9: Pharmacological properties of sea anemone toxins

Toxin[a]	Toxicity		Binding to synaptosomes, K_d (nM)	$^{22}Na^+$ uptake of neuroblastoma cells ED_{50} (nM)
	LD_{50} mice (g/kg)	LD_{100} crabs (g/kg)		
AS_I	>4000	4.4	7000	>10000
AS_{II}	100	3.7	150	200
AS_{III}	>18000	6.7	>10000	>10000
AS_V	19	10.4	50	15
AX_I	66	22	120	47
AX_{II}	8	78	35	7
SG_I	>2000	14	>10000	>10000
AP_I	>6000	32	>10000	>10000

[a] The abbreviations AS, AX, SG and AP correspond to toxins derived from *Anemonia sulcata*, *Anthopleura xanthogrammica*, *Stoichactis giganteus* and *Actinodendron plumosum* sea anemones, respectively.

Source: from Schweitz *et al.* (1981).

and in its binding properties to the Na^+ channel of rat brain synaptosomes (Barhanin *et al.*, 1981). There are two essential points which deserve attention. First, the fact that the described series of modifications have affected both the mice and crab toxicities in a similar manner suggests principally the same molecular mode of action in both organisms. Secondly, the striking result from these studies is that modification of three carboxylic functions of AS_{II} does not impair its ability to associate with the specific receptor but induces a complete loss of its toxic properties as well as effects on sodium conductance in both organisms. A reasonable interpretation given by Barhanin *et al.* (1981) was that carboxylate side-chains are not involved in the binding of the toxin to its specific receptor but play an important role in one of the subsequent steps that permit the expression of neurotoxicity (Barhanin *et al.*, 1981).

The amino acid sequences of five sea anemone toxins, among them two of a potent toxicity to crustaceans and of an extremely low toxicity to mammals (AS_I and RP_{II}), are presented in Figure 7.13. It is difficult, at the present stage, to draw any conclusion concerning a relation between the primary structure and the animal-group specificity of these toxins.

Crustacean-selective neurotoxins derived from nemertines

Toxic substances in nemertine secretions

The nemertines (Rhychocoela) are a small phylum of fewer than a thousand described species of carnivorous worms feeding mainly upon annelids, crustaceans, molluscs and even fish. The toxic substances produced by nemertines originate substantially from venom storage in the proboscis of the enoplan nemertines, and from mucous secretions from the body surface which occur in both anoplan and enoplan nemertines (Kem, 1973; Kem *et al.*, 1976). The proboscis and body surface of nemertine worms have been shown to possess two main groups of toxic substances, pyridine bases and toxic polypeptides (Kem, 1973). The latter are subdivided into two categories: the cytolytic 10000-Da polypeptides designated as A-toxins (Blumenthal and Kem, 1980b; Blumenthal, 1982), and the 6000-Da B-toxins. The short B-toxins are crustacean-selective toxins.

Crustacean selective neurotoxins

The purification of the B-toxins from the heteronemertine *Cerebratulus lacteus* was accomplished through the following sequence of manipulations (Kem, 1976):

(1) Stimulation of live worms to secrete mucus by 1 per cent solution of acetic acid.
(2) Batch absorption of the basic polypeptides upon CM-cellulose

300

Figure 7.13: Homologies and differences in the amino acid sequences from different sea anemone toxins: toxins AS$_I$, AS$_{II}$ and AS$_V$ from *Anemonia sulcata*, toxin AX$_I$ from *Anthopleura xanthogrammica*, and toxin RP$_{II}$ from *Radianthus paumotensis*.

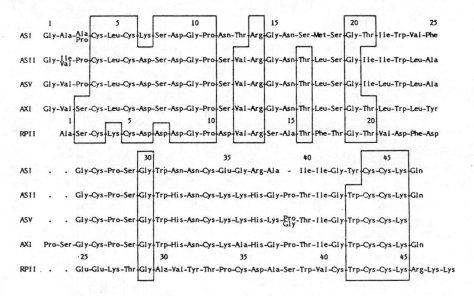

Source: from Schweitz *et al.* (1985).

(Whatman CM52);

(3) Desorption of the toxic components by stepwise elution with 0.1 and 1.0 M ammonium acetate pH 6.5 buffer.

(4) The concentrated, desalted and finally lyophilised CM-cellulose-desorbed material was separated by Sephadex G-50F, resulting in fraction II — containing the A-toxins and fraction III — containing the B toxins (Table 7.10).

(5) Further purification of the Sephadex G-50 fraction II on a CM-cellulose gradient chromatograph, resulting in the CM-cellulose toxic fractions VI, VII and VIII causing crayfish paralysis.

(6) The various B-toxins I, II, III and IV were obtained by rechromatography of the Sephadex III fraction on CM-cellulose. The purification process of the B toxins was monitored through a series of lethality and paralysis assays including crustacean (crabs *Procambarus*), insect (cockroaches *Neophytes*) and mammalian (laboratory mice) species. The specific toxicities of the four purified B-toxins differed over a 25- to 30-fold range. Toxin B-II was the most active (35 200 units/mg) (Kem, 1976).

Table 7.10: Amino acid compositions of purified *Cerebratulus* toxins (residues per molecule)

Amino acid	A-toxins			B-toxins			
	II	III	IV	I	II	III	IV
Lysine	15	14	18	7	8	8	10
Histidine	0	1	1	3	2	1	1
Arginine	3	3	4	3	2	3	3
CmCysteine	6	6	6	6	8	8	8
Aspartic acid	6	6	5	6	5	6	5
Threonine	3	3	2	1	3	1	1
Serine	7	7	6	1	3	2	1
Glutamic acid	4	4	5	3	5	5	4
Proline	4	2	5	1	0	0	0
Glycine	11	12	9	5	5	5	5
Alanine	12	13	11	4	6	7	7
Valine	5	7	5	3	1	0	0
Methionine	0	0	2	1	0	0	0
Isoleucine	7	7	6	1	1	3	3
Leucine	6	5	7	1	0	0	1
Tyrosine	1	1	2	1	2	2	2
Phenylalanine	2	2	2	2	0	0	0
Tryptophan	2	2	2	0	2	2	2
Total residues	94	95	98	49	53	54	53
Formula wt	9 797	9 835	10 533	5 453	5 857	5 898	5 927
Molecular wt (10% agarose)	10 400	10 500	10 400	5 380	5 930	5 970	6 000

Source: from Kem and Blumenthal (1978).

The selective toxicity of B-toxins to crustaceans is emphasised by their inability to affect mice as well as other arthropods. Very high doses (>50 mg/kg) of toxin B-IV were also completely inactive when injected into two species of teleost fishes, a frog, an annelid worm, a hoplonemertine and *Cerebratulus* itself. Furthermore, toxin B-IV induced spontaneous action potentials and prolonged the repolarisation phase of the action potential when applied to crustacean neurones, but failed to produce these effects when applied to garfish olfactory, squid giant axon and frog sciatic axon (Kem, 1976).

The amino acid analysis of the crustacean selective B-neurotoxins derived from heteronemertine *Cerebratulus* is presented in Table 7.10 and compared with the A-toxins derived from the same animal. The primary structures (Blumenthal and Kem, 1976) and the disulphide bonds (Blumenthal and Kem, 1977) of toxin B-IV and the primary structures of toxin B-II and B-IV (Blumenthal *et al.*, 1981) were studied and are presented in Figures 7.14 and 7.15. As shown (Figure 7.15) hydroxyproline, an amino acid not normally found in non-fibrous proteins, occupies positions 10 of both toxins. There exists a high degree of homology between these two polypeptides, particularly within the N-terminal region up to two-thirds of each chain (Blumenthal *et al.*, 1981). It was recently shown that toxin B-IV is completely inactivated by reduction of its four disulphide bonds. Reduction is also accompanied by loss of secondary structure. Reoxidation of the reduced protein may be catalysed by the oxidised forms of glutathione or dithiothreitol. Secondary structure and toxicity were shown to recover in parallel (Blumenthal, 1986). Nitration of a specific tyrosine residue 9 (Blumenthal and Kem, 1980a) and alkylation of tryptophan 30 (Blumenthal, 1980) were shown to abolish toxicity of the B-IV toxin without causing gross conformational changes in the protein. It was suggested that these residues might be directly involved in the interaction of the toxin with axonal receptor involved in the inactivation of voltage-sensitive Na^+ channels in crustacean nerves (Blumenthal and Kem, 1980b).

The binding of the *Cerebratulus* ^{125}I-labelled B-IV toxin to lobster axonal vesicles was studied through a series of equilibrium saturation, displacement and kinetic binding studies (Toth and Blumenthal, 1983). It has been shown that the toxin binds to a single class of non-interacting binding sites of high affinity (K_D 5–20 nM) and a relatively low capacity (6–9 pmol per milligram of protein) which closely resembles that of [3H]saxitoxin to the same preparation. The K_D value obtained through the kinetic studies (k_{-1}/k_1) was in good agreement with those performed in equilibrium saturation conditions. There is no competition with *Leiurus quinquestriatus* mammal toxin V, which is also known to affect the sodium inactivation mechanism (Catterall, 1980).

The synthesis of a radioactive bifunctional cross-linking reagent

Figure 7.14: Schematic diagram of the complete covalent structure of *Cerebratulus lacteus* toxin B–IV. The disulphide bond linking Cys-10 and Cys-47 is assigned by difference.

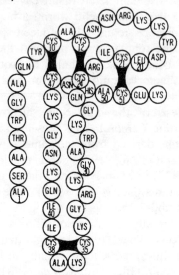

Source: from Blumenthal and Kem (1977).

Figure 7.15: Comparison of the sequences of *Cerebratulus* toxins B–II and B–IV. Residues occupying identical positions in the two proteins have been enclosed in boxes

```
                                        10
B-II    Ala-Ser┤Ser├Thr-Trp-Gly┤Gly-Ser├Tyr-Hyp-Ala-Cys-Glu-Asn-Asn-Cys-Arg-Lys┤
B-IV    Ala-Ser┤Ala├Thr-Trp-Gly┤Ala-Ala├Tyr-Hyp-Ala-Cys-Glu-Asn-Asn-Cys-Arg-Lys┤


            20                              30
B-II    Gln┤Tyr-Asp├Asp┤Cys-Ile├Lys┤Cys-Gln-Gly-Lys-Trp-Ala-Gly-Lys-Arg-Gly-Lys-
B-IV    Lys┤Tyr-Asp├Leu┤Cys-Ile├Arg┤Cys-Gln-Gly-Lys-Trp-Ala-Gly-Lys-Arg-Gly-Lys-


            40                              50
B-II    Cys-Ala-Ala-His-Cys┤Ala-Val├Gln┤Thr-Thr-Ser├Cys┤Asn-Asp├Lys-Cys-Lys-Lys┤His
B-IV    Cys-Ala-Ala-His-Cys┤Ile-Ile├Gln┤Lys-Asn-Asn├Cys┤Lys-Gly├Lys-Cys-Lys-Lys┤Glu
```

[^{125}I]azidosalicylic acid (ASA) B-IV toxin coupled with autoradiography of polyacrylamide gel electrophoresis has recently (Lieberman and Blumenthal, 1986) enabled the identification of lobster axonal proteins of molecular masses of 38 and 40 kDa which are specifically cross-linked by this

304

reagent. The $[^{125}I]$ASA–B–IV cross-linker was shown to interact with the lobster axonal vesicles with binding constants (K_D = 30 nM, capacity 7.5 pmol per milligram of membrane protein) closely resembling those previously obtained with the iodinated toxin (Toth and Blumenthal, 1983; see above). The cross-linking was prevented in the presence of micromolar concentration of unlabelled toxin, in the absence of photolysis and in the presence of 150 mM K^+. The authors (Lieberman and Blumenthal, 1986) claim that the significance of the inhibition by this high potassium concentration is not clear. In the previous study, however (Toth and Blumenthal, 1983), it was suggested that binding of $[^{125}I]$B–IV toxin to the lobster axonal residues might be dependent upon membrane potential and was decreased by high potassium concentrations assumed to depolarise the vesicle membrane. It was also suggested that the similarity of the molecular weights of the 40 and 38 kDa binding sites of the B-IV toxin to the 1 and 2 subunits of the rat brain sodium channel (39 and 37 kDa, respectively; Catterall, 1984) might indicate that the B-IV toxin binds to these (B1, B2) components of the Na^+ channel in lobster nerve (Lieberman and Blumenthal, 1986). However, this essential point demands further clarification.

DISCUSSION

The successful employment of neurotoxins in vertebrate neurobiology has strongly stimulated researchers to adopt the same approach in insect and arthropod neuropharmacology. The employment of α-bungarotoxin in the identification (Dudai and Amsterdam, 1977), characterisation (Dudai, 1978; Breer, 1981), allocation (Dudai and Amsterdam, 1977; Lees et al., 1983), function (Harrow et al., 1980), isolation (Harris et al., 1980; Mansour et al., 1980) and purification (Breer et al., 1985) of the cholinergic receptor in insect CNS, and the employment of saxitoxin for the quantitative determination (Gitschier et al., 1980), genetics (Jackson et al., 1984) and channel characterisation (Strichartz and Hansen-Bay, 1981) of the Na^+ channels in insects and crustaceans, may serve as examples.

The above toxicological approach to insect/arthropod neuropharmacology is directed by the concept of the uniformity and stability of the pharmacological properties of the nervous systems of various groups of multicellular organisms. When comparing arthropods with vertebrates, this concept is strongly supported by the fact that most of the chemical signals found in the nervous systems of vertebrates are also found in the central and peripheral nervous systems of insects (Usherwood, 1982), and that the majority of the neurotoxins studied in vertebrates are actually known to affect arthropodan systems. The latter aspect has obtained several expressions. It has been shown that a long series of various ion channel blockers (TTX, STX, 4-aminopyridine) and modifiers (DDT, pyrethroids, veratridine, atropine, and scorpion and sea anemone toxins) affect an insect axon in the same manner as they act on axonal preparations of other animals (Narahashi, 1974; Pichon, 1974, 1976; Pelhate and Sattelle, 1982; Sattelle, 1985). The presynaptic toxins latrotoxin and leptinotarsin discussed earlier were also shown to stimulate release of acetylcholine from preloaded insect synaptosomes (Breer, 1983b). With the postsynaptic toxins, in addition to α-bungarotoxin (see above) the glutaminergic blockers derived from araneid spider venoms were shown to affect both arthropod and vertebrate systems (Kawai *et al.*, 1982a,b, 1983a,b; Nagai *et al.*, 1984). It was recently shown that three toxins, representing three out of five classes of the Na^+ channel toxins considered earlier, have demonstrated high binding affinities to a fly-head synaptosomal preparation (Pauron *et al.*, 1985), this indicating a high degree of conservation of the Na^+ channels between insects and mammals.

To summarise, the employment of neurotoxic tools originally studied and developed in vertebrate systems is a common practice in insect neuropharmacology.

The phenomenon of the animal group selective neurotoxins, to which the present review is substantially dedicated, seems to disagree with the above concept of pharmacological uniformity. The animal group selective toxins are toxins that affect exclusively a given group of animals. The compounds presently reviewed that satisfy the above definition are the neurotoxins affecting insects derived from the venoms of scorpions, braconid wasps and black widow spider venoms and the toxins affecting crustaceans derived from nemertine secretions. The scorpion, sea anemones and nemertine toxins affect sodium conductance and thus can be defined as ion channel toxins whereas the toxins derived from braconid wasps and black widow spiders affect neurotransmitter release and can be defined as presynaptic toxins. The insect and crustacean toxins studied so far are proteins of medium weight in the braconid (about 50 kDa) and black widow spider (about 100 kDa) venoms, and are low molecular weight basic polypeptides (4–8 kDa with a relative abundance of disulphide bonds) in the case of the scorpion, sea anemone and nemertine toxins. According to some recent information (Zlotkin, 1985), insect-selective toxins may also occur in the venoms of reduviid bugs, tarantula spider, centipedes and social wasps. These aspects deserve attention in future studies.

Animal group selective toxins derived from the same venom share closely resembling action in the respective animals and their nervous systems. For example, the excitatory insect toxins derived from scorpion venoms resemble the scorpion venom β-mammalian toxins in their binding constants, voltage independence and the induction of repetitive activity in the respective neuronal preparations (Gordon *et al.*, 1984). It is also evident that the same essential effect is elicited, in the respective organisms, by the mammalian, insect and crustacean factors of the black widow spider venom concerning their presynaptic effects on transmitter release (Fritz *et al.*, 1980).

The animal group selectivity is expressed in the discrimination not only between major groups of animals such as vertebrates and arthropods but also between related groups of arthropods such as insects and crustaceans which are known to possess morphologically and physiologically identical nervous systems. This aspect receives an additional emphasis when dealing with the amazing selectivity in the action of the *Bracon* venom and its derived toxins. These toxins were shown to block exclusively the excitatory transmission in neuromuscular systems of a given group of insects. In other words, in addition to the distinction between the inhibitory and excitatory transmission in an insect, these toxins distinguish between insect and crustacean neuromuscular systems both known to employ *L*-glutamate as the excitatory neurotransmitter (Usherwood *et al.*, 1968; Gerschenfeld, 1973). An additional mystery is the specific susceptibility of lepidopterous insects for the *Bracon* venom.

In considering the pharmacological significance of animal group selectivity, let us consider for instance the various selective toxins that affect sodium conductance in insect and crustacean neuronal systems. The simplest conclusion from this phenomenon is that the animal group selectivity represents structural–functional differences in the voltage-activated sodium channels among the various groups of animals. The pharmacological diversity of Na^+ channels, as indicated by the various animal group selective toxins, is not surprising when we take into consideration some recent data concerning the diversity in their chemical composition in various mammalian systems. This is expressed in the occurrence of the smaller subunits in brain and skeletal muscle sodium channels (Hartshorne *et al.*, 1982; Barchi, 1983) in contrast to sodium channels isolated from electroplax or heart (Miller *et al.*, 1983; Catterall, 1986) which possess only the 260 kDa component. Furthermore a neurotoxic peptide isolated from a marine snail was recently shown to distinguish between mammalian muscle and brain sodium channels (Cruz *et al.*, 1985).

With this background it is not surprising that Na^+ channels corresponding to various groups of arthropods such as insects and crustaceans may possess similar diversities and peculiarities (presumably related to their regulatory mechanism) which are recognised and identified by the above animal group selective neurotoxins. The same basic considerations may apply to the insect or crustacean selective presynaptic toxins derived from black widow and braconid wasp venoms. They too may represent certain structural–functional peculiarities related to mechanisms of transmitter release and uptake in the above arthropods. The chemical–molecular basis of these peculiarities and diversities remains to be clarified. Whatever it may be, there is no doubt that these neuronal structural diversities among

307

the various animal groups are of critical functional significance, as indicated by the high toxic potency of the respective toxins. It may also be concluded that the above modified forms of the Na^+ channel are sufficiently stable and consistent in order to serve as a target for an evolutionary process of designing these selective neurotoxins.

To summarise, the phenomenon of the animal group selective neurotoxins represents a chemical adaptation of the venom of a venomous animal to its specific prey or natural enemy, or, more precisely, to certain peculiarities in the structure and function of their nervous systems. The potency and pharmacological specificity of these selective neurotoxins strongly suggest the clarification of their structure–function relationships and toxin–receptor interactions at a molecular level. This attitude is directed by the classical concept of the pharmacological tool. The fact that the selective neurotoxin interacts with a consistent, well conserved membranal structure related to a critical function enables the toxin to serve as a specific marker of the above structure and a key for clarifying the function. As such these toxins may serve as valuable tools in comparative neurochemistry and developmental neurobiology.

With regard to the insect-selective toxins, additional considerations may play an important role. These toxins are able first to distinguish between the nervous system of a non-insect and an insect, and secondly to identify in the latter a functionally critical site. As such these toxins may supply the rational approach and serve as models for the design of insect selective insecticides. The applicational significance of the insect selective peptide neurotoxins can be evaluated against the background of recent progress in the biotechnology of polypeptides with regard to their mass production, design and modification by techniques of recombinant DNA using either natural or artificial genes (Abelson, 1983) as well as the potential ability to mimic bioactive peptides by non-peptides (Farmer, 1980; Ariens, 1979). The latter point is exemplified in nature where a simple alkaloid such as morphine is mimicked at the receptor level by endorphin polypeptides or where a receptor of a complex polypeptide such as α-bungarotoxin may be recognised by a relatively simple alkaloid such as tubocurarine.

Briefly, the employment of the insect-selective neurotoxins for the future design of selective insecticides may follow one or a combination of the following approaches: their mimicry by synthetic non-peptidic substances; the design of modified metabolically stable neurotoxic polypeptides with oral toxicity; the insertion and association of the insect-toxin genes into microorganisms and/or viruses which possess the ability to penetrate and inoculate the insect body and thus to produce the insect toxins there. The data presented in the present review may contribute essential information in support of the above approaches.

ACKNOWLEDGEMENT

I am grateful to Mrs Zipora Spira for her assistance and particularly for her help in the collection of literature.

REFERENCES

Abe, T., Alema, S. and Miledi, R. (1977) Isolation and characterization of presynaptically acting neurotoxins from the venom of *Bungarus* snakes. *Eur. J. Biochem. 80*, 1

Abe, T., Kawai, N. and Niwa, A. (1983) Effects of spider toxin on the glutaminergic synapse of lobster muscle. *J. Physiol. 339*, 243

Abe, T., Limbrick, A.R. and Miledi, R. (1976) Acute muscle denervation induced by α-bungarotoxin. *Proc. R. Soc. Lond. B. Biol. Sci. 194*, 545

Abelson, P.H. (1983) Biotechnology: an overview. *Science 219*, 611

Abia, A., Lobaton, C.D., Moreno, A. and Garcia-Sancho, Y. (1986) *Leiurus quinquestriatus* venom inhibits different kinds of Ca-dependent K^+ channels. *Biochim. Biophys. Acta 856*, 403

Albuquerque, E.X. and Daly, J.W. (1976) Batrachotoxin, a selective probe for channels modulating sodium conductances in electrogenic membranes. In: Cuatrecasas, P. (ed.) *Receptors and recognition*, pp. 299-336. Chapman & Hall, London

Albuquerque, E.X. and Daly, J. (1977) Steroidal alkaloid toxin and ion transport in electrogenic membranes. In: Cuatrecasas, P. (ed.) *The specificity and action of animal, bacterial and plant toxin*, pp. 279-338. Chapman & Hall, London

Ariens, E.J. (1979) Receptors from fiction to fact. *Trends Pharm. Sci., inaugural issue*, 11

Baba, A. and Cooper, J.R. (1980) The action of black widow spider venom on cholinergic mechanisms in synaptosomes. *J. Neurochem. 34*, 1369

Babin, D.R., Watt, D.D., Goos, S.M. and Mlejnek, R.V. (1974) Amino acid sequences of neurotoxin protein variants from the venom of *Centruroides sculpturatus* Ewing. *Arch. Biochem. Biophys. 164*, 694

Babin, D.R., Watt, D.D., Goos, S.M. and Mlejnek, R.V. (1975) Amino acid sequences of neurotoxin I from *Centruroides sculpturatus* Ewing. *Arch. Biochem. Biophys. 166*, 125

Baguis, R., Chanteau, S., Chungue, E., Hartel, J.M., Yasumoto, T. and Inone, A. (1980) Origins of ciguatera fish poisoning: a new dinoflagellate, *Gambierdiscus toxicus* Adachi and Fukuyo, definitively involved as a causal agent. *Toxicon 18*, 199

Barchi, R.L. (1983) Protein components of the purified sodium channel from rat skeletal muscle sarcolemma. *J. Neurochem. 40*, 1377

Barhanin, J., Hugues, M., Schweitz, H., Vincent, J-P. and Lazdunski, M. (1981) Structure–function relationships of sea anemone toxin II from *Anemonia sulcata*. *J. Biol. Chem. 256*, 5764

Beard, R.L. (1952) The toxicology of *Habrobracon* venom: a study of natural insecticide. *Conn. Agric. Exp. Stn Bull. 562*

Beard, R.L. (1963) Insect toxins and venoms. *Ann. Rev. Ent. 8*, 1

Beard, R.L. (1978) Venoms of Braconidae. In: Bettini, S. (ed.) *Arthropod venoms*, pp. 773-800. Springer, Berlin and New York

Bearg, W.J. (1959) The black widow and five other venomous spiders in the United

States. *Agr. Exp. Stn. Arkansas Univ. Bull.* 608

Bearg, W.J. (1961) Scorpions biology and effect of the venom. *Univ. Kansas Agric. Exp. Stn. Bull.* 649

Beneski, D.A. and Catterall, W.A. (1980) Covalent labeling of protein components of the sodium channel with a photoactivable derivative of scorpion toxin. *Proc. Natl Acad. Sci. USA 77,* 639

Benzer, T.I. and Raftery, M.A. (1973) Solubilization and partial characterization of the tetrodotoxin binding component from nerve axons. *Biochem. Biophys. Res. Commun. 51,* 939

Beress, L. (1982) Biologically active compounds from coelenterates. *Pure and Appl. Chem. 54,* 1981

Beress, L., Wunderer, G. and Wachter, E. (1977) Amino acid seqence of toxin III from *Anemonia sulcata. Hoppe-Seyler's Z. Physiol. Chem. 358,* 985

Bergman, C., Dubois, J.M., Rojas, E. and Rathmayer, W. (1976) Decreased rate of sodium conductance inactivation in the node of Ranvier induced by a polypeptide toxin from sea anemone. *Biochim. Biophys. Acta 455,* 173

Bernheimer, A.W. and Avigad, L.S. (1981) New cybolysins in the sea anemones from the West Coast of the United·States. *Toxicon 19,* 529

Bernheimer, A.W. and Avigad, L.S. (1982) Toxins of the sea anemone *Epiactis prolifera. Arch. Biochem. Biophys. 217,* 174

Berwald-Netter, Y., Martin-Moutot, N., Kaoulakoff, A. and Couraud, F. (1981), Na⁺-channel associated scorpion toxin receptor sites as probes for neuronal evolution *in vivo* and *in vitro. Proc. Natl Acad. Sci. USA 78,* 1245

Bettini, S. and Maroli, M. (1978) Venoms of Theridiidae, genus *Latrodectus.* A. Systematics, distribution and biology of species: chemistry, pharmacology and mode of action of venom. In: Bettini, S. (ed.) *Arthropod venoms,* pp. 149-85. Springer, Berlin and New York

Bidard, J.N., Vijverberg, H.N.P., Frelin, Ch., Chungue, E., Legrand, A.M., Bagnis, R. and Lazdunski, M. (1984) Ciguatoxin is a novel type of Na⁺ channel toxin. *J. Biol. Chem. 259,* 8353

Blanquet, R. (1968) Properties and composition of the nematocyst toxin of the sea anemone, *Aiptasia pallida. Comp. Biochem. Physiol. 25,* 893

Blum, M.S. (1981) *Chemical defenses of arthropods.* Academic Press, New York and London

Blumenthal, K.M. (1980) Structure and action of heteronemertine polypeptide toxins: inactivation of *Cerebratulus lacteus* toxin B-IV concomitant with tryptophan alkylation. *Arch. Biochem. Biophys. 203,* 822

Blumenthal, K.M. (1982) Structure and action of heteronemertine polypeptide toxins. Membrane penetration by *Cerebratulus lacteus* toxin A-III. *Biochemistry 21,* 4229

Blumenthal, K.M. (1986) Renaturation of neurotoxin B-IV from the heteronemertine *Cerebratulus lacteus. Toxicon 24,* 63

Blumenthal, K.M. and Kem, W.R. (1976) Structure and action of heteronemertine polypeptide toxins. Primary structure of *Cerebratulus lacteus* toxin B-IV. *J. Biol. Chem. 251,* 6025

Blumenthal, K.M. and Kem, W.R. (1977) Structure and action of heteronemertine polypeptide toxins. Disulfide bonds of *Cerebratulus lacteus* toxin B-IV. *J. Biol. Chem. 252,* 3328

Blumenthal, K.M. and Kem, W.R. (1980a) Structure and action of heteronemertine polypeptide toxins: inactivation of *Cerebratulus lacteus* toxin B-IV by tyrosine nitration. *Arch. Biochem. Biophys. 203,* 816

Blumenthal, K.M. and Kem, W.R. (1980b) Structure–function relationships in *Cerebratulus* toxin B-IV. In: Eaker, D. and Wandstrom, T. (eds) *Natural toxins,* pp. 487-92. Pergamon Press, Oxford and New York

Blumenthal, K.M., Keim, P.S., Heinrikson, R.L. and Kem, W.R. (1981) Structure and action of heteronemertine polypeptide toxins. Amino acid sequence of *Cerebratulus lacteus* toxin B-II and revised structure of toxin B-IV. *J. Biol. Chem. 256*, 9063

Brazil, O.V., Neder, A.C. and Corrado, A.P. (1973) Effects and mechanism of action of *Tityus serrulatus* venom on skeletal muscle. *Pharmacol. Res. Commun. 5*, 137

Breer, H. (1981) Properties of putative nicotinic and muscarinic cholinergic receptors in the central nervous system of *Locusta migratoria. Neurochem. Int. 13*, 43

Breer, H. (1983a) Venoms and toxins in neurochemical research of insects. In: Hucho, F. and Ovchinnikov, Y.A. (eds) *Toxins as tools in neurochemistry*, pp. 115-25. De Gruyter, Berlin and New York

Breer, H. (1983b) Choline transport by synaptosomal membrane vesicles isolated from insect nervous tissue. *FEBS Lett. 153*, 345

Breer, H. and Jeserich, G. (1980) A microscale flotation technique for the isolation of synaptosomes from the nervous tissue of *Locusta migratoria. Insect Biochem. 10*, 457

Breer, H., Kleene, R. and Hinz, G. (1985) Molecular forms and subunit structure of the acetylcholine acceptor in the central nervous system of insects. *J. Neurosci. 5*, 3386

Brownell, P.H. (1984) Prey detection by the sand scorpion. *Sci. Amer. 251*, 94

Bucherl, W. (1971) Classification biology and venom extraction of scorpions. In: Bucherl, W. and Buckley, E. (eds) *Venomous animals and their venoms, Vol. 3, Venomous invertebrates*, pp. 317-46. Academic Press, New York

Cahalan, M.D. (1975) Modification of sodium channel gating of frog myelinated nerve fibers by *Centruroides sculpturatus* scorpion venom. *J. Physiol. (Lond.) 244*, 511

Carbone, E., Wanke, E., Prestipino, G., Possani, L.D. and Maelicke, A. (1982) Selective blockage of voltage-dependent K^+ channels by a novel scorpion toxin. *Nature (Lond.) 296*, 90

Catterall, W.A. (1975a) Activation of the action potential sodium ionophore of cultured neuroblastoma cells by veratridine and batrachotoxin. *J. Biol. Chem. 250*, 4053

Catterall, W.A. (1975b) Cooperative activation of the action potential Na^+ ionophore by neurotoxins. *Proc. Natl Acad. Sci. USA 72*, 1782

Catterall, W.A. (1977a) Membrane potential dependent binding of scorpion toxin to the action potential sodium ionophore. Studies with a toxin derivative prepared by lactoperoxidase-catalysed iodination. *J. Biol. Chem. 252*, 8660

Catterall, W.A. (1977b) Activation of the action potential Na^+ ionophore by neurotoxins. An allosteric model. *J. Biol. Chem. 252*, 8669

Catterall, W.A. (1979) Binding of scorpion toxin to receptor sites associated with sodium channels in frog muscle. Correlation with voltage-dependent activation. *J. Gen. Physiol. 74*, 375

Catterall, W.A. (1980) Neurotoxins that act on voltage-sensitive sodium channels in excitable membranes. *Am. Rev. Pharmacol. Toxicol. 20*, 15

Catterall, W.A. (1981) Localization of sodium channels in cultured neural cells. *J. Neurosci. 1*, 777

Catterall, W.A. (1984) The molecular basis of neuronal excitability. *Science 223*, 653

Catterall, W.A. (1985) The electroplax sodium channel revealed. *Trends Neurosci. 8*, 39

Catterall, W.A. (1986) Voltage-dependent gating of sodium channels: correlating structure and function. *Trends Neurosci. 9,* 7

Catterall, W.A. and Beress, L. (1978) Sea anemone toxin and scorpion toxin share a common receptor site associated with the action potential Na$^+$ ionophore. *J. Biol. Chem. 253,* 7393

Catterall, W.A. and Morrow, C.S (1978) Binding of saxitoxin to electrically excitable neuroblastoma cells. *Proc. Natl Acad. Sci. USA 75,* 218

Changeux, J.P., Kasai, M. and Lee, C.Y. (1970) Use of snake venom toxin to characterize the cholinergic receptor protein. *Proc. Natl Acad. Sci, USA 67,* 1241

Chibber, B.A., Martin, B.M., Walkinshaw, M.D., Saenger, W. and Maelicke, A. (1983) The sites of neurotoxicity in α-cobratoxin. In: Hucho, F. and Ovchinnikov, Y.A. (eds) *Toxins as tools in neurochemistry,* pp. 141-50. De Gruyter, Berlin and New York

Clark, A.W., Mauro, A., Longenecker, H. and Hurlbut, W.P. (1970) Effects of black widow spider venom on the frog neuromuscular junction. *Nature (Lond.) 225,* 703

Clark, R.B., Donaldson, P.L., Gration, K.A.F., Lambert, J.J., Piek, T., Spanjer, W. and Usherwood, P.N.R. (1980) Post-synaptic block at neuromuscular function on locust muscle by δ-philanthotoxin. *J. Physiol. (Lond.) 310,* 8P

Clark, R.B., Donaldson, P.L., Gration, K.A.F., Lambert, J.J., Piek, T., Ramsey, R., Spanjer, W. and Usherwood, P.N.R. (1982) Block of locust muscle glutamate receptor by δ-philanthotoxin occurs after receptor activations. *Brain Res. 241,* 105

Couraud, F. and Jover, E. (1984) Mechanism of action of scorpion toxins. In: Tu, A.T. (ed.) *Handbook of natural toxins,* vol. 2, pp. 659-78. Marcel Dekker, New York and Basle

Couraud, F., Jover, E., Dubois, J.M. and Rochat, H. (1982) Two types of scorpion toxin receptor sites, one related to the activation, the other to the inactivation of the action potential sodium channel. *Toxicon 20,* 9

Couraud, F., Rochat, H. and Lissitzky, S. (1978) Binding of scorpion and sea anemone neurotoxins to a common site related to the action potential Na$^+$ ionophore in neuroblastoma cells. *Biochem. Biophys. Res. Commun. 83,* 1525

Crosland, R.D., Hsiao, T.H. and McClure, W.O. (1984) Purification and characterization of α-leptinotarsin-h, an activator of presynaptic calcium channels. *Biochemistry 23,* 734

Cruz, L.J., Gray, W.R., Olivera, B.M., Zeikus, R.D., Kerr, L., Yoshikami, D. and Moczydlowski, E. (1985) *Conus geographicus* toxins that discriminate between neuronal and muscle sodium channels. *J. Biol. Chem. 260,* 9280

Cull-Candy, S.G., Neal, H. and Usherwood, P.N.R. (1973) Action of black widow spider venom on an aminergic synapse. *Nature (Lond.) 241,* 353

Curtis, D.R. and Johnston, G.A.R. (1974) Amino acid transmitters in the mammalian central nervous system. *Ergebn. Physiol. 69,* 97

D'Ajello, V., Magni, F. and Bettini, S. (1971) The effect of the venom of the black widow spider *Latrodectus mactans tredecimguttatus* on the giant neurones of *Periplaneta americana. Toxicon 9,* 103

Darbon, H., Jover, E., Couraud, F. and Rochat, H. (1983) Photoaffinity labeling of α- and β-scorpion toxin receptors associated with rat brain sodium channel. *Biochem. Biophys. Res. Commun. 115,* 415

Darbon, H., Zlotkin, E., Kopeyan, C., Van Rietschoten, J. and Rochat, H. (1982) Covalent structure of the insect toxin of the North African scorpion *Androctonus australis hector. Int. J. Peptide Protein Res. 20,* 320

Deitmer, J.W. (1973) Die Wirkung des Giftes der Schlupfwespe *Habrobracon hebetor* (Say) auf die Neuromuskulare Ubertragung am Sartoriusmuskel des

Frosches. Diplomarbeit, Universität Bonn

Drenth, D. (1974a) Stability of *Microbracon hebetor* (Say) venom preparation. *Toxicon 12*, 541

Drenth, D. (1974b) Susceptibility of different species of insects to an extract of the venom gland of the wasp *Microbracon hebetor* (Say). *Toxicon 12*, 189

Dudai, Y. (1978) Properties of an α-bungarotoxin binding to cholinergic nicotinic receptor from *Drosophila melanogaster. Biochim. Biophys. Acta 539*, 505

Dudai, Y. and Amsterdam, A. (1977)Nicotinic receptors in the brain of *Drosophila melanogaster* demonstrated by autoradiography with [^{125}I]α-bungarotoxin. *Brain Res. 130*, 551

Dunbar, S.J. and Piek, T. (1983) The action of iontophoretically applied *L*-glutamate on an insect visceral muscle. *Arch. Ins. Biochem. Physiol. 1*, 93

Eaker, D. and Wandstrom, T. (eds) (1980) *Natural toxins.* Pergamon Press, Oxford and New York

Farmer, P.S. (1980) Bridging the gap between bioactive peptides and nonpeptides: some perspectives in design. In: Ariens, E.J. (ed.) *Drug design*, Vol. 10, pp. 119-43. Academic Press, New York and London

Finkelstein, A., Rubin, L.L. and Tzeng, M.C. (1976) Black widow spider venom: effect of the purified toxin on lipid bilayer membranes, *Science 193*, 1009

Fosset, M., Schmid-Antomarchi, H., Hugues, M., Romey, G. and Lazdunski, M. (1984) The presence in pig brain of an endogenous equivalent of apamin, the bee venom peptide that specifically blocks Ca^{2+} dependent K^+ channels. *Proc. Natl Acad. Sci. USA 81*, 7228

Fritz, L.C., Tzeng, M.C. and Mauro, A. (1980) Different components of black widow spider venom mediate transmitter release at vertebrate and lobster neuromuscular junctions. *Nature (Lond.) 283*, 486

Frontali, N., Ceccarelli, B., Gorio, A., Mauro, A., Siekevitz, P., Tzeng, M.C. and Hurlbut, W.P. (1976) Purification from black widow spider venom of a protein factor causing the depletion of synaptic vesicles at neuromuscular junctions. *J. Cell Biol. 68*, 462

Frontali, N. and Grasso, A. (1964) Separation of three toxicologically different protein components from the venom of the spider *Latrodectus tredecimguttatus. Arch. Biochem. Biophys. 106*, 213

Gerschenfeld, H.M. (1973) Chemical transmission in invertebrate central nervous system and neuromuscular junctions. *Physiol. Rev. 53*, 1

Gitschier, J., Strichartz, G.R. and Hall, L.M. (1980) Saxitoxin binding to sodium channels in head extracts from wild-type and tetrodotoxin sensitive strains of *Drosophila melanogaster. Biochim. Biophys. Acta 595*, 291

Glatz, L. (1969) Correlations entre la capture de la proie et les structures des pieces buccales chez les Uloboridae. *Bull. Mus. Nat. Hist. Natur. 41*, 65

Gordon, D., Jover, E., Couraud, F. and Zlotkin, E. (1984) The binding of the insect selective neurotoxin (AaIT) from scorpion venom to locust synaptosomal membranes. *Biochim. Biophys. Acta 778*, 349

Gordon, D., Zlotkin, E. and Catterall, W.A. (1985) The binding of an insect selective neurotoxin and saxitoxin to insect neuronal membranes. *Biochim. Biophys. Acta 821*, 130

Gordon, D., Zlotkin, E. and Kanner, B. (1982) Functional membrane vesicles from the nervous system of insects. I. Sodium and chloride dependent α-aminobutyric acid transport. *Biochim. Biophys. Acta 688*, 229

Goyffon, M. and Kovoor, J. (1978) Chactoid venoms. In: Bettini, S. (ed.) *Arthropod venoms*, pp. 395-418. Springer, Berlin and New York

Grasso, A. (1976) Preparation and properties of a neurotoxin from the venom of black widow spider *(Latrodectus mactans tridecimguttatus). Biochim. Biophys.*

Acta 439, 406

Grasso, A. and Paggi, P. (1967) Effect of *Latrodectus mactans tredecimguttatus* venom on the crayfish stretch receptor neurone. *Toxicon 5*, 1

Grasso, A. and Senni, M.I. (1979) A toxin purified from the venom of black widow spider affects uptake and release of radioactive γ-aminobutyrate and N-epinephrine from rat brain synaptosomes. *Eur. J. Biochem. 102*, 337

Gray, W.R., Lugue, A. and Olivera, B.M. (1981) Peptide toxins from *Conus geographicus* venom. *J. Biol. Chem. 256*, 4734

Gregoire, J. and Rochat, H. (1983) Covalent structure of toxin I and II from the scorpion *Buthus occitanus tunetanus*. *Toxins 21*, 153

Griffiths, D.J.G. and Smyth, T., Jr (1973) Action of black widow spider venom of insect neuromuscular junctions. *Toxicon 11*, 369

Grishin, E.V., Sukhikh, A.P., Adamovich, T.B., Zhdanova, L.N., Soldatov, N.M., Ovchinnikov, Y.A., Atakusiev, B.U. and Tashmukhamedov, B.A. (1976) Symposium on the chemistry of peptides and proteins. *Dushanbe*, 91

Grishin, E.V., Volkova, T.M. and Soldatova, L.N. (1982) A study of the toxic component of the venom of the Caucasian subspecies of the scorpion *Buthus epeus. Biorg. Khim (USSR) 8*, 155

Gruener, R. (1973) Excitability blockage of the squid giant axon by the venom of *Latrodectus mactans* (black widow spider). *Toxicon 11*, 155

Habermann, E. (1971) Chemistry, pharmacology and toxicology of bee, wasp and hornet venom. In: Bucherl, W. and Buckley, E. (eds) *Venomous invertebrates*, pp. 61-89. Academic Press, New York

Harris, R., Cattell, K.J. and Donnellan, J.F. (1980) Characterization of α-bungarotoxin binding to homogenates of housefly brain. In: *Insect neurobiology and pesticide action (Neurotox '79)*, pp. 209-12. Society of Chemical Industry, London

Harrow, I.D., Hue, B., Gepner, J.I., Hall, L.M. and Sattelle, D.B. (1980) An α-bungarotoxin sensitive acetylcholine receptor in the central nervous system of the cockroach (*Periplaneta americana L.*). In: *Insect neurobiology and pesticide action (Neurotox '79)*, pp. 137-44. Society of Chemical Industry, London

Hartshorne, R.P. and Catterall, W.A. (1984) The sodium channel from rat brain: purification and subunit composition. *J. Biol. Chem. 259*, 1667

Hartshorne, R.P., Messner, D.J., Coppersmith, J.C. and Catterall, W.A. (1982) The saxitoxin receptor of the sodium channel from rat brain. Evidence for two non-identical subunits. *J. Biol. Chem. 257*, 13888

Henderson, R. and Wang, J.H. (1972) Solubilization of a specific tetrodotoxin binding component from garfish olfactory nerve membranes. *Biochemistry 11*, 4565

Herzog, W.H., Feibel, R.M. and Bryant, S.H. (1974) Effect of aconitine on the giant axon of the squid. *J. Gen. Physiol. 47*, 719

Hessinger, D.A. and Lenhoff, H.M. (1973) Assay and properties of hemolysis activity of pure venom from the nematocysts of the acontia of the sea anemone *Aiptasia pallida. Arch. Biochem. 159*, 629

Hille, B. (1968) Pharmacological modifications of the sodium channels of frog nerve. *J. Gen. Physiol. 51*, 199

Hille, B. (1971) The permeability of the sodium channel to organic cations in myelinated nerve. *J. Gen. Physiol. 58*, 599

Hille, B. (1975) The receptor for tetrodotoxin and saxitoxin. A structural hypothesis. *Biophys. J. 15*, 615

Howard, B.D. and Gundersen, C.B. (1980) Effects and mechanisms of polypeptide neurotoxins that act presynaptically. *Ann. Rev. Pharmacol. Toxicol. 20*, 307

Hu, S.L. and Kao, G.Y. (1986) The pH dependence of the tetrodotoxin-blockade of the sodium channel and implications for toxin binding. *Toxicon 24,* 25

Hucho, F. and Ovchinnikov, Y.A. (eds) (1983) *Toxins as tools in neurochemistry.* De Gruyter, Berlin and New York

Hugues, M., Romey, G., Duval, D., Vincent, J.P. and Lazdunski, M. (1982a) Apamin as a selective blocker of the calcium dependent potassium channel in neuroblastoma cells; voltage clamp and biochemical characterization of the toxin receptor. *Proc. Natl Acad. Sci. USA 79,* 1308

Hugues, M., Schmid, H. and Lazdunski, M. (1982b) Identification of a protein component of the Ca^{2+} dependent K^+ channel by affinity labelling with apamin. *Biochem. Biophys. Res. Commun. 107,* 1577

Jackson, F.R., Wilson, S.D., Strichartz, G.R. and Hall, L.M. (1984) Two types of mutants affecting voltage-sensitive sodium channels in *Drosophila melanogaster. Nature (Lond.) 308,* 189

Jacques, Y., Fosset, M. and Lazdunski, M. (1978) Molecular properties of the action potential Na^+ ionophore in neuroblastoma cells. *J. Biol. Chem. 253,* 7383

Jaimovich, E., Ildefonse, M., Barhanin, J., Rougier, O. and Lazdunski, M. (1982) *Centruroides* toxin, a selective blocker of surface Na^+ channels in skeletal muscle voltage-clamp analysis and biochemical characterization of the receptor. *Proc. Natl Acad. Sci. USA 79,* 3896

Jover, E., Couraud, F. and Rochat, H. (1980) Two types of scorpion neurotoxins characterized by their binding to two separate receptor sites on rat brain synaptosomes. *Biochem. Biophys. Res. Commun. 95,* 1607

Kagan, B.L., Pollard, H.B. and Hanna, R.B. (1982) Induction of ion-permeable channels by the venom of the fanged bloodworm *Glycera dibranchiata. Toxicon 20,* 887

Katz, N.L. and Edwards, C.H. (1972) The effect of scorpion venom on the neuromuscular junction of the frog. *Toxicon 10,* 133

Kawai, N., Mauro, A. and Grundfest, H. (1972) Effect of black widow spider venom on the lobster neuromuscular junctions. *J. Gen. Physiol. 60,* 650

Kawai, N., Niwa, A. and Abe, T. (1982a) Spider venom contains specific receptor blocker of glutaminergic synapses. *Brain Res. 247,* 169

Kawai, N., Niwa, A. and Abe, T. (1982b) Effect of a spider toxin on glutaminergic synapses in the mammalian brain. *Biomed. Res. 3,* 353

Kawai, N., Niwa, A. and Abe, T. (1983a) Specific antagonism of the glutamate receptor by an extract from the venom of the spider *Araneus ventricosus. Toxicon 21,* 438

Kawai, N., Niwa, A. and Abe, T. (1983b) Block of glutamate receptors by a spider toxin. In: Mandel, P. and De Fendis, F.V. (eds) *CNS receptors — from molecular pharmacology to behavior,* pp. 221-7. Raven Press, New York

Kem, W.R. (1973) Biochemistry of nemertine toxins. In: Martin, D.F. and Padilla, G.M. (eds) *Marine pharmacognosy,* pp. 38-84. Academic Press, New York

Kem, W.R. (1976) Purification and characterization of a new family of polypeptide neurotoxins from the heteronemertine *Cerebratulus lacteus* (Leidy). *J. Biol. Chem. 251,* 4189

Kem, W.R. and Blumenthal, K.M. (1978b) Polypeptide cytolysins and neurotoxins isolated from the mucus secretions of the heteronemertine *Cerebratulus lacteus* (Leidy). In: Rosenberg, P. (ed.) *Toxins — animal, plant and microbial,* pp. 509-16. Pergamon Press, Oxford and New York

Kem, W.R., Scott, K.N. and Duncan, J.H. (1976) Hoplonemertine worms — a new source of pyridine neurotoxins. *Experientia 32,* 684

Kerr, L. and Yoshikami, D. (1984) A venom peptide with a novel presynaptic blocking action. *Nature (Lond.) 308,* 282

Kobayashi, M., Miyakoda, G., Nakamura, T. and Ohizumi, Y. (1985) Ca^{2+}-

dependent arrhythmogenic effects of maitotoxin, the most potent marine toxin known, on isolated rat cardiac muscle cells. *Eur. J. Pharmacol. 111*, 121

Kobayashi, J., Nakamura, H., Hirata, Y. and Ohizumi, Y. (1983) Tessulatoxin, the vasoactive protein from the venom of the marine snail *Conus tessulatus. Comp. Biochem. Physiol. 74B, 381*

Kopeyan, C., Martinez, G., Lissitzky, S., Miranda, F. and Rochat, H. (1974) Disulfide bonds of toxin II of the scorpion *Androctonus australis* Hector. *Eur. J. Biochem. 47*, 483

Koppenhofer, E. and Schmidt, R. (1968) Die Wirkung von Skorpiongift auf die Ionenstrome des Ranvier'schen Schnurrings. I. Die Permeabilitaten von P_{Na+} und p_{K+}. *Pflügers Arch. 303*, 133

Krasilnikov, O.V., Ternovskii, V.I. and Tashmukhamedov, B.A. (1982) Investigation of the channel-forming properties of black widow spider venom. *Biophysics 27*, 71

Krnjevic, K. (1974) Chemical nature of synaptic transmission in vertebrates. *Physiol. Rev. 54*, 418

Lafranconi, W.M., Ferlan, I., Russell, F.E. and Huxtalele, R.J. (1984) The action of equinatoxin, a peptide from the venom of the sea anemone, *Actinia equina*, on the isolated lung. *Toxicon 22*, 347

Lazarovici, P., Menashe, M. and Zlotkin, E. (1984) Toxicity to Crustacea due to polypeptide–phospholipase interaction in the venom of a chactoid scorpion. *Arch. Biochem. Biophys. 229*, 270

Lazarovici, P. and Zlotkin, E. (1982) A mammal toxin derived from the venom of a chactoid scorpion. *Comp. Biochem. Physiol. 71C*, 177

Lazarovici, P., Yanai, P., Pelhate, M. and Zlotkin, E. (1982) Insect toxic components from the venom of a chactoid scorpion *Scorpio maurus palmatus* (Scorpionidae). *J. Biol. Chem. 257*, 8397

Lee, C.Y. (1972) Chemistry and pharmacology of polypeptide toxins in snake venoms. *Ann. Rev. Pharmacol. 12*, 265

Lee, C.Y. (1979a) Recent advances in chemistry and pharamacology of snake toxins. In: Ceccarelli, B. and Clementi, F. (eds) *Advances in cytopharmacology*, Vol. 3, pp. 1-16. Raven Press, New York

Lee, C.Y. (1979b) *Snake venoms.* Springer, Berlin and New York

Lees, G., Beadle, D.J. and Botham, R.P. (1983) Cholinergic receptors on cultured neurones from the central nervous system of embryonic cockroaches. *Brain Res. 288*, 49

Lester, D., Lazarovici, P., Pelhate, M. and Zlotkin, E. (1982) Two insect toxins from the venom of the scorpion *Buthotus judaicus.* Purification, characterization and action. *Biochim. Biophys. Acta 701*, 370

Lieberman, D.L. and Blumenthal, K.M. (1986) Structure and action of heteronemertine polypeptide toxins. Specific cross-linking of *Cerebratulus lacteus* toxin B-IV to lobster axon membrane vesicles. *Biochim. Biophys. Acta 855*, 41

Llados, F., Matteson, D.R. and Kriebel, M.F. (1980) α-Bungarotoxin preferentially blocks one class of miniature endplate potentials. *Brain Res. 192*, 598

Longenecker, H.E. Jr, Hurlbut, W.P,, Mauro, A. and Clark, A.W. (1970) Effects of black widow spider venom on the frog neuromuscular junction. Effects on end plate potential, miniature end plate potential and nerve terminal spike. *Nature (Lond.) 225*, 701

Low, P.A., Wu, C.H. and Narahashi, T. (1979) Effect of anthopleurin A on crayfish giant axon. *J. Pharmacol. Exp. Ther. 210*, 417

Manaranche, R., Thieffry, M. and Israel, M. (1980) Effect of the venom of *Glycera convoluta* on the spontaneous quantal release of transmitter. *J. Cell. Biol. 85*, 446

316

Mansour, N.A., Pessah, I.N. and Eldefrawi, A.T. (1980) Binding of [^{125}I] α-bungar-otoxin and reversible cholinergic ligands to proteins in house fly brains. In: *Insect neurobiology and pesticide action (Neurotox '79)*, pp. 201-8. Society of Chemical Industry, London

Marlas, G. and Bon, C. (1982) Relationship between the pharmacological action of crotoxin and its phospholipase activity. *Eur. J. Biochem. 125*, 157

Maroli, M., Bettini, S. and Parrfili, B. (1973) Toxicity of *Latrodectus mactans trede-cimguttatus* venom on frog and birds. *Toxicon 11*, 203

May, T.E. and Piek, T. (1979) Neuromuscular block in locust skeletal muscle caused by a venom preparation made from the digger wasp *Philanthus triangu-lum F.* from Egypt. *J. Insect. Physiol. 25*, 685

McClure, W.O., Abbott, B.C., Baxter, D.E., Hsiao, T.H., Satin, L.S., Siger, A. and Yosino, J.E. (1980) Leptinotarsin: a presynaptic neurotoxin that stimulates release of acetylcholine. *Proc. Natl Acad. Sci. USA 77*, 1219

McIntosh, M., Cruz, L.J., Hunkapiller, M.W., Gray, W.R. and Olivera, B.M. (1982) Isolation and structure of a peptide toxin from the marine snail *Conus magus*. *Arch. Biochem. Biophys. 218*, 329

McIntosh, M.E. and Watt, D.D. (1972) Purification of toxins from the North American scorpion *Centruroides sculpturatus*. In: de Vries, A. and Kochva, E. (eds) *Toxins of animal and plant origin*, Vol. 2, pp. 529-44. Gordon & Breach, London

Mebs, D., Liebrich, M., Reul, A. and Samejima, Y. (1983) Hemolysins and protein-ase inhibitors from sea anemones of the Gulf of Aqaba. *Toxicon 21*, 257

Mebs, D., Narita, K., Iwanaga, S., Samejima, Y. and Lee, C.Y. (1972) Purification, properties and amino acid sequence of α-bungarotoxin from the venom of *Bungarus multicinctus*. *Hoppe-Seyler's Z. Physiol. Chem. 353*, 243

Meldolesi, J. (1982) Studies on α-latrotoxin receptors in rat brain synaptosomes. Correlation between toxin binding and stimulation of transmitter release. *J. Neurochem. 38*, 1559

Meldolesi, J., Madeddu, L., Gatti, G. and Watanabe, O. (1983) Studies on α-latro-toxin of black widow spider venom and its receptor in presynaptic membranes. *Period. Biol. 85*, 107

Michaelis, E.K., Galton, N. and Early, S. (1984) Spider venoms inhibit *L*-glutamate binding to brain synaptic membrane receptors. *Proc. Natl Acad. Sci. USA 81*, 5571

Miller, J.A., Agnew, W.S. and Levinson, S.R. (1983) Principal glycopeptide of the tetrodotoxin/saxitoxin binding protein from *Electrophorus electricus*: isolation and partial chemical and physical characterization. *Biochemistry 22*, 462

Miller, Ch., Moczydlowski, E., Latorre, R. and Phillips, M. (1985) Charybdotoxin, a protein inhibitor of single Ca^{2+}-activated K$^+$ channels from mammalian skeletal muscle. *Nature (Lond.) 313*, 316

Miranda, F., Kopeyan, C., Rochat, C., Rochat, H. and Lissitsky, S. (1970) Purifi-cation of animal neurotoxins. Isolation and characterization of eleven neuro-toxins from the venom of the scorpions *Androctonus australis* Hector, *Buthus occitanus tunetatus* and *Leiurus quinquestriatus*. *Eur. J. Biochem. 16*, 514

Miyamoto, T., Ohizumi, O., Washio, H. and Yasumoto, T. (1984) Potent excitatory effect of maitotoxin on Ca channels in the insect skeletal muscle. *Pflügers Arch. 400*, 439

Morel, N., Theiffry, M. and Manaranche, R. (1983) Binding of a *Glycera convoluta* neurotoxin to cholinergic nerve terminal plasma membranes. *J. Cell. Biol. 97*, 1737

Nagai, T., Obara, S. and Kawai, N. (1984) Differential blocking effects of a spider toxin on synaptic and glutamate responses in the afferent synapse of the acous-

317

tico lateralis receptor of *Plotosus*. *Brain Res. 300*, 183

Nakamura, Y., Nakajima, S. and Grundfest, H. (1965) The action of tetrodotoxin on electrogenic components of squid giant axons. *J. Gen. Physiol. 48*, 985

Narahashi, T. (1974) Chemicals as tools in the study of excitable membranes. *Physiol. Rev. 54*, 813

Narahashi, T., Anderson, N.C. and Moore, J.W. (1966) Tetrodotoxin does not block excitation from inside the nerve membrane. *Science 153*, 765

Narahashi, T., Haas, H.G. and Terrien, E.F. (1967) Saxitoxin and tetrodotoxin: comparison of nerve blocking mechanism. *Science 157*, 1441

Narahashi, T., Moore, J.W. and Scott, W.R. (1964) Tetrodotoxin blockage of sodium conductance increase in lobster giant axon. *J. Gen. Physiol. 47*, 965

Narahashi, T., Shapiro, B.I., Deguchi, T., Scuka, M. and Wang, C.M. (1972) Effects of scorpion venom on squid axon membranes. *Am. J. Physiol. 222*, 850

Nicholls, D.G., Rugolo, M., Scott, I.G. and Meldolesi, J. (1982) α-Latrotoxin of black widow spider depolarizes the plasma membrane, induces massive calcium influx and stimulates transmitter release in guinea pig synaptosomes. *Proc. Natl Acad. Sci. USA 79*, 7924

Oblas, B., Boyd, N.D. and Singer, R.H. (1983) Analysis of receptor–ligand interactions using nitrocellulose gel transfer: application to *Torpedo* acetylcholine receptor and alpha-bungarotoxin. *Anal. Biochem. 130*, 1

Ohizumi, Y., Kajiwara, A. and Yasumoto, T. (1983) Excitatory effect of the most potent marine toxin, maitotoxin, on the guinea-pig vas deferens. *J. Pharmacol. Exp. Ther. 227*, 199

Ohizumi, Y., Nakamura, H. and Kobayashi, J. (1986) Presynaptic inhibitory effect of geographitoxin II, a new peptide toxin from *Conus geographicus* venom in the guinea pig vas deferens. *Eur. J. Pharmacol.* 245

Ohizumi, Y. and Yasumoto, T. (1983) Contractile response of the rabbit aorta to maitotoxin, the most potent marine toxin. *J. Physiol. 337*, 711

Ohta, M., Narahashi, T. and Keeler, R.F. (1973) Effects of veratrum alkaloids on membrane potential and conductance of squid and crayfish giant axons. *J. Pharmacol. Exp. Ther. 184*, 143

Ornberg, R.L., Smyth, T. and Benton, A.W. (1976) Isolation of neurotoxin with a presynaptic action from the venom of the black widow spider (*Latrodectus mactans*, Fabr.). *Toxicon 14*, 329

Ovchinnikov, Yu. A. and Grishin, E.V. (1982) Scorpion neurotoxins as tools for studying fast sodium channels. *Trends Biochem. Sci. 7*, 26

Paggi, P. and Toschi, G. (1977) Effects of denervation and lack of calcium on the action of *Latrodectus* venom on rat sympathetic ganglion. *Life Sci. 11*, 413

Pansa, M.C., Migliori Natalizi, G. and Bettini, S. (1973) Effect of scorpion venom and its fractions on the crayfish stretch receptor organ. *Toxicon 11*, 283

Parnas, I., Avgar, D. and Shulov, A. (1970) Physiological effects of venom of *Leiurus quinquestriatus* on neuromuscular systems of locust and crab. *Toxicon 8*, 67

Parnas, I. and Russell, F.E. (1967) Effects of venom on nerve muscle and neuromuscular junction. In: Russell, F.E. and Saunders, P.R. (eds) *Animal toxins*, pp. 401-27. Pergamon Press, Oxford and New York

Pauron, D., Barhanin, J. and Lazdunski, M. (1985) The voltage-dependent Na^+ channel of insect nervous system identified by receptor sites for tetrodotoxin and scorpion and sea anemone toxins. *Biochem. Biophys. Res. Commun. 131*, 1226

Pelhate, M. and Sattelle, D.B. (1982) Pharmacological properties of insect axons: a review. *J. Insect Physiol. 28*, 889

Pelhate, M. and Zlotkin, E. (1981) Voltage dependent slowing of the turn off of Na^+ current in the cockroach giant axon induced by the scorpion venom 'insect toxin'. *J. Physiol. (Lond.) 319*, 30

Pelhate, M. and Zlotkin, E. (1982) Actions of insect toxin and other toxins derived from the venom of the scorpion *Androctonus australis* in isolated giant axons of the cockroach (*Periplaneta americana*). *J. Exp. Biol. 97*, 67

Pichon, Y. (1974) The pharmacology of the insect nervous system. In: Rockstein, M. (ed.) *The physiology of Insecta*, Vol. 4, pp. 101-74. Academic Press, New York

Pichon, Y. (1976) Pharmacological properties of the ionic channels in insect axons. In: Spencer Davies, P. (ed.) *Periplaneta in experimental biology*, Vol. 1, pp. 297-312. Pergamon Press, Oxford

Piek, T. (1982) δ-Philanthotoxin, a semi-irreversible blocker of ion channels. *Comp. Biochem. Physiol. 72C*, 311

Piek, T., Buitenhuis, A., Veldsema-Currie, R.D. and Mantel, P. (1983) Smooth muscle contracting factors in the venoms of sphecid wasps (Hym: Sphecidae) *Comp. Biochem. Physiol. 75C*, 153

Piek, T., Dunbar, S.J., Kits, K.S., Van Marle, J. and Van Wilgenburg, H. (1985) Philanthotoxins: a review of the diversity of actions on synaptic transmission. *Pest. Sci. 16*, 488

Piek, T., Hue, B., Pelhate, M., David, J.A., Spanjer, W. and Veldsema-Currie, R.D. (1984) Effects of the venom of *Philanthus triangulum* F. (Hym. Sphecidae) and β- and δ-philanthotoxin on axonal excitability and synaptic transmission in the cockroach CNS. *Arch. Ins. Biochem. Physiol. 1*, 297

Piel, T., Mantel, P. and Engels, E. (1971) Neuromuscular block in insects caused by the venom of the digger wasp *Philanthus triangulum. F. Comp. Gen. Pharmacol. 2*, 317

Piek, T., Mantel, P. and Jas, H. (1980a) Ion-channel block in insect muscle fibre membrane by the venom of the digger wasp *Philanthus triangulum F. J. Insect Physiol. 26*, 345

Piek, T., May, T.E. and Spanjer, W. (1980b) Paralysis of locomotion in insects by the venom of the digger wasp *Philanthus triangulum*. In: *Insect neurobiology and pesticide action (Neurotox '79)*, pp. 219-26. Society of Chemical Industry, London.

Piek, T. and Simon-Thomas, R.T. (1969) Paralysing venoms of solitary wasps. *Comp. Biochem. Physiol. 30*, 13

Piek, T. and Spanjer, W. (1986) Chemistry and pharmacology of solitary wasp venoms. In: Piek, T. (ed.) *Venoms of the Hymenoptera*, pp. 161-308. Academic Press, London and New York

Piek, T., Spanjer, W., Veldsema-Currie, R.D., Van Groen, T., De Hanna, W. and Mantel, P. (1982a) Effect of venom of the digger wasp *Philanthus triangulum* F. on the sixth abdominal ganglion of the cockroach. *Comp. Biochem. Physiol. 71C*, 159

Piek, T., Veenendaal, R.L. and Mantel, P. (1982b) The pharmacology of *Microbracon* venom. *Comp. Biochem. Physiol. 72C*, 303

Possani, L.D., Dent, M.A.R., Martin, B.M., Maelicke, A. and Svendsen, I. (1981) The amino terminal sequence of several toxins from the venom of the Mexican scorpion *Centruroides noxius* Hoffman. *Carlsberg Res. Commun. 46*, 207

Quicke, D.L.J. (1985) Antagonism of locust muscle glutamate receptor-channel complexes by fractions of orb-web spider venoms. *Neurotox '85*, University of Bath, England, Abstracts

Raftery, M.A., Hunkapiller, M.W., Strader, C.D. and Hood, L.E. (1980) Acetylcholine receptor: complex of homologous subunits. *Science 208*, 1454

Rathmayer, W. (1962a) Paralysis caused by the digger wasp *Philanthus*. *Nature (Lond.) 196*, 1148

Rathmayer, W. (1962b) Das Paralysierungsproblem beim Bienenwolf *Philanthus triangulum* F. (Hym. Sphec). *Z. Vergl. Physiol.* 45, 413

Rathmayer, W. (1966) The effect of the poison of spider and digger wasps on the prey. *Mem. Inst. Butantan Simp. Inst.* 33, 651

Rathmayer, W. (1978) Venoms of Sphecidae, Pompilidae, Multilidae and Bethylidae. In: Bettini, S. (ed.) *Arthropod venoms*, pp. 661-90. Springer, Berlin and New York

Rathmayer, W., Ruhland, M., Tintpulver, M., Walther, Ch. and Zlotkin, E. (1978) The effect of toxins derived from the venom of the scorpion *Androctonus australis* Hector on neuromuscular transmission. In: Rosenberg, P. (ed.) *Toxins: animal, plant and microbial*, pp. 629-37. Pergamon Press, Oxford and New York

Rathmayer, W. and Walther, C. (1976) Mode of action and specificity of *Habrobracon* venom. In: Ohsaka, A., Hayashi, K. and Sawai, Y. (eds) *Animal, plant and microbial toxins*, Vol. 2, pp. 299-307. Plenum Press, New York

Rathmayer, W., Walther, Ch. and Zlotkin, E. (1977) The effect of different toxins from scorpion venom on neuromuscular transmission and nerve action potentials in the crayfish. *Comp. Biochem. Physiol.* 56C, 35

Ray, R., Morrow, C.S. and Catterall, W.A. (1978) Binding of scorpion toxin to receptor sites associated with voltage-sensitive sodium channels in synaptic nerve ending particles. *J. Biol. Chem.* 253, 7307

Ritchie, J.M., Rogart, R.B. and Strichartz, G.R. (1976) A new method for labelling saxitoxin and its binding to nonmyelinated fibers of the rabbit vagus, lobster walking leg and garfish olfactory nerves. *J. Physiol. (Lond.)* 261, 472

Rochat, H., Bernard, P. and Couraud, F. (1979) Scorpion toxins: chemistry and mode of action. In: Ceccarelli, B. and Clementi, F. (eds) *Advances in cytopharmacology*, Vol. 3, pp. 325-34. Raven Press, New York

Rochat, H., Rochat, C., Kopeyan, C., Miranda, F., Lissitzky, S. and Edman, P. (1970) Scorpion neurotoxins — a family of homologous proteins. *FEBS Lett.* 10, 349

Rochat, C., Rochat, H., Miranda, F. and Lissitzky, S. (1967) Purification and some properties of the neurotoxins of *Androctonus australis* Hector. *Biochemistry* 6, 578

Romey, G., Abita, J.P., Schweitz, H., Wunderer, G. and Lazdunski, M. (1976) Sea anemone toxin: a tool to study molecular mechanism of nerve conduction and excitation-contraction coupling. *Proc. Natl Acad. Sci. USA* 73, 4055

Romey, G., Chicheportiche, R., Lazdunski, M., Rochat, H., Miranda, F. and Lissitzky, S. (1975) Scorpion neurotoxin — a presynaptic toxin which affects both Na^+ and K^+ channels in axons. *Biochem. Biophys. Res. Commun.* 64, 115

Romey, G., Renaud, J.F., Fosset, M. and Lazdunski, M. (1980) Pharmacological properties of the interaction of the sea anemone polypeptide toxin with cardiac cells in culture. *J. Pharmacol. Exp. Ther.* 213, 607

Rosenberg, P. (ed.) (1978) *Toxins: animal, plant and microbial.* Pergamon Press, Oxford and New York

Rosin, R. and Shulov, A. (1963) Studies on the scorpion *Nebo* hierochonticus. *Proc. Zool. Soc. Lond.* 140, 547

Ruhland, M., Zlotkin, E. and Rathmayer, W. (1977) The effect of toxins from the venom of the scorpion *Androctonus australis* on a spider nerve—muscle preparation. *Toxicon* 15, 157

Saenger, W., Walkinshaw, M.D. and Maelicke, A. (1983) α-Cobratoxin and α-bungarotoxin, two members of the 'long' neurotoxin family — a structural comparison. In: Hucho, F. and Ovchinnikov, Yu. A. (eds) *Toxins as tools in neurochemistry*, pp. 151-7. De Gruyter, Berlin and New York

Salikhov, S.I., Tashmukhamedov, M.S., Adylbekov, M.T., Abdurakhmanova, Ya., Korneev, A.S. and Sadykov, A.S. (1982a) Isolation and structural studies of neurotoxin from the venom of spider *Latrodectus tredecimguttatus. Chem. Pept. Proteins Proc. USSR–FRG Symp. 3rd. 1980*, 109

Salikhov, S.I., Tashmukhamedov, M.S., Adylbekov, M.T., Korneev, A.S. and Sadykov, A.S. (1982b) Isolation and quaternary structure of neurotoxin from *Latrodectus tredecimguttatus* spider venom. *Dokl. Acad. Nauk SSR 262*, 485

Sattelle, D.B. (1985) Acetylcholine receptors. In: Kerkut, G.A. and Gilbert, L.I. (eds) *Comprehensive insect physiology biochemistry and pharmacology*, Vol. II, pp. 395-434. Pergamon Press, Oxford and New York

Schmidt, H. and Schmitt, O. (1974) Effect of aconitine on the sodium permeability of the node of Ranvier. *Pflügers Arch. 349*, 133

Schmitt, O. and Schmidt, H. (1972) Influence of calcium ions on the ionic currents of nodes of Ranvier treated with scorpion venom. *Pflügers Arch. 333*, 51

Schweitz, H., Bidard, J.N., Frelin, Ch., Pauron, D., Vijverberg, H.P.M., Mahasneh, D.M. and Lazdunski, M. (1985) Purification, sequence and pharmacological properties of sea anemone toxins from *Radianthus paumotensis*. A new class of sea anemone toxins acting on the sodium channel. *Biochemistry 24*, 3554

Schweitz, H., Vincent, J.P., Barhanin, J., Frelin, Gh., Linden, G., Hugues, M. and Lazdunski M. (1981) Purification and pharmacological properties of eight sea anemone toxins from *Anemonia sulcata, Anthopleura xanthogrammica, Stoichactis giganteus* and *Actinodendron plumosum. Biochemistry 20*, 5245

Sharkey, R.G., Beneski, D.A. and Catterall, W.A. (1984) Differential labeling of the α- and β_1 subunits of the sodium channel by photoreactive derivatives of scorpion toxin. *Biochemistry 23*, 78

Spanjer, W., Grosu, L. and Piek, T. (1977) Two different paralysing preparations obtained from a homogenate of the wasp *Microbracon hebetor* (Say). *Toxicon 15*, 413

Spanjer, W., May, T.E., Piek, T. and De Hann, W. (1982) Partial purification of components from the paralyzing venom of the digger wasp *Philanthus triangulum* F. (Hym. Sphec.) and their action on neuromuscular transmission in the locust. *Comp. Biochem. Physiol. 71C*, 149

Stahnke, H.L. (1966) Some aspects of scorpions behavior. *Bull. South Calif. Acad. Sci. 65*, 65

Steiner, A.L. (1962) Etude du comportement predateur d'un hymenoptere sphegien: *Liris nigra* V.d.L. (= *Notogonia pompiliformis* Pz). *Ann. Sci. Nat. Zool. (Ser. 12)* 41

Steiner, A.L. (1986) Stinging behavior of solitary wasps. In: Piek, T. (ed.) *Venoms of the Hymenoptera*, pp. 63-160. Academic Press, London and New York

Strichartz, G.R. and Hansen-Bay, C.M. (1981) Saxitoxin binding in nerves from walking legs of lobster *Homarus americanus*. Two classes of receptors. *J. Gen. Physiol. 77*, 205

Strichartz, G.R., Rogart, R.B. and Ritchie, J.M. (1979) The binding of radioactively labelled saxitoxin to the squid giant axon. *J. Memb. Biol. 48*, 357

Strong, P.N., Georke, J., Oberg, S.G. and Kelly, R.B. (1976) β-Bungarotoxin, a pre-synaptic toxin with enzymatic activity. *Proc. Natl Acad. Sci. USA 73*, 178

Takahashi, M., Ohizumi, Y. and Yasumoto, T. (1982) Maitotoxin: a Ca^{2+} channel activator candidate. *J. Biol. Chem. 257*, 7287

Takahashi, M., Tatsumi, M., Ohizumi, Y. and Yasumoto, T. (1983) Ca^{2+} channel activating function of maitotoxin, the most potent marine toxin known, in clonal rat pheochromocytoma cells. *J. Biol. Chem. 258*, 10944

Tamashiro, M. (1971) A biological study of venoms of two species of *Bracon. Tech. Bull. Hawaii Agric. Exp. Stn 70*

Tanaka, J.C., Eccleston, J.F. and Barchi, R.L. (1983) Cation selectivity characteristics of the reconstituted voltage-dependent sodium channel purified from rat skeletal muscle sarcolemma. *J. Biol. Chem. 258*, 7519

Teitelbaum, Z., Lazarovici, P. and Zlotkin, E. (1979) Selective binding of the scorpion venom insect toxin to insect nervous tissue. *Insect Biochem. 9*, 343

Tintpulver, M., Zerachia, T. and Zlotkin, E. (1976) The action of toxins derived from scorpion venom on the ileal smooth muscle preparation. *Toxicon 14*, 371

Toth, G.P. and Blumenthal, K.M. (1983) Structure and action of heteronemertine polypeptide toxins. Binding of *Cerebratulus lacteus* toxin B-IV to axon membrane vesicles. *Biochim. Biophys. Acta 732*, 160

Tu, A.T. (1974) Sea snake venom and neurotoxins. *J. Agric. Food Chem. 22*, 36

Tu, A.T. (1977) *Venoms: chemistry and molecular biology.* John Wiley, New York

Tzeng, M.C., Cohen, R.S. and Siekevitz, P. (1978) Release of neurotransmitters and depletion of synaptic vesicles in cerebral cortex slices by α-latrotoxin from black widow spider venom. *Proc. Natl Acad. Sci. USA 75*, 4016

Ulbricht, W. (1969) The effect of veratridine on excitable membranes of nerve and muscle. *Ergeb. Physiol. Biol. Chem. Exp. Pharmakol. 61*, 18

Usherwood, P.N.R., Machili, P. and Leaf, G. (1968) *L*-Glutamate at insect excitatory nerve muscle synapses. *Nature (Lond.) 219*, 1169

Usherwood, P.N.R. (1982) Review of symposium. In: *Ciba Foundation Symposium 88. Neuropharmacology of insects*, pp. 307-17. Pitman, London

Usmanov, P.B., Kalikulov, D., Shadyeva, N.G., Nenilin, A.B. and Tashmukhamedov, B.A. (1985) Postsynaptic blocking of glutaminergic and cholinergic synapses as a common property of Araneidae spider venoms. *Toxicon 23*, 528

Vachon, M. (1952) *Etudes sur les scorpiones.* Institute Pasteur de Algerie

Vachon, M. (1953) The biology of scorpions. *Endeavour 12*, 80

Van Marle, J., Piek, T., Lind, A. and Van Weeren-Kramer, J. (1985) Specificity of two insect toxins as inhibitors of high affinity transmitter uptake. *Comp. Biochem. Physiol. 28C*, 435

Van Marle, J., Piek, T., Lind, A. and Van Weeren-Kramer, J. (1986) Reduction of high affinity glutamate uptake in rat hippocampus by two polyamine-like toxins isolated from the venom of the predatory wasp *Philanthus triangulum* F. *Experientia 42*, 157

Van Rietschoten, J., Granier, C., Rochat, H., Lissitzky, S. and Miranda, F. (1975) Synthesis of apamin, a neurotoxic peptide from bee venom. *Eur. J. Biochem. 56*, 35

Van Wilgenburg, H., Piek, T. and Mantel, P. (1984) Ion channel block in rat diaphragm by the venom of the digger wasp *Philanthus triangulum*. *Comp. Biochem. Physiol. 79C*, 205

Vincentini, L.M. and Meldolesi, J. (1984) α-Latrotoxin of black widow spider venom binds to a specific receptor coupled to phosphoinositide breakdown in PC12 cells. *Biochem. Biophys. Res. Commun. 121*, 538

Visser, B.J., Labruyere, W.T., Spanjer, W. and Piek, T. (1983) Characterization of two paralysing protein toxins (A-MTX and B-MTX), isolated from a homogenate of the wasp *Microbracon hebetor* (Say). *Comp. Biochem. Physiol. 75B*, 523

Walther, C. and Rathmayer, W. (1974) The effect of *Habrobracon* venom on excitatory neuromuscular transmission in insects. *J. Comp. Physiol. 89*, 23

Walther, C. and Reinecke, M. (1983) Block of synaptic vesicle exocytosis without block of Ca^{2+} influx. An ultrastructural analysis of the paralyzing action of *Habrobracon* venom on locust motor nerve terminals. *Neuroscience 9*, 213

Walther, C., Zlotkin, E. and Rathmayer, W. (1976) Action of different toxins from the scorpion *Androctonus australis* on a locust nerve–muscle preparation. *J. Insect Physiol. 22*, 1187

322

Watt, D.D., Simard, J.M., Babin, D.R. and Mlejnek, R.V. (1978) Physiological characterization of toxins isolated from scorpion venom. In: Rosenberg, P. (ed.) *Toxins: animal, plant and microbial,* pp. 647-60. Pergamon Press, Oxford

Weiland, G.A. and Molinoff, P.B. (1981) Quantitative analysis of drug–receptor interactions. I. Determination of kinetic and equilibrium properties. *Life Sci. 29,* 313

Wernike, J.F., Vanker, A.D. and Howard, B.D. (1975) The mechanism of action of β-bungarotoxin. *J. Neurochem. 25,* 483

Witkop, B. and Brossi, A. (1984) Natural toxins and drug development. In: Krogsgaard-Larsen, P., Christensen, S.B. and Kofod, H. (eds) *Natural products and drug development,* pp. 283-300. Munksgaard, Copenhagen

Wunderer, G. and Eulitz, M. (1978) Amino acid sequence of toxin I from *Anemonia sulcata. Eur. J. Biochem. 89,* 11

Wunderer, G., Fritz, H., Wachter, E. and Machleidt, W. (1976) Amino acid sequence of a coelenterate toxin: Toxin II from *Anemonia sulcata. Eur. J. Biochem. 68,* 193

Yasumoto, T., Nakajima, I., Oshima, Y. and Bagnis, R. (1979) A new toxic dinoflagellate found in association with ciguatera. In: Taylor, D.L. and Seligen, H. (eds) *Toxic dinoflagellate blooms,* pp. 65-70 Elsevier, North Holland, New York

Zhdanova, L.N., Adamovich, T.B., Nazimov, I.V., Grishin, E.V. and Ovchinnikov, Yu. A. (1977) Amino acid sequence of insectotoxin I, from the venom of the central Asian scorpion *Buthus epeus. Biorg. Kim.* (USSR) *3,* 485

Zlotkin, E. (1973) Chemistry of animal venoms. *Experientia 29,* 1453

Zlotkin, E. (1985) Toxins derived from arthropod venoms specifically affecting insects. In: Kerkut, G.A. and Gilbert, L.I. (eds) *Comprehensive insect physiology, biochemistry and pharmacology,* Vol. 10, pp. 499-546. Pergamon Press, Oxford

Zlotkin, E., Fraenkel, G., Miranda, F. and Lissitzky, S. (1971a) The effect of scorpion venom on blowfly larvae; a new method for the evaluation of scorpion venom potency. *Toxicon 9,* 1

Zlotkin, E. and Gordon, D. (1985) Detection, purification and receptor binding assays of insect selective neurotoxins derived from scorpion venom. In: Breer, H. and Miller, T.A. (eds) *Neurochemical techniques in insect research,* pp. 243-95. Springer, Berlin and New York

Zlotkin, E., Kadouri, D., Gordon, D., Pelhate, M., Martin, M.F. and Rochat, H. (1985) An excitatory and a depressant insect toxin from scorpion venom — both affect sodium conductance and possess a common binding site. *Arch. Biochem. Biophys. 240,* 877

Zlotkin, E., Lebovits, N. and Shulov, A. (1972c) Toxic effects of the venom of the scorpion *Scorpio maurus palmatus* (Scorpionidae). *Riv. Parassit. 33,* 237

Zlotkin, E., Lebovits, N. and Shulov, A. (1973) Hemolytic action of the venom of the scorpion *Scorpio maurus palmatus* (Scorpionidae). In: Kaiser, E. (ed.) *Animal and plant toxins,* pp. 67-72. Goldman, Munich

Zlotkin, E., Martinez, G., Rochat, H. and Miranda, F. (1975) A protein toxic to Crustacea from the venom of the scorpion *Androctonus australis. Insect Biochem. 5,* 243

Zlotkin, E., Miranda, F., Kupeyan, G. and Lissitzky, S. (1971b) A new toxic protein in the venom of the scorpion *Androctonus australis* Hector. *Toxicon 9,* 9

Zlotkin, E., Miranda, F. and Lissitzky, S. (1972a) A factor toxic to crustaceans in the venom of the scorpion *Androctonus australis* Hector. *Toxicon 10,* 211

Zlotkin, E., Miranda, F. and Lissitzky, S. (1972b) Proteins in scorpion venoms toxic to mammals and insects. *Toxicon 10,* 207

Zlotkin, E., Miranda, F. and Rochat, H. (1978) Chemistry and pharmacology of

Buthinae scorpion venoms. In: Bettini, S. (ed.) *Arthropod venoms*, pp. 317-69. Springer, Berlin and New York

Zlotkin, E., Rochat, H., Kupeyan, C., Miranda, F. and Lissitzky, S. (1971c) Purification and properties of the insect toxin from the venom of the scorpion *Androctonus australis* Hector. *Biochimie (Paris) 53*, 1073

Zlotkin, E., Teitelbaum, Z., Rochat, H. and Miranda, F. (1979) The insect toxin from the venom of the scorpion *Androctonus mauretanicus*. Purification, characterization and specificity. *Insect Biochem. 9*, 347

Index